Research
Methods

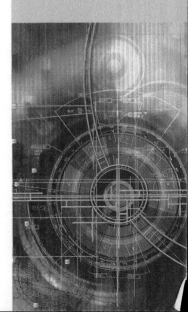

Research Methods

THE ESSENTIAL KNOWLEDGE BASE

William M. Trochim
Cornell University

James P. Donnelly
Canisius College

Kanika Arora
Syracuse University

CENGAGE
Learning·

Australia • Brazil • Japan • Korea • Mexico • Singapore • Spain • United Kingdom • United States

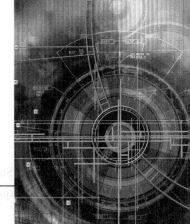

Research Methods: The Essential Knowledge Base, Second Edition
William M. Trochim, James P. Donnelly, and Kanika Arora

Product Director: Jon-David Hague

Product Manager: Tim Matray

Content Developer: Gary O'Brien

Associate Content Developer:
  Jessica Alderman

Product Assistant: Nicole Richards

Media Developer: Kyra Kane

Content Project Manager: Samen Iqbal

Art Director: Vernon Boes

Art Editor: Precision Graphics

Manufacturing Planner: Karen Hunt

IP Analyst: Deanna Ettinger

IP Project Manager: Brittani Hall

Production Service/Project Manager:
  MPS Limited, Teresa Christie

Image Researcher: Nazveena Begum Syed

Text Researcher: Sharmila Srinivasan

Text and Cover Designer: Lisa Delgado

Cover Image Credit: Data Funk/Digital
  Vision

Compositor: MPS Limited

For product information and technology assistance, contact us at
Cengage Learning Customer & Sales Support, 1-800-354-9706.

For permission to use material from this text or product, submit all requests online at **www.cengage.com/permissions.**
Further permissions questions can be e-mailed to
permissionrequest@cengage.com.

Library of Congress Control Number: 2014940179

Student Edition:

ISBN: 978-1-133-95477-4

Cengage Learning
20 Channel Center Street
Boston, MA 02210
USA

Cengage Learning is a leading provider of customized learning solutions with office locations around the globe, including Singapore, the United Kingdom, Australia, Mexico, Brazil, and Japan. Locate your local office at **www.cengage.com/global.**

Cengage Learning products are represented in Canada by Nelson Education, Ltd.

To learn more about Cengage Learning Solutions, visit **www.cengage .com.** Purchase any of our products at your local college store or at our preferred online store **www.cengagebrain.com.**

Printed in the United States of America
Print Number: 01   Print Year: 2014

About the Authors

WILLIAM M. TROCHIM, PH.D. William M. Trochim, Ph.D., Cornell University. William M. Trochim is a Professor in the Department of Policy Analysis and Management at Cornell University and a Professor of Public Health in the Department of Healthcare Policy and Research at the Weill Cornell Medical College (WCMC). He is the Director of the Cornell Office for Research on Evaluation, Director of Evaluation for Extension and Outreach at Cornell, and the Director of Evaluation for the WCMC's Clinical and Translational Science Center. He has taught both undergraduate and graduate required courses in applied social research methods since joining the faculty at Cornell in 1980. He received his Ph.D. in 1980 from the program in Methodology and Evaluation Research of the Department of Psychology at Northwestern University. Trochim's research interests include the theory and practice of research, conceptualization methods (including concept mapping, pattern matching, logic and pathway modeling), strategic and operational planning methods, performance management and measurement, and change management. His current research is primarily in the areas of translational research, research-practice integration, evidence-based practice, and evaluation policy.

JAMES P. DONNELLY James P. Donnelly, Ph.D., Canisius College. Dr. Donnelly is a licensed psychologist and an Associate Professor affiliated with the Institute for Autism Research and the Department of Counseling & Human Services. He completed his undergraduate degree at Allegheny College, his masters at Claremont Graduate University, and his doctorate at the University at Buffalo. He teaches courses related to research methods, health, and counseling psychology at the graduate level. His research and clinical interests are in quality-of-life issues related to chronic and life-limiting conditions. He lives in Clarence, New York, with his wife Kerry and sons Seamus and Paddy.

KANIKA ARORA, MPA Kanika Arora, MPA, Syracuse University. Kanika Arora is a Ph.D candidate in the Department of Public Administration and International Affairs at Syracuse University. She received her MPA from Cornell University in 2007. Kanika's research focuses on long-term care in the United States, including the provision of intergenerational support by adult children. She is also interested in topics related to performance management and measurement. In particular, she studies tools that facilitate the link between program planning and evaluation. Previously, she worked as a Monitoring and Evaluation Specialist for Orbis—an international nonprofit in the field of blindness prevention. Kanika lives in Syracuse, New York, with her husband Vikas.

Brief Contents

Contents

Preface

We shall not cease from exploration
And the end of all our exploring
Will be to arrive where we started
And know the place for the first time.
—*T. S. Eliot*

How is it that we can look at the familiar things that are around us and see them in a new way? The three of us who have co-authored this text have certainly been dealing in familiar territory for us. Together we have decades of experience in research methods, as students, teachers, and practitioners. Every Monday morning at 9 a.m. for the past several years we have gotten on the phone to talk over how the text was coming, to discuss some arcane aspect of research methods, to divvy up responsibilities for next steps and, okay, we'll admit it, to have some fun just playing with ideas and coming up with new ways to present this material. For us this has been an exploration of very familiar territory. But, as T. S. Eliot suggests, the end is that we have arrived here, at this preface, at the beginning of this text, writing the last few lines that will finish our journey, and we feel like we know the place for the first time.

Throughout, we've imagined you, the reader, and have tried to put ourselves in your place. We've tried to think about what it must be like to experience this unfamiliar territory of research methods for the first time. We've tried to sense the panic, the feeling of being overwhelmed, and your desperation as a test approaches. We tried to be there with you in spirit as you hit the college town bars last Saturday night knowing you had a whole chapter on measurement to digest before the quiz at 8 a.m. on Monday morning. In order to feel what you went through, we even went so far as to simulate the experience ourselves a few times—the bars, that is. And in the end, we tried to write this text with one overarching principle in mind—you have to get a grip! We know that if this is really your first time in a course like this, the material can be daunting. We know you probably put this course off until the last possible semester (even though it would have been much better if you had taken this stuff earlier so you could have understood the research in your other courses). We can sense that many of you will feel disoriented by the strangeness of research thinking. And so we have done our best to try to calm you down.

Learning about research methods is a lot like learning about a new culture. You're going to meet a lot of strange people along the way. You're not going to understand the language. You're going to have a hard time communicating. You're going to have trouble even reading the menu. You're going to feel foolish at times and, yes, maybe you'll actually say some foolish things. You will make mistakes. But like all new cultural experiences, once you immerse yourself in the context you'll begin to get your bearings. You'll pick up a useful phrase here and there and actually use it properly in a sentence. You'll get the lay of the land and begin to move around more comfortably. And one day you'll suddenly find yourself feeling that sense of mastery that comes from having stayed with it. All right, maybe not everyone who reads this text will feel that way. But we're confident that you will come away from this a better person for having experienced this new culture. So, let's set out on this exploration and come to "know the place for the first time."

The Road to Research

When you come to a fork in the road—take it.
—*Yogi Berra*

Remember when you were a little kid, piling into the family car and setting off on a trip? It might have been to Grandma's house, or it might have been a cross-country vacation, but there was the thrill of the journey to come, the unexpected, perhaps even something exciting. Or maybe you didn't do the family-car thing. Perhaps for you it was setting off on the subway for the museum on a Saturday afternoon. Or getting on a plane to fly off to new places. Never traveled when you were a kid? Okay, this metaphor won't work—skip down to the next section, and we'll try again. But if you did any traveling, you know how exciting and mysterious setting out can be. Research is a lot like setting out on a new trip. No, really. You're going to have fun. Honest.

When you start out on a trip it's useful to take a map. We're not talking about Google maps on an iPhone, we're talking about a real map, crinkled at the edges, a marked-up and well-worn map that shows the terrain you'll move through on your journey. You're going to take your trip via this map, following a path. We and your instructor will guide you in moving down the road—let's call it the Road to Research. Figure 1 shows what this road might look like and, not incidentally, depicts the high-level contents of this text in a way that suggests that the research process is a practical sequence of events, a type of trip down the road. As with all maps, the actual trip down the research road is a little more exciting than Figure 1 suggests! The map shows a territory that looks a lot like Middle Earth in the Tolkien's *Hobbit* and the *Lord of the Rings* trilogy. And, even though the map itself looks relatively benign, you know that as you and your friends move down this road, stopping off at the major research methods destinations, you can't anticipate all of the challenges along the way, how you will be constantly avoiding dangers and defying death while trying to get the ring into the fiery volcano. Okay, maybe it's not that exciting. Maybe we're overstating the metaphor a bit.

But research is like a journey in that it typically involves a set of steps. Every research project needs to start with a clear problem formulation. As you develop your project, you will find critical junctions where you will make choices about how to proceed, where you will consider issues of sampling, measurement, design, and analysis, as well as the theories of validity that underlie each step. In the end, you will need to think about the whole picture and write up your findings. You might even find yourself backtracking from time to time and reassessing your previous decisions. You might get waylaid by dwarves in the land of measurement or be set upon by strange orcs and trolls when doing statistics in the land of analysis. Really, it's been known to happen. Especially the orcs and trolls who seem especially prone to hanging around statistics. And it's important to know that this is a two-way road; planning and reflection—looking forward and backward—are critical and interdependent. You can take a step back on the way to making two steps forward. You might spend time in the Northern Waste before finally making it to Eriador. Think of the hard surface of the road as the foundation of research philosophy and practice. Without consideration of the basics in research, you'll find yourself bogged down in the mud of Dunland! And if you really want to go nuts, you might think of your teacher as the kids in the back seat of the car (Okay, perhaps to keep the metaphor straight, it should be a cart), constantly needling

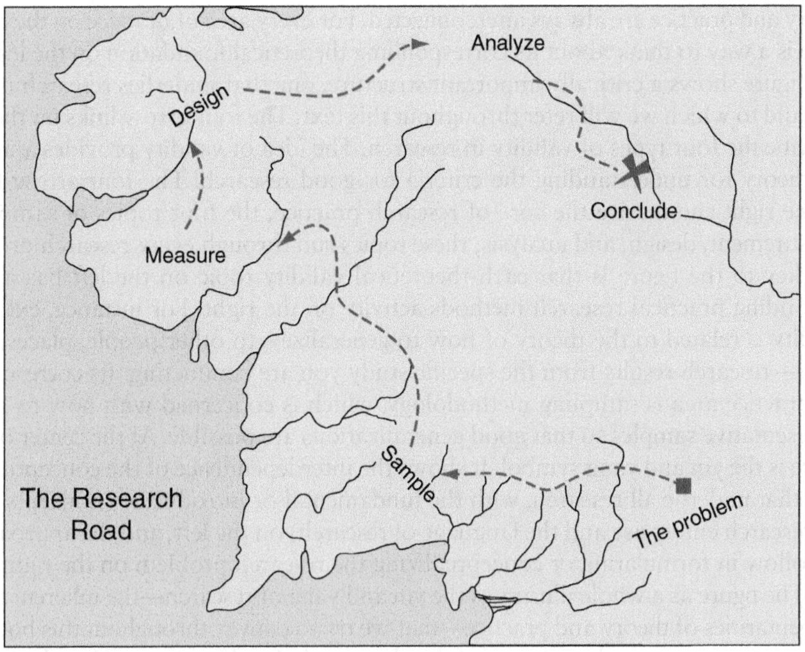

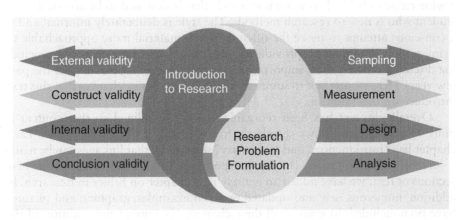

Analyze

Design

Conclude

Measure

Sample

The Research
Road

The problem

Figure 1 The Research
Road Map.

you with, "Are we there yet?" But with all of the twists and turns, the map is useful because it reminds us of the general course we are on. It reminds us that research follows well-known paths, and that even if sometimes you feel like you are lost, the map is always there to guide you.

The Yin and the Yang of Research

For this second metaphor of the research process, imagine that you're a Buddhist. You might want to sit cross-legged on the floor, burn some incense, and turn up your sitar music. To the Buddhist, everything in the universe is connected to everything else. To the Buddhist researcher, if you can imagine such a person, all parts of the research process are interconnected. The Buddhist view of research might be something like that shown in Figure 2. The left side of the figure refers to the theory of research. The right side of the figure refers to the practice of research. The yin-yang figure (okay, so that's more Daoist than Buddhist) in the center shows you that

Figure 2 The Yin and
Yang of Research

External validity

Introduction
to Research

Sampling

Construct validity

Measurement

Internal validity

Design

Research
Problem
Formulation

Conclusion validity

Analysis

theory and practice are always interconnected. For every area of practice on the right, there is a way to think about its corresponding theoretical foundation on the left. The figure shows a critically important structure, one that underlies research methods, and to which we will refer throughout this text. The four arrow links on the left describe the four types of validity in research. The idea of validity provides a unifying theory for understanding the criteria for good research. The four arrow links on the right summarize the core of research practice, the four topics of sampling, measurement, design, and analysis; these topics run through every research project. The key to the figure is that each theoretical validity topic on the left has a corresponding practical research methods activity on the right. For instance, external validity is related to the theory of how to generalize—to other people, places, and times—research results from the specific study you are conducting. Its corresponding practice area is sampling methodology, which is concerned with how to draw representative samples so that good generalizations are possible. At the center of the figure is the yin and yang symbol. It shows the interdependence of the conceptual issues that underlie all research, with the fundamental or introductory concepts (like the research enterprise and the language of research) on the left, and the approaches we follow in formulating or conceptualizing the research problem on the right.

The figure as a whole illustrates the yin and yang of research—the inherent complementarities of theory and practice—that we try to convey throughout this book. If you can come to an understanding of this deeper relationship, you will be a better researcher, one who is able to create research processes, rather than to simply use them.

Okay, it's time for you to sit cross-legged and meditate on the yin and yang of it all, as we start down the road to research.

What's New in This Text

This volume is the latest in a long line of writing about research methodology that began in the late 1990s with the Research Methods Knowledge Base website (http://www.socialresearchmethods.net/kb/), which was essentially the translation of the first author's class lectures to the Internet. This was followed by the publication of revised content in several significant prior textbook publications, including *The Research Methods Knowledge Base* and *Research Methods: The Concise Knowledge Base*. The current text constitutes the next step forward in this decades-long tradition. It was designed for a broad, applied social research readership, a text that could be used in any standard research methods courses in a wide range of fields. It is written in a style that is designed to be accessible to a student who is new to research methods. The style is deliberately informal and is a conscious attempt to make the often-daunting material more approachable to many readers. And this text provides significant updates of the prior texts, including discussions of the most important changes in research methods over the past few years. Here we attempt to summarize some of the major changes that this text introduces to this tradition.

Overall, the text has been reorganized and streamlined so that content is more tightly knit and flows seamlessly from "broad" to "specific" topics. Each chapter has "Introduction" and "Summary" sections so that linkages can be made to preceding and following chapters, respectively. In addition to numerous new sections of text, we have added an entirely new chapter on Ethics in Research. In addition, numerous new and updated research examples, graphics, and pictures have been included. In spite of all these changes, the core of the tradition of the original Research Methods Knowledge Base remains intact. Readers of earlier texts will recognize the fundamentals of research methods that have not changed in several decades.

The text begins with an introductory chapter that describes the growing awareness of the field of research methods—an awareness that there is a large and complex research enterprise in contemporary society. This is evident in some of the most important movements in contemporary research: the idea of translational research; the notion of evidence-based practice; the growing importance of research syntheses (such as systematic reviews and meta-analysis) and practice guidelines; and the continuing reverberation of the implications of the theory of evolution in our views of how research evolves.

Chapter 2 introduces the increasingly important topic of ethics in research. We placed this chapter immediately after the introduction to signal to the reader that ethical issues permeate the entire research endeavor. This is the first edition of "The Knowledge Base" series that has a separate chapter on ethics. The topic now receives a complete treatment that includes a detailed history as well as the key principles and practices that all researchers need to know. The discussion is framed in terms of defining the meaning of "good" in research. We suggest that a thorough understanding of the historic and current context of research ethics is essential to good research—every bit as important as the technical or methodological aspects. The review of key events in the history of research ethics includes both world events outside the normal boundaries of research (e.g., the Nazi crimes conducted under the guise of experimentation) and legitimate but ethically problematic research programs (e.g., Stanley Milgram's obedience studies). Our discussion then moves to the key events that occurred in response to the ethical issues that became known in the problematic studies. This includes the National Research Act and the *Belmont Report*, which established the key principles for our modern research ethics system: Respect for Persons, Beneficence, and Justice. We also discuss the rights of research participants, the responsibilities of investigators, and the role of Institutional Review Boards (IRBs) in the oversight of research. The chapter then discusses the integrity of the research enterprise itself. In particular, we focus on the matter of research ethics in the production and publication of scholarly work. We cite key principles such as honesty in reporting, as well as several cases of scientific misconduct that have undermined the integrity of research. We conclude by emphasizing that research ethics is now defined by formal principles and practices, but will always depend on the ethical compass that resides in each member of the research community.

The third chapter on Qualitative Approaches to research is now included earlier in the book as part of the Foundations section. This was done, as with the chapter on ethics, to signal to the reader that these approaches are in some way foundational to all research. Unobtrusive measures relating to the qualitative tradition are integrated within the discussion of Qualitative Measures—they are no longer treated as separate from Qualitative Measures, as they were in previous editions. Unobtrusive measures relating to "Secondary Analysis of Data" are discussed in later chapters of the book. The chapter now begins more generally by introducing Qualitative Research. The section on "When are qualitative research methods most appropriate to use" has been expanded and the section on Qualitative Traditions is discussed earlier in the chapter, in order to provide context for the subsequent discussion on qualitative measures. Research examples are now integrated in the discussion of each qualitative tradition. The section on "Qualitative Methods" is expanded to include "focus groups," and the section on "Indirect

Measures" now discusses technological innovation in such measures. The discussion on "Qualitative Data" has also been expanded, and the discussion on differences between qualitative and quantitative data is now integrated within this section. The "Summary" emphasizes the appropriateness of qualitative research methods in the context of specific research questions.

The next section of the book, Chapters 4 through 12, constitutes the heart of the steps in research methodology—sampling, measurement, design, and analysis. While much of the discussion remains true to the Knowledge Base tradition, each chapter has been significantly revised and updated. Chapter 4 on Sampling has more detail and includes research-based examples for each type of sampling method. The organization of the chapter is more intuitive and logical, with added sections summarizing probability and nonprobability sampling methods, and the subsection on "How big should the sample be?" was also included. Chapter 5 on Measurement has been reorganized to begin more generally with "Theory of Measurement" and "Levels of Measurement." In an effort to provide context, the concepts of "Reliability" and "Validity" are discussed under the larger topic of "Quality of Measurement." For consistency purposes, we conclude the chapter with a big-picture discussion about integrating "Validity" and "Reliability." The previously disparate sections on construct validity throughout the chapter are better integrated. We also include a new subsection on "Construct Validity of What?" A discussion on Cohen's kappa is included under the subsection on "Inter-Rater Reliability," and the section on "Discriminant Validity" has an entirely new example. Chapter 6 on Scales, Tests and Indices now comes ahead of Survey Research. The section on "Scales" now leads off the chapter and there is an entirely new section on "Tests" that includes: Validity, Reliability and Test Construction, Standardized Tests, Test Fairness, and Finding Good Tests. Chapter 7 on Survey Research begins broadly by defining surveys, the different ways in which surveys are administered, and what factors to consider when selecting a particular survey method. There is an expanded discussion on different types of questionnaires and interviews, and the topic of "Point of Experience Surveys" is now included. The chapter also has an expanded discussion on "Selecting the Survey Method" and "Survey Construction" and updated examples in the subsection on "Structured Response Formats."

Chapter 8 introduces the critically important topic of research design. It begins with the tricky issue of how to establish causality, using the new example of the Aurora, Colorado, shooting and the issue of whether movie violence causes real violence. The discussion then shifts to the topic of internal validity and the different threats to internal validity, especially in two-group comparative designs. A considerable amount of the discussion is devoted to the issue of selection threats. The chapter concludes with a discussion of the logic of how to design a design. Chapter 9 introduces the idea of experimental designs, particularly the randomized experiment. The chapter begins with a new introduction that provides a history of the evolution of the randomized experiment. Throughout the chapter there is a consistent effort to provide a balanced view of both the strengths and weaknesses of randomized experiments, especially considering their importance in the evidence-based practice debate. The chapter covers the basic two-group experimental design, introduces the design notation, and discusses two ways to address the signal-noise problem in experiments: factorial designs and blocking strategies. The chapter concludes with some important variations on experimental designs and a discussion of the limitations of the randomized experiment. Chapter 10 introduces quasi-experimental designs and begins with the basic two-group, pre-post nonequivalent groups design, including how to interpret the major outcomes and the major threats to internal validity that might result. It then moves on to a design that has taken on increasing importance in the evidence-based practice debate as an alternative to the randomized experiment—the regression-discontinuity design. The chapter concludes with several important quasi-experimental designs that illustrate critical principles for addressing threats to internal validity. Chapters 8, 9, and 10 incorporate numerous changes and updates that reflect the evolving nature of research design.

The next two chapters of the book, Chapters 11 and 12, deal with the topic of data analysis. Chapter 11 is an introduction to the topic, and it covers everything from data preparation to descriptive statistics. The discussion of conclusion validity, a central idea in this chapter, has been expanded. We attempt to connect every step in the management and analysis of data to the credibility and confidence we can obtain in our analysis. For example, we added encouragement to consider research context in the interpretation of data. This discussion also introduces effect sizes as an important part of conclusion validity. The discussion of p values has been revised to present a tighter and more restrictive conceptualization of what p values are and what they are not. Chapter 12 addresses inferential statistics. The chapter now adds to the conceptual and procedural understanding of conclusion validity with a discussion of the correct interpretation of p values, effect sizes, confidence intervals, and practical significance and their relationship to conclusion validity. And we have added a data-based example of signal to noise ratio in the section on "What does difference mean?"

The final chapter of the book deals with the general topic of research communication. It revisits the idea of the research–practice continuum introduced in Chapter 1 and shows the critical role that research write-ups have in translational research and evidence-based practice. A new section on oral presentation has been added. This includes guidelines for giving a talk as well as a sample conference poster. The poster is based on the sample paper. The presentation is simple and straightforward but compliant with current reporting recommendations. The sample paper is new to the book and is consistent with current standards of analysis and reporting, including the APA 6th Edition and the recently announced requirements of the American Psychological Society (Cumming, 2013). These include a statement regarding IRB review, statistical power, a CONSORT-type flow diagram, effect sizes, and confidence intervals.

Acknowledgments

This work, as is true for all significant efforts in life, is a collaborative achievement. It is also one that has evolved in unexpected ways since the publication of the original website on which it is based. One happy discovery in the creation of this volume is the excellence of the Cengage team. They have been wonderful to work with, except for their annoying habit of paying attention to the passage of time, deadlines, and such. Seriously, though, the team lead by Tim Matray has provided continuous support, responsive listening, and a very clear commitment to high standards of teaching and learning. In addition to Tim's guidance, we have had the great benefit of working with Gary O'Brien (yes, the handsome and debonair one), who has shared his experience, knowledge, kindness, and sense of humor from the beginning. We are also very grateful to the many special people on the Cengage team who have made cheerful suggestions in every phase, from debating the frequency of contractions to helping with artwork, graphics, photos, and cartoons. These wonderful people include Jon-David Hague, Jessica Alderman, Nicole Richards, Kyra Kane, Samen Iqbal, Vernon Boes, Karen Hunt, Deanna Ettinger, Brittani Hall, Teresa Christie, Nazveena Begum Syed, Sharmila Srinivasan, Lisa Delgado, Charlene Carpentier, and the folks at Precision Graphics.

Finally, we acknowledge the thoughtful, constructive comments and suggestions of the following reviewers: Veanne Anderson, Indiana State University; Steven Branstetter, Pennsylvania State University; Michael Cassens, Irvine Valley College; Tom Copeland, Geneva College; Bob Dubois, Marquette University, Waukesha County Technical College; Jonathan Gore, Eastern Kentucky University; Gary King, Pennsylvania State University; Christine Lofgren, University Of California, Irvine; Edward Maguire, American University; Charlotte Markey, Rutgers University—Camden; Kristine Olson, Dixie State College Of Utah; Leigh Alison Phillips, The George Washington University; Janice Thorpe, University of Colorado at Colorado Springs; Kari Tucker, Irvine Valley College; and Alyssa Zucker, The George Washington University.

Bill's acknowledgments: There are lots of popular aphorisms about the value of working with others. The phrases "many hands make light work" and "misery loves company" come immediately to mind. So it should be no surprise that I want to begin by thanking my two incredible co-authors who did make the work lighter and whose company I do indeed love. We met by phone every Monday morning at 9 a.m. for the past several years. Those calls started our week. They were times to check in on the progress on this book, but they were also an opportunity to check in on each other's lives. We would touch base about big events, things we were struggling with, travel plans, and much more. Jim is remarkable for many, many things, but I think most immediately of his infectious enthusiasm and his invariably positive cheerful manner. Kanika joined our collaboration when Jim and I realized we needed a bright graduate student to work through the details of the text with us. She quickly demonstrated that she was more than a student assistant. Through her notable intelligence and ability to cut to the heart of complex issues, she rapidly became an indispensable part of the team until Jim and I realized that she was in fact a fully functioning co-author. Meeting with the two of them was the best way I could imagine to get my week off to a positive start. And this book simply would not be in your hands today (or on your screen) if the two of them had not become involved. So I can't thank them enough for their dedication and hard work, and I am proud and honored to have my name next to theirs on this book. So that's who I wrote this with.

Equally important is who inspired this book. I have been fortunate throughout my career to have encountered so many incredible students, graduate and undergraduate alike. Whenever I was lost in the details of this text, not able to see the forest for the trees, I found my mind wandering to the many experiences I have had the privilege to observe with my students, as they wrestled with this material. They were my teachers and the primary way I learned what little I know about how to learn about research methods. Much of this text would have been impossible without them, their struggles, and ultimately, their triumphs on the road to research. Finally, I reserve my deepest gratitude for the most special of those former students who, like all the others, graduated and seemingly left Cornell to go lead her life. We didn't know then that circumstances would bring us together over twenty-five years later. So, my deepest thanks, Gwenn, to you and your three wonderful children, Andrew, Alix, and Rebecca, for the encouragement and support you always provide me that helped make this book possible and that has so greatly enhanced and enriched my life.

And finally, there is who I wrote this book for. There are two precious souls, those of my grandchildren Evangeline and Curran, who represent their generation and those still to come. I can only dream that this book might contribute in some small way to the stream of human thought that is still our best chance to enhance the world they will inhabit. My hopes are invested in them, and my deepest pride is with their parents, my daughter Nora and my son-in-law Derek, without whom the present world and the prospects for that future one would not nearly be so bright.

Jim's acknowledgments: Students are the reason that Professor Trochim launched the original "Knowledge Base," and students remain the main reason for this edition's development. I still feel like a student most of the time. Maybe that is because I have had the wonderful opportunity to keep learning from Professor Trochim since he graciously replied to my first email (or was it a an actual letter?) way back in 1990. He has allowed me to join him on the research road all these years as collaborator, friend, and, more than he knows, student. Most recently, he has introduced me to Kanika, who has made our team and this volume immeasurably better and even more fun. Thank you both! I've also had the good fortune to meet hundreds of students and researchers on the research road and I have certainly been their student as well. Thanks to all of you, I still look forward to every semester and every new class. As ever, my deepest gratitude goes to Kerry, Seamus, and Paddy, who make life good in every way.

Kanika's acknowledgments: I would like to dedicate this book to my co-authors. This collaboration has been enormously rewarding in more ways than one, and I am very fortunate to have both Bill and Jim as mentors. I am also grateful to Vikas for making this book a "team effort," even at home.

MindTap

MindTap for Research Methods: The Essential Knowledge Base engages and empowers students to produce their best work—consistently. By seamlessly integrating course material with videos, activities, apps, and much more, MindTap creates a unique learning path that fosters increased comprehension and efficiency.

For students:

- MindTap delivers real-world relevance with activities and assignments that help students build critical thinking and analytic skills that will transfer to other courses and their professional lives.
- MindTap helps students stay organized and efficient with a single destination that reflects what's important to the instructor, along with the tools students need to master the content.
- MindTap empowers and motivates students with information that shows where they stand at all times—both individually and compared to the highest performers in class.

Additionally, for instructors, MindTap allows you to:

- Control what content students see and when they see it with a learning path that can be used as-is or matched to your syllabus exactly.
- Create a unique learning path of relevant readings and multimedia and activities that move students up the learning taxonomy from basic knowledge and comprehension to analysis, application, and critical thinking.

- Integrate your own content into the MindTap Reader using your own documents or pulling from sources like RSS feeds, YouTube videos, websites, Googledocs, and more.
- Use powerful analytics and reports that provide a snapshot of class progress, time in course, engagement, and completion.

In addition to the benefits of the platform, MindTap for Research Methods: The Essential Knowledge Base includes:

- Formative assessments following each section and summative assessments at the conclusion of each chapter.
- SPSS video tutorials on the most commonly taught procedures that provide students with a foundation in SPSS.
- Small experimental data sets that provide students practice in SPSS/SAS/Excel etc.
- Research Tutor, a project management tool that helps students stay on task with the research assignment that is often included in the behavioral sciences research methods course. Research Tutor breaks the process down into 10 assignable modules that help manage timelines and turn research ideas into well-constructed research proposals, research papers, or presentations. It's the only interactive tool that helps students evaluate and choose an appropriate topic early in the course and stay on task as they move through their study.

Supplements

Cengage Learning Testing, powered by Cognero Instant Access (ISBN-13: 978-1-305-57716-9) Cognero is a flexible, online system that allows you to author, edit, and manage test bank content as well as create multiple test versions in an instant. You can deliver tests from your school's learning management system, your classroom, or wherever you want.

Online Instructor's Manual (ISBN-13: 978-1-305-57710-7) The Instructor's Manual (IM) contains a variety of resources to aid instructors in preparing and presenting text material in a manner that meets their personal preferences and course needs. It presents chapter-by-chapter suggestions and resources to enhance and facilitate learning.

Online PowerPoint® (ISBN-13: 978-1-305-57711-4) These vibrant Microsoft® PowerPoint® lecture slides for each chapter assist you with your lecture by providing concept coverage using images, figures, and tables directly from the textbook.

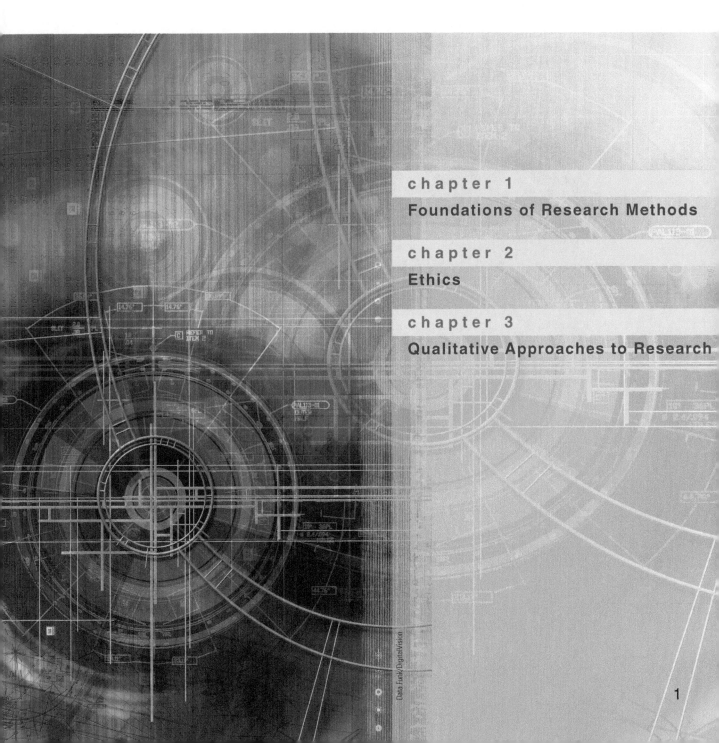

PART 1 Foundations

Data Funk/DigitalVision

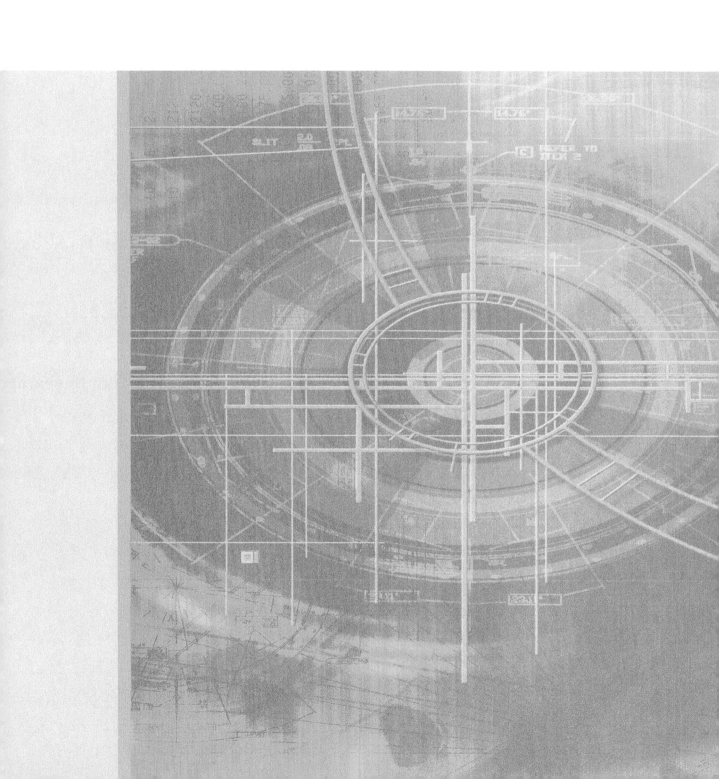

1

Foundations of Research Methods

Data Funk/DigitalVision

3

The Greek philosopher Aristotle reportedly said: "Well begun is half done." This chapter is designed to get you "well begun" on the often-daunting topic of research methods. The good news—if you believe this saying—is that when you're done with this chapter you will be "half done." The bad news of course is that it's not literally true. You have an entire text yet to complete. But, we were looking for something you might find consoling as you start out on this journey into research methods.

This chapter begins with consideration of the big picture, what we term the "research enterprise." It describes how the tens of thousands of research projects conducted around the world over time are increasingly being integrated to provide a more empirical knowledge base for humanity. Some of the most exciting and important developments in research are occurring at this more macro level as we collectively get a more global view of how we have evolved into a research-based society. Next, we explore where research ideas come from, the task of conceptualizing a research project. We consider both the inspirational and insightful aspects of conceptualizing and the ways we attempt to build new ideas on the research literatures that preceded our work. Following that, we begin to provide you with the basic tools you will need to navigate the research terrain, the beginnings of a vocabulary that will enable you to understand and speak the language of research. Here we consider some of the most basic terms and concepts that you will need throughout the text as you acquire vocabulary that is even more advanced. Then, we consider the idea that every research project has a structure: a beginning, a middle, and an end. We introduce the basic components that make up the typical research project and describe how they fit together. Finally, since research is concerned with learning about the world around and within us, we end by introducing the basic idea of validity in research, how we judge the degree to which the research we conduct is an accurate depiction of our world.

That ought to be enough to get you started. At least it ought to be enough to get you thoroughly confused. But don't worry, there's stuff that's far more confusing than this yet to come!

1.1 The Research Enterprise

It is amazing, when you think about it, how much our modern society relies upon research. Virtually everything we do, see or come into contact with is the subject of research. Humans carry out research that takes us into the universe, that explores our environment, that leads to all of the technologies that we use, that investigates what we eat and drink and how to produce these things better, that looks at our cities and towns, that assesses our interactions and relationships, and that explores our physical and mental selves. There is almost no aspect of our world that is not the subject of considerable research. Sometimes it is difficult to believe that all of this human activity is a relatively recent invention. Prior to the 1600s there was virtually nothing that resembled what we would today call research. In the space of less than 400 years, humanity has created the idea of research and has seen it permeate into almost every aspect of our lives.

1.1a What Is Research?

So, let's begin with a few simple definitions. The obvious place to start is with the term research itself. Virtually every field of study involves some form of research. But the term means many different things across different fields. Research in the field of history doesn't look much like research in medicine. Research in costume design typically won't resemble research in meteorology. So what is it that is common to all of these definitions? Perhaps the most important thing that holds all definitions together is that research is *systematic investigation*. In our everyday life we think about the world around us. We consider options and make choices. But much of this thinking is done dynamically, changing and adapting to the circumstances as they unfold. Research is different. It is a conscious effort to concentrate our thinking, to do it in a rational, careful manner. This is the key to the systematic nature of research.

Research also involves collecting data. It is an *empirical* endeavor. When you go to the store or to a market to buy something and you just browse through the aisles seeing what catches your eye, you may be gathering information, but you are not doing so systematically. On the other hand, when you systematically compare products, collecting comparable information about their features, quality, service history, and so on, you are engaged in an empirical effort, an effort that is based upon systematic observation that yields data that you can use in your decision making.

Research is also typically a *public* effort. While you might collect lots of information systematically so that you can make a better decision, researchers typically conduct research so that it can contribute to a broader base of knowledge than just their own. Consequently, it is important that research procedures are described in a way that enables other people to understand them, duplicate them and make judgments about their quality.

So, we might put this together into a simple definition:

> **Research** is a type of systematic investigation that is empirical in nature and is designed to contribute to public knowledge.

In this volume, we focus on a particular subclass of research known as *social* research. The topics that are investigated in social research have to do with our societies, the things we do, how we interact, how we live, how we feel, and how we see ourselves. It encompasses much of the research that is done in fields like sociology, education, public health, criminology, housing, public welfare, applied and social psychology, and many more. While much of what this text talks about is relevant to other fields like biomedical research or engineering, our focus here is on the social aspects of phenomena.

Typically, when we conduct research, we do a research *project* or study that addresses one or more specific questions, collects specific data, involves conducting specific analyses, and so on. We virtually never do a research project in a vacuum. Every research project is undertaken with the realization that there was prior research that addressed some aspect of what we are looking at. Even if no one has previously looked at exactly the question we are investigating in a project, it's still likely that someone has previously looked at something similar, used similar data, or done similar analyses. We also know that every research project will have flaws and that no research project on its own is likely to provide a definitive answer to any truly important question. In every field, we conduct multiple research projects, addressing important issues, each project fallible and imperfect, and each one contributing to the broader accumulating knowledge base.

This text concentrates most on how you learn to conduct a research project, a specific investigation of a question of interest. But it is important that you understand the broader effort that each research project contributes to. We refer to that broader effort as the research enterprise. The **research enterprise** is the macro-level effort to accumulate knowledge across multiple empirical systematic public research

research A type of systematic investigation that is empirical in nature and is designed to contribute to public knowledge.

research enterprise The macro-level effort to accumulate knowledge across multiple empirical systematic public research projects.

projects (Sung et al., 2003). In the past few decades, as more research projects and studies have been done, we have become much more aware of this cross-project endeavor. This makes sense. After hundreds of years of conducting individual research studies and then series of studies, we are now finally turning our attention to the broader environment within which all this activity takes place. In the next few sections, we consider some of the most important aspects of this larger research enterprise, in order to provide a foundation for understanding how to conduct individual research projects—the central focus of the remainder of this text.

1.1b Translational Research

So, what are we doing all these research projects for? The traditional answer has been that we do research studies in order to contribute to our knowledge. This, of course, leads to the next obvious question: What are we accumulating knowledge for? Some would argue that we accumulate knowledge for its own sake. They would claim that not all knowledge has to be useful or lead to something. Sometimes when we learn something we cannot possibly anticipate how that knowledge could be used. A classic example is of the Post-It notes that are in almost every office. The creators of the Post-It note did not set out to create such objects. They were discovered at the 3M research laboratories in the 1970s when chemists were trying to create a new glue. The glue they created, however, didn't work as they had hoped. It stuck things like two pieces of paper together, but they could be pulled apart again, with the glue remaining only on the original sheet. It seemed like a totally useless type of glue until one of the researchers hit upon the idea that there are times when you want to be able to unstick two pieces of paper without doing any damage to either. The result was the Post-It note. Many of the major discoveries in research—penicillin, the telephone, Velcro—happened by accident. The research that led to them contributed to knowledge that was subsequently used in unanticipated ways. So, we accumulate knowledge with the idea that it may contribute some day to something we can use. In this sense, we are the toolmakers. Our research contributes to instrumental knowledge that we hope can make our lives or our world better. That is, knowledge gained from research may at some point be able to be put into practice.

When we move research from discovery to practice (and to the effects of that practice on our lives) we can say we are translating research into practice. **Translational research** is the systematic effort to move research from initial discovery to practice and ultimately to impacts on our lives. There are a wide variety of clever phrases that are used in various fields to convey the idea of translational research concisely, such as: from "bench to bedside"; from "bench to behavior"; from "the mind to the marketplace"; from "brain to vein"; and from "bench-to-practice-to-community," to name but a few. There are lots of different models of translational research that divide the process into stages in different ways (Dougherty & Conway, 2008; Khoury et al., 2007; Sung et al., 2003; Trochim, Kane, Graham, & Pincus, 2011; Westfall, Mold, & Fagnan, 2007), but all of them convey the central agenda of translational research: to move research from discovery to impact in the research enterprise.

We can think of the research enterprise as encompassing a **research-practice continuum** within which translation occurs. In the course of moving through this continuum it is likely that many individual research projects will be conducted. Some of these are what might be called **basic research** and are designed to generate discoveries and to understand their mechanisms better. For discoveries that relate to humans, this is usually followed by a series of **applied research** projects where the discovery is tested under increasingly controlled conditions with humans. If a

translational research The systematic effort to move research from initial discovery to practice and ultimately to impacts on our lives.

research-practice continuum The process of moving from an initial research idea or discovery to practice, and the potential for the idea to influence our lives or world.

basic research Research that is designed to generate discoveries and to understand how the discoveries work.

applied research Research where a discovery is tested under increasingly controlled conditions in real-world contexts.

Figure 1.1 Translational Research.

discovery survives this applied research testing, there is usually a process of seeing how well it can be implemented in and disseminated to a broad range of contexts that extend beyond the original controlled studies. This is sometimes referred to as **implementation and dissemination research**. Ultimately, many such discoveries are assessed for the impacts they have broadly on society, what might be termed **impact research**. Sometimes discoveries lead to the development of new policies that are investigated with **policy research** in the broader population. The research-practice continuum might be depicted as shown in **Figure 1.1**. It is assumed that different discoveries take different pathways through this continuum. Some take longer to go through one stage or another. The bidirectional arrow in the figure is meant to convey that the translational process works in both directions. Sometimes insights from practitioners and policy makers can inform basic and applied researchers and improve their ability to transform their discoveries to better anticipate the real-world contexts that they will eventually need to be implemented in.

1.1c Research Syntheses and Guidelines

Typically, during the testing of a new discovery during the basic and applied research period a number of separate research projects are likely to be conducted. In the past, it was assumed that implementers and practitioners of new discoveries would read the research journals to find new things that they could do to address their problem or issue of interest. But the research literature has become voluminous and is often very technical, making it a barrier for practitioners that reduces the rate of adoption of new discoveries.

To address this challenge over the past several decades the research enterprise has evolved a system for synthesizing the large numbers of research studies in different topical areas. In the next several decades, we expect that this system will increasingly become the normative way that research about new discoveries moves from the basic-applied stage to implementation and dissemination in broader contexts. For example, a recent Institute of Medicine report calls for the U.S. government to develop a national system for managing systematic reviews of research in health and biomedicine (Institute of Medicine, 2008).

A **research synthesis** is a systematic study of multiple prior research projects that address the same research question or topic and summarize the results in a manner that can be used by practitioners. There are two major types of research syntheses. A **meta-analysis** uses statistical methods to combine the results of similar studies quantitatively in order to allow general conclusions to be made. A **systematic review** is a research synthesis approach that focuses on a specific question or issue and uses specific preplanned methods to identify, select, assess, and summarize the findings of multiple research studies. It may or may not include a meta-analysis (a quantitative synthesis of results). Often, a systematic review involves a panel of experts who discuss the research literature and reach conclusions about how well a discovery works to address a problem or issue. So, while

implementation and dissemination research Research that assesses how well an innovation or discovery can be distributed in and carried out in a broad range of contexts that extend beyond the original controlled studies.

impact research Research that assesses the broader effects of a discovery or innovation on society.

policy research Research that is designed to investigate existing policies or develop and test new ones.

research synthesis A systematic study of multiple prior research projects that address the same research question or topic and that summarizes the results in a manner that can be used by practitioners.

meta-analysis A type of research synthesis that uses statistical methods to combine the results of similar studies quantitatively in order to allow general conclusions to be made.

systematic review A type of research synthesis that focuses on a specific question or issue and uses preplanned methods to identify, select, assess, and summarize the findings of multiple research studies.

a meta-analysis is always a quantitative synthesis, a systematic review may be a judgmental expert-driven synthesis, a meta-analysis, or both.

It turns out that even meta-analyses and systematic reviews are sometimes not by themselves sufficient to be used by practitioners as guides for how they might change what they implement. Both of these types of reviews can be somewhat technical and are written in a scientific style that typically is cautious about making formal recommendations for action. To help address this problem, the research enterprise has increasingly developed a mechanism called a practice guideline. A **guideline** is the result of a systematic process that leads to a specific set of research-based recommendations for practice that usually includes some estimates of how strong the evidence is for each recommendation.

It's important to recognize what a major shift this move to research syntheses and guidelines represents. Throughout the several hundred years of the evolution of research in our societies, the primary unit of a research project's results was the scientific publication. But this system has become unwieldy and has not been as useful as desired to those who might implement the results of research. Practitioners are flooded with new research studies and find it hard to keep up with the technical literature. So, research syntheses and guidelines represent a major effort of the research enterprise to deal with the results of research at a macro or systems level. This involves a major shift in the unit of what we are considering when moving from discovery to impact. During the basic and applied research phases, we are still focusing on the individual research project (this text is designed to introduce you to how to conduct such a project). But, the research synthesis stage introduces a major phase transition. After conducting a research synthesis, the unit of what practitioners will look at is the multistudy synthesis result, rather than the results of individual research projects. It is as if the research synthesis stage on the research-practice continuum acts like a sieve or strainer that combines previous multiple research projects and distills the core results that are needed to guide practice, as reflected in guidelines. We can add this idea into the earlier figure on translational research, as shown in **Figure 1.2**.

guideline A systematic process that leads to a specific set of research-based recommendations for practice that usually includes some estimates of how strong the evidence is for each recommendation.

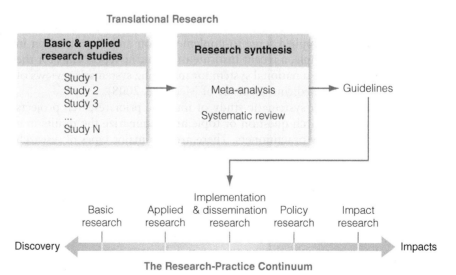

Figure 1.2 Translational research and the research-practice continuum with a system of research syntheses and guidelines included.

The figure shows the overarching idea of translational research across the research-practice continuum with the addition of a system for research synthesis and the development of practice guidelines interposed between basic and applied research and its subsequent implementation and dissemination. This represents a major change in the research enterprise in the first part of the twenty-first century.

1.1d Evidence-Based Practice

The interpositioning of a synthesis and guideline process in the middle of the research-practice continuum has transformed virtually every area of applied social research practice in our society. The term that is most associated with this change is evidence-based practice. It originated first in medicine as evidence-based medicine (Sackett, 1997) and then rapidly moved into other fields (Gibbs, 2003). Virtually every area of social practice today has an effort to integrate research and practice in this type of way. **Evidence-based practice (EBP)** is a movement designed to encourage or require practitioners to employ practices that are based on research evidence as reflected in research syntheses or practice guidelines. The EBP movement represents a major attempt of the research enterprise to achieve a better integration of research and practice. Primary examples of EBP efforts based on research syntheses include the Cochrane Collaboration in medicine (http://www.cochrane.org/), the Community Guide in public health (http://www.thecommunityguide.org/index.html), and the Campbell Collaboration (http://www.campbellcollaboration.org/) in education. Good examples of guideline clearinghouses include the National Guidelines Clearinghouse (http://www.guideline.gov/) in medicine and the What Works Clearinghouse in education (http://ies.ed.gov/ncee/wwc/). You may want to take a look at some of these to see how they are organized and what they address.

The EBP movement has not been without controversy. There are significant debates about what kinds of research projects should be allowed in research syntheses and guidelines, with different researchers and practitioners lining up in favor of or opposed to research studies that use certain types of methods (as described in this text). For instance, there are some who would only allow studies based on randomized experimental or strong quasi-experimental designs into the evidence base, while others would allow a broader range of studies to qualify as evidence. This introductory research methods text will provide you with a basic understanding of the strengths and weaknesses of different research designs, a necessary foundation for understanding these debates. In addition, some practitioners have questioned the primacy of the term "evidence" in EBP. They argue that it should not only be evidence that drives practice; it should also be that practice helps drive the development of evidence (Urban & Trochim, 2009). That is, they are arguing for something like a practice-based evidence movement (McDonald & Viehbeck, 2007). When all of these debates finally work their way through various fields, it is likely that we will have a transformed research enterprise with considerably stronger research-practice integration than in the past.

1.1e An Evolutionary Perspective on the Research Enterprise

As our societies have gradually become more aware of the dominance of research in all aspects of our lives, we have seen a corresponding shift in how we think about the research enterprise. Increasingly we view research as an evolutionary system. This view is based upon the idea of **evolutionary epistemology** which is the

evidence-based practice (EBP) A movement designed to encourage or require practitioners to employ practices that are based on research evidence as reflected in research syntheses or practice guidelines.

evolutionary epistemology The branch of philosophy that holds that ideas evolve through the process of natural selection.

branch of philosophy that holds that ideas evolve through the process of natural selection (Campbell, 1988; Popper, 1985). In this view, an individual discovery or idea is like an organism in biology. It competes with other established and emerging discoveries. It is selected for (or not) through a complex selection mechanism that involves the communities engaged in assessing the constantly emerging research projects in the research enterprise. Ultimately, any discovery has survival value; it either survives or it doesn't.

In this view, when you conduct a research study or project of the type described in this text, you are entering into this broader evolutionary system that constitutes the research enterprise. What you do may contribute ultimately to our knowledge base and influence the degree to which discoveries are taken up as part of the evidence base for practice. Our research-based knowledge evolves, just like everything else in nature, often in unpredictable and surprising ways. This broader perspective, taken together with the ideas of translational research, the research-practice continuum, and evidence-based practice, helps unify our thinking about the emerging research enterprise in the twenty-first century.

In the remainder of this chapter, we begin to focus on the primary topic of this book: how to develop a specific research project. But it is important to keep in mind throughout this text that we now view the kind of specific research project you'll learn how to do as one of many such projects that collectively act as sources of evidence within a larger and evolving research enterprise.

1.2 Conceptualizing Research

One of the most difficult aspects of research—and, surprisingly, one of the least discussed—is how to develop the idea for a research project in the first place. In training students, most faculty members simply assume that if students read enough of the research in an area of interest, they will somehow magically be able to produce sensible ideas for further research. Now, that may be true. And heaven knows that's the way researchers have been doing this higher education thing for some time now; but we probably could do a better job of helping our students learn how to formulate good research problems. One thing we can do (and some texts at least cover this at a surface level) is to give students a better idea of how professional researchers typically generate research ideas. Some of this is introduced in the discussion of problem formulation that follows.

1.2a Where Research Topics Come From

So how do researchers come up with the idea for a research project? Probably one of the most common sources of research ideas is the experience of *practical problems in the field*. Many researchers are directly engaged in social, health, or human service program implementation (or know practitioners who are) and come up with their ideas based on what they see happening around them. Others aren't directly involved in service contexts, but work with (or survey) people to learn what needs to be better understood. Many of the ideas would strike the outsider as silly or worse. For instance, in health services areas, there is great interest in the problem of back injuries among nursing staff. It's not necessarily the thing that comes first to mind when you think about the health care field; but if you reflect on it for a minute, it should be obvious that nurses and nursing staff do an awful lot of lifting while performing their jobs. They lift and push heavy equipment, and they lift and push heavy patients! If five or ten out of every 100 nursing staff were to strain their backs on average over the period of one year, the costs would be enormous and that's pretty much what's happening. Even minor injuries can result in increased absenteeism. Major ones can result in lost jobs and expensive medical bills. The nursing industry figures this problem costs tens of millions of dollars annually in increased health care. Additionally, the health care industry has developed a number of approaches, many of them educational, to try to reduce the scope and cost of the problem. So, even though it might seem trivial at first, many of the practical problems that arise in practice can lead to extensive research efforts.

Another source for research ideas is the *literature in your specific field*. Certainly, many researchers get ideas for research by reading the literature and thinking of ways to extend or refine previous research. Most journal articles include suggestions for further study in their conclusion or discussion sections. Another type of literature that acts as a source of good research ideas is the **requests for proposals (RFPs)** that are published by government agencies and some foundations and companies. These RFPs describe some problem that the agency would like researchers to address; they are virtually handing the researcher an idea. Typically, the RFP describes the problem that needs addressing, the contexts in which it operates, the approach they would like you to take to investigate the problem, and the amount they would be willing to pay for such research. Clearly, there's nothing like potential research funding to get researchers to focus on a particular research topic!

Finally, let's not forget the fact that many researchers simply *think up their research* topic on their own. Of course, no one lives in a vacuum, so you would expect that even the ideas you come up with on your own are influenced by your background, culture, education, and experiences.

1.2b The Literature Review

One of the most important early steps in a research project is the conducting of the literature review. The **literature review** is a systematic compilation and written summary of all of the literature published in scientific journals that is related to a research topic of interest. This is also one of the most humbling experiences you're likely to have. Why? Because you're likely to find out that just about any worthwhile idea you will have has been thought of before, at least to some degree. We frequently have students who come to us complaining that they couldn't find anything in the literature that was related to their topic. And virtually every time they have said that, we were able to show them that was only true because they only looked for articles that were *exactly* the same as their research topic. A literature review is designed to identify *related* research, to set the current research project within a conceptual and theoretical context. When looked at that way, almost no topic is so new or unique that you can't locate relevant and informative related research done previously.

Here are some tips about conducting the literature review. First, *concentrate your efforts on the research literature*. Try to determine what the most credible research journals are in your topical area and start with those. Put the greatest emphasis on research journals that use a blind or juried peer review system. In a **peer review,** authors submit potential articles to a journal editor who solicits several reviewers who agree to give a critical review of the paper. The paper is sent to these reviewers with no identification of the author so that there will be no personal bias (either for or against the author). Based on the reviewers' recommendations, the editor can accept the article, reject it, or recommend that the author revise and resubmit it. Articles in journals with peer review processes are likely to have a fairly high level of credibility. Second, *do the review early* in the research process. You are likely to learn a lot in the literature review that will help you determine what the necessary trade-offs are. After all, previous researchers also had to face trade-off decisions.

What should you look for in the literature review? First, you might be able to find a study that is quite similar to the one you are thinking of doing. Since all credible research studies have to review the literature themselves, you can check their literature review to get a quick start on your own. Second, prior research will help ensure that you include all of the major relevant constructs in your study. You may find that other similar studies routinely look at an outcome that you might not have included. Your study might not be judged credible if it ignored such a major construct. Third, the literature review will help you find and select appropriate measurement instruments. You will readily see what measurement instruments researchers used themselves in contexts similar to yours. Finally, the literature review will help you anticipate common problems in your research context. You can use the prior experiences of others to avoid common traps and pitfalls. Chapter 13 shows a sample research article with a brief literature review included.

1.2c Feasibility Issues

Soon after you get an idea for a study, reality begins to kick in and you begin to think about whether the study is feasible at all. Several major considerations come into play. Many of these involve making *trade-offs between rigor and practicality*. Performing a scientific study may force you to do things you wouldn't do normally. You might want to ask everyone who used an agency in the past year to fill in your evaluation survey only to find that there were thousands of people and it would be prohibitively expensive. Or, you might want to conduct an in-depth interview on your subject of interest only to learn that the typical participant in your study won't willingly take the hour that your interview requires. If you had unlimited resources and unbridled control over the circumstances, you would always be able to do the best-quality research; but those ideal circumstances seldom exist, and researchers are almost always forced to look for the best trade-offs they can find to get the rigor they desire.

When you are determining a research project's feasibility, you usually need to bear in mind several practical considerations. First, you have to think about *how long the research will take* to accomplish. Second, you have to question whether any important *ethical constraints* require consideration (see Chapter 2). Third, you must determine whether you can acquire the *cooperation* needed to take the project to its successful conclusion. And finally, you must determine the degree to which the costs will be manageable. Failure to consider any of these factors can mean disaster later.

1.3 The Language of Research

Learning about research is a lot like learning about anything else. To start, you need to learn the jargon people use, the big controversies they fight over, and the different factions that define the major players. Research blends an enormous range of skills and activities. Learning about research is a lot like learning a new language. You need to develop a specialized vocabulary of terms that describe the different types of research, the methods used, and the issues and problems that arise. You need to learn how to use those words correctly in a sentence. You need to understand the local idioms of this language. Just as in any language, if you aren't aware of the subtle way words are used, you run the risk of embarrassing yourself. To begin, we'll introduce some basic ideas like the types of studies you can perform, the role of time in research, and the different types of relationships you can learn about. Then we define some basic vocabulary terms like hypothesis, variable, data, and unit of analysis. Finally, we introduce several basic terms that describe different types of thinking in research.

1.3a Research Vocabulary

Just to get you warmed up to the idea that learning about research is in many ways like learning a new language, we want to introduce you to four terms that we think help describe some of the key aspects of contemporary social research. This list is far from exhaustive. It's really just the first four terms that came into our minds when we were thinking about research language. Think of these like a "word of the day" for the next four days. You might even try to work one term into your conversation each day (go ahead, we dare you), to become more at ease with this language.

We present the first two terms—theoretical and empirical—together because they are often contrasted with each other. Social research is **theoretical**, meaning that much of it is concerned with developing, exploring, or testing the theories or ideas that social researchers have about how the world operates. It is also **empirical**, meaning that it is based on observations and measurements of reality—on what you perceive of the world around you. You can even think of most research as a blending of these two terms—a comparison of theories about how the world operates with observations of its operation.

In the old days, many scientists thought that one major purpose of science was to measure what was "really there," and some believed that we could develop measuring instruments that were perfectly accurate. Alas, experience has shown that even the most accurate of our instruments and measurement procedures inevitably have some inaccuracy in them. Whether measuring the movement of subatomic particles or the height or weight of a person, there is some error in all measurement. Thus, the third big word that describes much contemporary social research is **probabilistic**, or based on probabilities. The inferences made in social research have probabilities associated with them; they are seldom if ever intended as covering laws that pertain to all cases with certainty. Part of the reason statistics has become so dominant in social research is that it enables the estimation of the probabilities for the situations being studied.

The last term we want to introduce is **causal**. You have to be careful with this term. Note that it is spelled *causal* not *casual*. You'll really be embarrassed if you write about the "casual hypothesis" in your study! (Beware the automatic spell checker). The term causal has to do with the idea of cause-and-effect (Cook & Campbell, 1979). A lot of social researchers are interested (at some point) in looking at a cause-effect or **causal relationship**. For instance, we might want to know whether a new program causes improved outcomes or performance. Now, don't get us wrong. There are lots of studies that don't look at cause-and-effect relationships. Some studies simply observe; for instance, a survey might be used to describe the percentage of people holding a particular opinion. Many studies explore relationships—for example, a study may attempt to determine whether

theoretical Pertaining to theory. Social research is theoretical, meaning that much of it is concerned with developing, exploring, or testing the theories or ideas that social researchers have about how the world operates.

empirical Based on direct observations and measurements of reality.

probabilistic Based on probabilities.

causal Pertaining to a cause-effect relationship, hypothesis, or relationship. Something is causal if it leads to an outcome or makes an outcome happen.

causal relationship A cause-effect relationship. For example, when you evaluate whether your treatment or program causes an outcome to occur, you are examining a causal relationship.

there is a relationship between gender and salary. Probably the vast majority of applied social research consists of these descriptive and correlational studies. So why are we talking about causal studies? Because for most social sciences, it is important to go beyond just passively observing the world or looking at relationships. You might like to be able to change the world, to improve it and help address some of its major problems. If you want to change the world (especially if you want to do this in an organized, scientific way), you are automatically interested in causal relationships—ones that tell how causes (for example, programs and treatments) affect the outcomes of interest.

1.3b Types of Studies

Research projects usually can be classified into one of three basic forms:

1. **Descriptive studies** are designed primarily to document what is going on or what exists. Public opinion polls that seek to describe the proportion of people who hold various opinions are primarily descriptive in nature. For instance, if you want to know what percentage of the population would vote for a Democrat or a Republican in the next presidential election, you are simply interested in describing something.
2. **Relational studies** look at the relationships between two or more variables. A public opinion poll that compares the proportion of males and females who say they would vote for a Democratic or a Republican candidate in the next presidential election is essentially studying the relationship between gender and voting preference.
3. **Causal studies** are designed to determine whether one or more variables (for example, a program or treatment variable) causes or affects one or more outcome variables. If you performed a public opinion poll to try to determine whether a recent political advertising campaign changed voter preferences, you would essentially be studying whether the campaign (cause) changed the proportion of voters who would vote Democratic or Republican (effect).

The three study types can be viewed as cumulative. That is, a relational study generally assumes that you can first describe (by measuring or observing) each of the variables you are trying to relate. A causal study generally assumes that you can describe both the cause-and-effect variables and that you can show that they are related to each other.

1.3c Time in Research

Time is an important element of any research design, and here we want to introduce one of the most fundamental distinctions in research design nomenclature: cross-sectional versus longitudinal studies. **Cross-sectional studies** take place at a single point in time. In effect, you are taking a slice or cross-section of whatever it is you're observing or measuring. **Longitudinal studies** take place over multiple points in time. In a longitudinal study, you measure your research participants on at least two separate occasions or at least two points in time. When you measure at different time points, we often say that you are measuring multiple waves of measurement. Just as with the repeated motion of the waves in the ocean or of waving with your hand, multiple waves of measurement refers to taking measurements on a variable several times.

descriptive studies A study that documents what is going on or what exists.

relational studies A study that investigates the connection between two or more variables.

causal studies A study that investigates a causal relationship between two variables.

cross-sectional study A study that takes place at a single point in time.

longitudinal A study that takes place over time.

A further distinction is made between two types of longitudinal designs: repeated measures and time series. There is no universally agreed-upon rule for distinguishing between these two terms; but in general, if you have two or a few waves of measurement, you are using a **repeated measures** design. If you have many waves of measurement over time, you have a **time series**. How many is many? Usually, you wouldn't use the term time series unless you had at least twenty waves of measurement. With fewer waves than that, you would usually call it a repeated measures design.

1.3d Types of Relationships

A relationship refers to the correspondence between two variables (see the next section on variables in this chapter). When you talk about types of relationships, you can mean that in at least two ways: the *nature* of the relationship or the *pattern* of it.

The Nature of a Relationship

We start by making a distinction between two types of relationships: a correlational relationship and a causal relationship. A correlational **relationship** simply says that two things perform in a synchronized manner. For instance, economists often talk of a correlation between inflation and unemployment. When inflation is high, unemployment also tends to be high. When inflation is low, unemployment also tends to be low. The two variables are correlated; but knowing that two variables are correlated does not tell whether one *causes* the other. For instance, there is a correlation between the number of roads built in Europe and the number of children born in the United States. Does that mean that if fewer children are desired in the United States there should be a cessation of road building in Europe? Or, does it mean that if there aren't enough roads in Europe, U.S. citizens should be encouraged to have more babies? Of course not. (At least, we hope not.) While there is a relationship between the number of roads built and the number of babies, it's not likely that the relationship is a causal one. A causal relationship is a synchronized relationship between two variables just as a correlational relationship is, but in a causal relationship we say that one variable *causes* the other to occur.

This leads to consideration of what is often termed the **third variable or missing variable problem**. In the example considered above, it may be that a third variable is causing both the building of roads and the birthrate and leading to the correlation that is observed. For instance, perhaps the general world economy is responsible for both. When the economy is good, more roads are built in Europe and more children are born in the United States. The key lesson here is that you have to be careful when you interpret correlations. If you observe a correlation between the number of hours students use the computer to study and their grade-point averages (with high computer users getting higher grades), you *cannot* assume that the relationship is causal—that computer use improves grades. In this case, the third variable might be socioeconomic status—richer students, who have greater resources at their disposal, tend to both use computers more and make better grades. Resources, the third variable, may drive both use and grades; computer use doesn't cause the change in the grade-point averages.

Patterns of Relationships

Several terms describe the major different types of patterns one might find in a relationship. First, there is the case of *no relationship* at all. If you know the values

repeated measures Two or more waves of measurement over time.

time series Many waves of measurement over time.

relationship An association between two variables such that, in general, the level on one variable is related to the level on the other. Technically, the term "correlational relationship" is redundant: a correlation by definition always refers to a relationship. However the term correlational relationship is used to distinguish it from the specific type of association called a causal relationship.

third variable or missing variable problem An unobserved variable that accounts for a correlation between two variables.

Figure 1.3 Graphs of the different types of relationships between two variables.

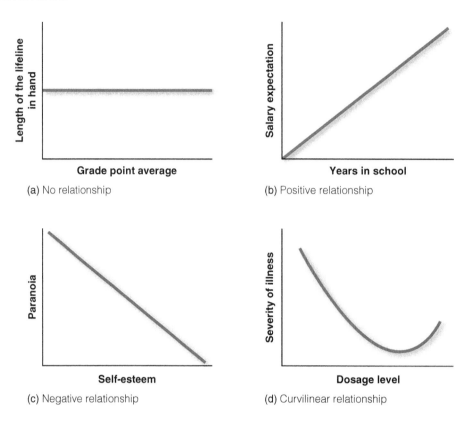

(a) No relationship

(b) Positive relationship

(c) Negative relationship

(d) Curvilinear relationship

on one variable, you don't know anything about the values on the other. For instance, we suspect that there is no relationship between the length of the lifeline on your hand and your grade-point average. If we know your GPA, we don't have any idea how long your lifeline is. Graph "a" in the upper left of **Figure 1.3** shows the case where there is no relationship.

Then, there is the **positive relationship**. In a positive relationship, high values on one variable are associated with high values on the other, and low values on one are associated with low values on the other. Graph "b" in the upper right of Figure 1.3 shows an idealized positive relationship between years of education and the salary one might expect to be making.

On the other hand, a **negative relationship** implies that high values on one variable are associated with low values on the other. This is also sometimes termed an inverse relationship. Graph "c" in the lower left of Figure 1.3 shows an idealized negative relationship between a measure of self-esteem and a measure of paranoia in psychiatric patients.

These are the simplest patterns of relationships that might typically be estimated in research. However, the pattern of a relationship can be more complex than this. For instance, Graph "d" in the lower right of Figure 1.3 shows a relationship that changes over the range of both variables, a curvilinear relationship. In this example, the horizontal axis represents the dosage of a drug for an illness and the vertical axis represents a severity of illness measure. As the dosage rises, the severity of illness goes down; but at some point, the patient begins to experience negative side effects associated with too high a dosage, and the severity of illness begins to increase again.

positive relationship A relationship between variables in which high values for one variable are associated with high values on another variable, and low values are associated with low values on the other variable.

negative relationship A relationship between variables in which high values for one variable are associated with low values on another variable.

1.3e Hypotheses

An **hypothesis** is a specific statement of prediction. It describes in concrete (rather than theoretical) terms what you expect to happen in your study. Not all studies have hypotheses. Sometimes a study is designed to be exploratory (see the section, Deduction and Induction, later in this chapter). There is no formal hypothesis, and perhaps the purpose of the study is to explore some area more thoroughly to develop some specific hypothesis or prediction that can be tested in future research. A single study may have one or many hypotheses.

Actually, whenever we talk about an hypothesis, we are really thinking simultaneously about *two* hypotheses. Let's say that you predict that there will be a relationship between two variables in your study. The way to set up the hypothesis test is to formulate two hypothesis statements: one that describes your prediction and one that describes all the other possible outcomes with respect to the hypothesized relationship. Your prediction might be that variable A and variable B will be related (in this example you don't care whether it's a positive or negative relationship). Then the only other possible outcome would be that variable A and variable B are *not* related. Usually, the hypothesis that you support (your prediction) is called the **alternative hypothesis**, and the hypothesis that describes the remaining possible outcomes is termed the **null hypothesis**. Sometimes a notation like H_A or H_1 is used to represent the alternative hypothesis or your prediction, and H_O or H_0 to represent the null case. You have to be careful here, though. In some studies, your prediction might well be that there will be no difference or change. In this case, you are essentially trying to find support for the null hypothesis and you are opposed to the alternative (Marriott, 1990).

If your prediction specifies a direction, the null hypothesis automatically includes both the no-difference prediction *and* the prediction that would be opposite in direction to yours. This is called a **one-tailed hypothesis**. For instance, let's imagine that you are investigating the effects of a new treatment for depression and that you believe one of the outcomes will be that there will be *less* depression. Your two hypotheses might be stated something like this:

The null hypothesis for this study is

H_O: As a result of the new program, there will either be no significant difference in depression or there will be a significant *increase,*

which is tested against the alternative hypothesis:

H_A: As a result of the new program, there will be a significant *decrease* in depression.

In **Figure 1.4**, this situation is illustrated graphically. The alternative hypothesis—your prediction that the program will decrease depression—is shown there. The null must account for the other two possible conditions: no difference, or an increase in depression. The figure shows a hypothetical distribution of depression difference scores. That is, a value of zero means that there has been no difference in depression observed, a positive value means that depression has increased, and a negative value means it has decreased. The term one-tailed refers to the tail of the distribution on the outcome variable.

hypothesis A specific statement of prediction.

alternative hypothesis A specific statement of prediction that usually states what you expect will happen in your study.

null hypothesis The hypothesis that describes the possible outcomes other than the alternative hypothesis. Usually, the null hypothesis predicts there will be no effect of a program or treatment you are studying.

one-tailed hypothesis A hypothesis that specifies a direction; for example, when your hypothesis predicts that your program will increase the outcome.

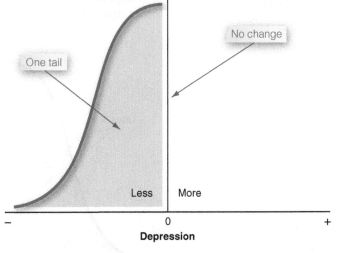

Figure 1.4 One-tailed hypothesis test.

When your prediction does *not* specify a direction, you have a **two-tailed hypothesis**. For instance, let's assume that you are studying a new drug treatment for depression. The drug has gone through some initial animal trials, but has not yet been tested on humans. You believe (based on theory and the previous research) that the drug will have an effect, but you are not confident enough to hypothesize a direction and say the drug will reduce depression. After all, you've seen more than enough promising drug treatments come along that eventually were shown to have severe side effects that actually worsened symptoms. In this case, you might state the two hypotheses like this:

The null hypothesis for this study is:

H_O: As a result of 300 mg/day of the ABC drug, there will be no significant difference in depression,

which is tested against the alternative hypothesis:

H_A: As a result of 300 mg/day of the ABC drug, there will be a significant difference in depression.

Figure 1.5 illustrates this two-tailed prediction for this case. Again, notice that the term two-tailed refers to the tails of the distribution for your outcome variable.

The important thing to remember about stating hypotheses is that you formulate your prediction (directional or not), and then you formulate a second hypothesis that is mutually exclusive of the first and incorporates all possible alternative outcomes for that case. When your study analysis is completed, the idea is that you will have to choose between the two hypotheses. If your prediction was correct, you would (usually) reject the null hypothesis and accept the alternative. If your original prediction was not supported in the data, you will accept the null hypothesis and reject the alternative. The logic of hypothesis testing (Marriott, 1990) is based on these two basic principles:

- Two mutually exclusive hypothesis statements that, together, exhaust all possible outcomes, need to be developed.
- The hypotheses must be tested so that one is necessarily accepted and the other rejected.

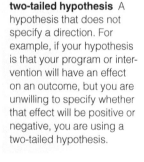

two-tailed hypothesis A hypothesis that does not specify a direction. For example, if your hypothesis is that your program or intervention will have an effect on an outcome, but you are unwilling to specify whether that effect will be positive or negative, you are using a two-tailed hypothesis.

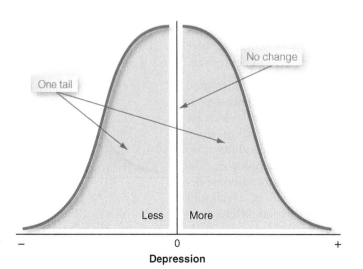

Figure 1.5 Two-tailed hypothesis test.

One tail

No change

Less   More

$-$ 0 $+$

Depression

Okay, we know it's a convoluted, awkward, and formalistic way to ask research questions, but it encompasses a long tradition in science and statistics called the **hypothetico-deductive model** (Nagel, 1979; Popper, 1959), and sometimes things are just done because they're traditions. And anyway, if all of this hypothesis testing was easy enough that anybody could understand it, how do you think statisticians and methodologists would stay employed?

1.3f Variables

You won't be able to do much in research unless you know how to talk about variables. A **variable** is any entity that can take on different values. Okay, so what does that mean? Anything that can vary can be considered a variable. For instance, *age* can be considered a variable because age can take different values for different people or for the same person at different times. Similarly, *country* can be considered a variable because a person's country can differ from another's, and each can be assigned a value.

Variables aren't always **quantitative** or numerical. The variable *gender* consists of two values expressed in words: *male* and *female*. These values of the variable gender can be called "text values" to differentiate them from numeric values. However, if it is useful, quantitative values like "1" for female and "2" for male can be assigned instead of (or in place of) the words. But it's not necessary to assign numbers for something to be a variable. It's also important to realize that variables aren't the only things that are measured in the traditional sense. For instance, in much social research and in program evaluation, the treatment or program (i.e., the "cause") is considered to be a variable. An educational program can have varying amounts of time on task, classroom settings, student-teacher ratios, and so on. Therefore, even the program can be considered a variable, which can be made up of a number of subvariables.

An **attribute** is a specific value on a variable. For instance, the variable *sex* or *gender* has two attributes: male and female, or, the variable *agreement* might be defined in a particular study as having five attributes:

1 = strongly disagree
2 = disagree
3 = neutral
4 = agree
5 = strongly agree

Another important distinction having to do with the term variable is the distinction between an independent and dependent variable. This distinction is particularly relevant when you are investigating cause-effect relationships. This can often be a tricky distinction to understand. (Are you someone who gets confused about the signs for arrivals and departures at airports—do you go to arrivals because you're arriving at the airport or does the person you're picking up go to arrivals because they're arriving on the plane?) Some people mistakenly think that an independent variable is one that would be free to vary or respond to some program or treatment, and that a dependent variable must be one that *depends* on your efforts (that is, it's the *treatment*). However, this is entirely backwards! In fact *the* **independent variable** *is what you (or nature) manipulates*—a treatment or program or cause. The **dependent variable** *is what you presume to be affected by the independent variable*—your effects or outcomes. For example, if you are studying the effects of a new educational program on student achievement, the program is the independent variable and your measures of achievement are the dependent ones. Or,

hypothetico-deductive model A model in which two mutually exclusive hypotheses that together exhaust all possible outcomes are tested, such that if one hypothesis is accepted, the second must therefore be rejected.

variable Any entity that can take on different values. For instance, age can be considered a variable because age can take on different values for different people at different times.

quantitative The numerical representation of some object. A quantitative variable is any variable that is measured using numbers.

attribute A specific value of a variable. For instance, the variable *sex* or *gender* has two attributes: male and female.

independent variable The variable that you manipulate. For instance, a program or treatment is typically an independent variable.

dependent variable The variable affected by the independent variable; for example, the outcome.

if you are looking at the effects of a new surgical treatment for cancer on rates of mortality for that cancer, the independent variable would be the surgical treatment and the dependent variable would be the mortality rates. The independent variable is what you (or nature) do, and the dependent variable is what results from that.

Finally, the attributes of a variable should be both exhaustive and mutually exclusive. Each variable's attributes should be **exhaustive**, meaning that they should include all possible answerable responses. For instance, if the variable is *religion* and the only options are *Protestant*, *Jewish*, and *Muslim*, there are quite a few religions we can think of that haven't been included. The list does not exhaust all possibilities. On the other hand, if you exhaust all the possibilities with some variables—religion being one of them—you would simply have too many responses. The way to deal with this is to list the most common attributes and then use a general category like *Other* to account for all remaining ones.

In addition to being exhaustive, the attributes of a variable should be **mutually exclusive**, meaning that no respondent should be able to have two attributes simultaneously. While this might seem obvious, it is often rather tricky in practice. For instance, you might be tempted to represent the variable *Educational Status* by asking the respondent to check one of the following response attributes: *High School Degree, Some College, Two-Year College Degree, Four-Year College Degree,* and *Graduate Degree*. However, these attributes are not mutually exclusive—a person who has a two-year or four-year college degree also could correctly check *some college*! In fact, if someone went to college, got a two-year degree, then got a four-year degree, they could check all three. The problem here is that you are asking the respondent to provide a single response to a set of attributes that are not mutually exclusive. But don't researchers often use questions on surveys that ask the respondent to check all that apply and then list a series of categories? Yes, but technically speaking, each of the categories in a question like that is its own variable and is treated dichotomously as either checked or unchecked—attributes that *are* mutually exclusive.

1.3g Types of Data

Data will be discussed in lots of places in this text, but here we just want to make a fundamental distinction between two types of data: qualitative and quantitative. Typically data are called quantitative if they are in numerical form and qualitative if they are not. Note that **qualitative data** could be much more than just words or text. Photographs, videos, sound recordings, and so on, can be considered qualitative data.

Personally, while we find the distinction between qualitative and **quantitative data** to have some utility, we think most people focus too much on the differences between them, and that can lead to all sorts of confusion. In some areas of social research, the qualitative-quantitative distinction has led to protracted arguments, with the proponents of each claiming the superiority of their kind of data over the other. The quantitative types argue that their data are hard, rigorous, credible, and scientific. The qualitative proponents counter that their data are sensitive, nuanced, detailed, and contextual.

For many of us in social research, this kind of polarized debate has become less than productive. Additionally, it obscures the fact that qualitative and quantitative data are intimately related to each other. *All quantitative data are based upon qualitative judgments; and all qualitative data can be summarized and manipulated numerically.* For instance, think about a common quantitative

exhaustive The property of a variable that occurs when you include all possible answerable responses.

mutually exclusive The property of a variable that ensures that the respondent is not able to assign two attributes simultaneously. For example, gender is a variable with mutually exclusive options if it is impossible for the respondents to simultaneously claim to be both male and female.

qualitative data Data in which the variables are not in a numerical form, but are in the form of text, photographs, sound bites, and so on.

quantitative data Data that appear in numerical form.

measure in social research—a self-esteem scale. The most common such scales use simple text items like "I feel good about myself" and have respondents rate them on a 1-to-5 scale where 1 = strongly disagree and 5 = strongly agree. We add up the responses to all of these items to get a total self-esteem score. Because the measure ultimately yields a number, it is considered a quantitative measure. But the researchers who developed such instruments had to make countless qualitative judgments in constructing them: how to define self-esteem; how to distinguish it from other related concepts; how to word potential scale items; how to make sure the items would be understandable to the intended respondents; what kinds of contexts they could be used in; what kinds of cultural and language constraints might be present, and so on. Researchers who decide to use such a scale in their studies have to make another set of judgments: how well the scale measures the intended concept; how reliable or consistent it is; how appropriate it is for the research context and intended respondents, and so on. Believe it or not, even the respondents make many judgments when filling out such a scale: what various terms and phrases mean; why the researcher is giving this scale to them; how much energy and effort they want to expend to complete it, and so on. Even the consumers and readers of the research make judgments about the self-esteem measure and its appropriateness in that research context. What may look like a simple, straightforward, cut-and-dried quantitative measure is actually based on lots of qualitative judgments made by many different people.

On the other hand, all qualitative information can be easily converted into quantitative, and many times doing so would add considerable value to your research. The simplest way to do this is to divide the qualitative information into categories and number them! We know that sounds trivial, but even simply assigning a number to each category can often enable you to organize and process qualitative information more efficiently. Perhaps a more typical example of converting qualitative data into quantitative is when we do a simple content coding. For example, imagine that you have a written survey and as the last question you ask the respondent to provide any additional written comments they might wish to make. What do you do with such data? A straightforward approach would be to read through all of the comments from all respondents and, as you're doing so, develop a list of categories into which they can be classified. Once you have a simple classification or "coding" scheme you can go back through the statements and assign the best code to each specific comment. If you use a computer to analyze these comments, you might summarize the results by counting the number of comments in each category. Or, you might use percentages to describe what percent of all comments each category constitutes. This is a simple example of coding qualitative data so that they can be summarized quantitatively. There are more sophisticated approaches for analyzing qualitative data quantitatively (see, for example, the discussion of content analysis in Chapter 3), but the essential point should be clear—qualitative and quantitative data are intimately related and we often move from one form to another in the course of a research project.

1.3h The Unit of Analysis

One of the most important ideas in a research project is the **unit of analysis**. The unit of analysis is whatever entity you are analyzing in your study. For instance, any of the following could be a unit of analysis in a study:

- Individuals
- Groups

unit of analysis The entity that you are analyzing in your analysis; for example, individuals, groups, or social interactions.

- Artifacts (books, photos, newspapers)
- Geographical units (town, census tract, state)
- Social interactions (dyadic relations, divorces, arrests)

For instance, if you conduct a survey where you ask individuals to tell you their opinions about something, and you combine their responses to get some idea of what the "typical" individual thinks, your unit of analysis is the individual. On the other hand, if you collect data about crime rates in major cities in the country, your unit of analysis would be the city. Why is it called the unit of analysis and not something else (like, the unit of sampling)? Because *it is the analysis you do in your study that determines what the unit is*. For instance, if you are comparing the children in two classrooms on achievement test scores, the unit is the individual child because you have a score for each child. On the other hand, if you are comparing the two classes on noise levels, your unit of analysis is the group, in this case the classroom, because you will measure noise for the class as a whole, not separately for each individual student.

For different analyses in the same study, you may have different units of analysis. If you decide to base an analysis on student scores, the individual student is the unit. However, you might decide to compare average achievement test performance for the students with a classroom climate score. In this case, the data that go into the analysis include a variable (achievement) where the student is the unit of analysis and a variable (classroom climate) where the classroom is. In many areas of social research, these hierarchies of analysis units have become particularly important and have spawned a whole area of statistical analysis referred to as **hierarchical modeling**. In education, for instance, where a researcher might want to compare classroom climate data with individual student-level achievement data, hierarchical modeling allows you to include data at these two different levels within the same analysis without averaging the individual student data first.

hierarchical modeling A statistical model that allows for the inclusion of data at different levels, where the unit of analysis at some levels is nested within the unit of analysis at others (e.g., student within class within school within school district)

deductive Top-down reasoning that works from the more general to the more specific.

1.3i Deduction and Induction

In logic, a distinction is often made between two broad methods of reasoning known as the deductive and inductive approaches.

Deductive reasoning works from the more general to the more specific (see **Figure 1.6**). Sometimes this is informally called a top-down approach. You might

Figure 1.6 Deductive reasoning.

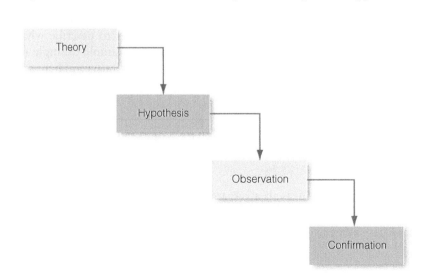

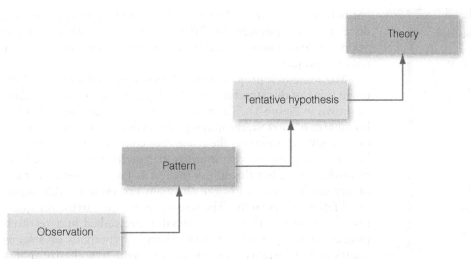

Figure 1.7 Inductive reasoning.

begin with a *theory* about your topic of interest. You then narrow that down into more specific hypotheses that you can test. You narrow down even further when you collect *observations* to address the hypotheses. This ultimately leads you to be able to test the hypotheses with specific data—a *confirmation* (or not) of your original theories.

Inductive reasoning works the other way, moving from specific observations to broader generalizations and theories (see **Figure 1.7**). Informally, this is sometimes called a bottom-up approach. (Please note that it's bottom-up and *not* "Bottoms up!", which is the kind of thing the bartender says to customers when he's trying to close for the night!) In inductive reasoning, you begin with specific observations and measures, detect initial patterns and regularities, formulate some tentative hypotheses that you can explore, and finally end up developing some general conclusions or theories.

These two methods of reasoning have a different feel to them when you're conducting research. Inductive reasoning, by its nature, is more open-ended and exploratory, especially at the beginning. Deductive reasoning is narrower in nature and is concerned with testing or confirming hypotheses. Even though a particular study may look like it's purely deductive (for example, an experiment designed to test the hypothesized effects of some treatment on some outcome), most social research involves both inductive and deductive reasoning processes at some time in the project. In fact, it doesn't take a rocket scientist to see that you could assemble the two graphs from Figures 1.5 and 1.6 into a single circular one that continually cycles from theories down to observations and back up again to theories. Even in the most constrained experiment, the researchers might observe patterns in the data that lead them to develop new theories.

inductive Bottom-up reasoning that begins with specific observations and measures and ends up as general conclusion or theory.

1.4 The Structure of Research

You probably think of research as something abstract and complicated. It can be, but you'll see (we hope) that if you understand the basic logic or rationale that underlies research, it's not nearly as complicated as it might seem at first glance.

A research project has a well-known structure: a beginning, middle, and end. We introduce the basic stages of a research project in the following section titled

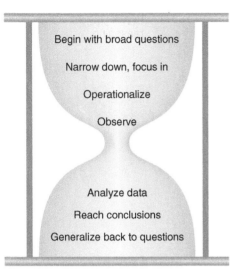

Begin with broad questions

Narrow down, focus in

Operationalize

Observe

Analyze data

Reach conclusions

Generalize back to questions

Figure 1.8 The "hourglass" structure of research.

"Structure of Research." Here, we also introduce some important distinctions in research: the different types of questions you can ask in a research project; and the major components or parts of a research project.

Most research projects share the same general structure. You might think of this structure as following the shape of an hourglass as shown in **Figure 1.8**. The research process usually starts with a broad area of interest, the initial problem that the researcher wishes to study. For instance, the researcher could want to investigate how to use computers to improve the performance of students in mathematics; but this initial interest is far too broad to study in any single research project. (It might not even be addressable in a lifetime of research.) The researcher has to narrow the question down to one that can reasonably be studied in a research project. This might involve formulating a hypothesis or a focus question. For instance, the researcher might hypothesize that a particular method of computer instruction in math will improve the ability of elementary school students in a specific district. At the narrowest point of the research hourglass, the researcher is engaged in direct measurement or observation of the question of interest. This is the point at which the "rubber hits the road" and the researcher is most directly involved in interacting with the environment within which the research is being conducted.

Once the basic data are collected, the researcher begins trying to understand it, usually by analyzing it in a variety of ways. Even for a single hypothesis, there are a number of analyses a researcher might typically conduct. At this point, the researcher begins to formulate some initial conclusions about what happened as a result of the computerized math program. Finally, the researcher often will attempt to address the original broad question of interest by generalizing from the results of this specific study to other related situations. For instance, on the basis of strong results indicating that the math program had a positive effect on student performance, the researcher might suggest that other school districts similar to the one in the study might expect similar results.

Notice that both ends of the hourglass represent the realm of ideas and the research questions that guide the project. The hourglass center is the most concrete or specific part of the process. The parts in between show how we translate the research questions into procedures for measurement (top part of the hourglass) and how we translate the data we observe into conclusions and new or revised questions (bottom part).

1.4a Components of a Research Study

What are the basic components or parts of a research study? Here, we describe the basic components involved in a causal study. Remember that earlier in the chapter (see the section, Types of Studies), you learned that causal studies build on descriptive and relational questions; therefore, many of the components of causal studies will also be found in descriptive and relational studies.

Most social research originates from some general problem or question. You might, for instance, be interested in examining which programs help to prevent and reduce childhood obesity (Foster et al., 2008). Usually, the problem is broad enough that you could not hope to address it adequately in a single research study. Consequently, the problem is typically narrowed down to a more specific **research question** that can be addressed. The research question is often stated in

research question The central issue being addressed in the study, typically phrased in the language of theory.

the context of some theory that has been advanced to address the problem. For instance, you might have a theory that school-based interventions can lead students to make healthier food choices. The research question is the central issue being addressed in the study and is often phrased in the language of theory. The research question might at this point be:

> Are school-based interventions more effective (as compared to no such interventions) in reducing childhood obesity?

The problem with such a question is that it is still too general to be studied directly. Consequently, in much research, an even more specific statement, called a hypothesis is developed that describes in *operational* terms exactly what you think will happen in the study (see the section, Hypotheses, earlier in this chapter). For instance, the hypothesis for your study might be something like the following:

> Schools that integrate nutritional education in their curriculum will see a significant decrease in the proportion of overweight children as compared with schools that do not adopt such a program.

Notice that this hypothesis is specific enough that a reader can understand quite well what the study is trying to assess.

In causal studies, there are at least two major variables of interest: the cause and the effect. Usually the cause is some type of event, program, or treatment (the cause is also sometimes called the independent variable, as mentioned in the section on "variables"). A distinction is made between causes that the researcher can control (such as a program) versus causes that occur naturally or outside the researcher's influence (such as a change in interest rates, recessions, or the occurrence of an earthquake). The effect (or dependent variable) is the outcome that you wish to study.

For both the cause and effect, a distinction is made between the idea of the cause or effect (the constructs) and how they are actually manifested in reality. For instance, when you think about school-based interventions for preventing and reducing childhood obesity, you are thinking of the construct. On the other hand, the real world is not always what you think it is. In research, a distinction is made between your view of an entity (the construct) and the entity as it exists (the **operationalization**). Ideally, the two should agree, but in most situations, the reality falls short of your ideal.

Social research is always conducted in a social context. Researchers ask people questions, observe families interacting, or measure the opinions of people in a city. The units that participate in the project are important components of any research project. Units are directly related to sampling. Note that there is a distinction between units (the participants in your study) and the units of analysis (as described earlier) in any particular analysis. In most projects, it's not possible to involve everyone it might be desirable to involve. For instance, in studying school-based interventions for childhood obesity, you can't possibly include in your study every student in the world, or even in the country. Instead, you have to try to obtain a representative sample of such people. When sampling, a distinction is made between the theoretical population of interest and the final sample that is actually included in the study. Usually the term *units* refers to the *people* that are sampled and from whom information is gathered; but for some projects the units are organizations, groups, or geographical entities like cities or towns. Sometimes the sampling strategy is multilevel; a number of cities are selected and within them families are sampled. Sampling will be discussed in greater detail in Chapter 4.

operationalization The act of translating a construct into its manifestation—for example, translating the idea of your treatment or program into the actual program, or translating the idea of what you want to measure into the real measure. The result is also referred to as an *operationalization;* that is, you might describe your actual program as an *operationalized program.*

In causal studies, the interest is in the effects of some cause on one or more *outcomes*. The outcomes are directly related to the research problem; usually the greatest interest is in outcomes that are most reflective of the problem. In the hypothetical childhood obesity study, you would probably be most interested in measures like the student's Body Mass Index, dietary intake, physical activity levels, and so on. Finally, in a causal study, the effects of the cause of interest (for example, the program) are usually compared to other conditions (for example, another program or no program at all). Thus, a key component in a causal study concerns how you decide which units (people) receive the program and which are placed in an alternative condition. This issue is directly related to the research design that you use in the study. One of the central themes in research design is determining how people wind up in or are placed in various programs or treatments that you are comparing. Different types of research designs will be explored in Chapters 8 through 10.

These, then, are the major components in a causal study:

- The research problem
- The research question
- The program (cause)
- The units
- The outcomes (effect)
- The design

1.5 The Validity of Research

Quality is one of the most important issues in research. **Validity** is a term that we use to discuss the quality of various conclusions you might reach based on a research project. Here's where we have to give you the pitch about validity. When we mention validity, most students roll their eyes, curl up into a fetal position, or go to sleep. They think validity is just something abstract and philosophical (and at some level it is). But we think if you can understand *validity*—the principles that are used to judge the quality of research—you'll be able to do much more than just complete a research project. You'll be able to be a virtuoso at research because you'll have an understanding of *why* you need to do certain things to ensure quality. You won't just be plugging in standard procedures you learned in school—sampling method X, measurement tool Y—you'll be able to help create the next generation of research methodologies.

Validity is technically defined as "the best available approximation to the truth or falsity of propositions, including propositions about cause" (Cook & Campbell, 1979, p. 37). What does this mean? The first thing to ask is: "validity of *what?*" When people think about validity in research, they typically think in terms of research components. You might say that a measure is a valid one, or that a valid sample was drawn, or that the design had strong validity; but all of those statements are technically incorrect. Measures, samples, and designs don't *have* validity—only propositions can be said to be valid. Technically, you should say that a measure leads to valid conclusions or that a sample enables valid inferences, and so on. Validity is relevant to a proposition, inference, or conclusion.

Researchers make lots of different inferences or conclusions while conducting research. Many of these are related to the process of doing research and are not the major questions or hypotheses of the study. Nevertheless, like the bricks

validity The best available approximation of the truth of a given proposition, inference, or conclusion.

that go into building a wall, these intermediate processes and methodological propositions provide the foundation for the overall conclusions that they wish to address. For instance, virtually all social research involves measurement or observation, and, no matter what researchers measure or observe, they are concerned with whether they are measuring what they intend to measure or with how their observations are influenced by the circumstances in which they are made. They reach conclusions about the quality of their measures—conclusions that will play an important role in addressing the broader substantive issues of their study. When researchers talk about the validity of research, they are often referring to the many conclusions they reach about the quality of different parts of their research methodology.

Validity is typically subdivided into four types (Cook & Campbell, 1979). Each type addresses a specific methodological question. To understand the types of validity, you have to know something about how researchers investigate a research question. We will use a causal study as an example because only causal questions involve all four validity types.

Figure 1.9 shows that two realms are involved in research. The first, on the top, is the land of theory. It is what goes on inside your head. It is where you keep your theories about how the world operates. The second, on the bottom, is the land of observations. It is the real world into which you translate your ideas: your programs, treatments, measures, and observations. When you conduct research, you are continually moving back and forth between these two realms, between what you think about the world and what is going on in it. When you are investigating a cause-effect relationship, you have a theory (implicit or otherwise) of what the cause is (the **cause construct**). For instance, if you are testing a new educational program, you have an idea of what it would look like ideally. Similarly, on the effect side, you have an idea of what you are ideally trying to affect and measure (the **effect construct**). But each of these—the cause and the effect—have to be translated into real things, into a program or treatment and a measure or observational method. The term operationalization is used to describe the act of translating a construct into its manifestation. In effect, you take your idea and describe it as a series of operations or procedures. Now, instead of it only being an idea in your mind, it becomes a public entity that others can look at and examine for themselves. It is one thing, for instance, for you to say that you would like to measure self-esteem (a construct). But when you show a ten-item, paper-and-pencil self-esteem measure that you developed for that purpose, others can look at it and understand more clearly what you mean by the term self-esteem.

Now, back to explaining the four validity types. They build on one another, with two of them (conclusion and internal) referring to the land of observation on the bottom of Figure 1.9, one of them (construct) emphasizing the linkages between the bottom and the top, and the last (external) being primarily concerned about the range of the theory on the top.

Imagine that you want to examine whether use of an Internet virtual classroom improves student understanding of course material. Assume that you took

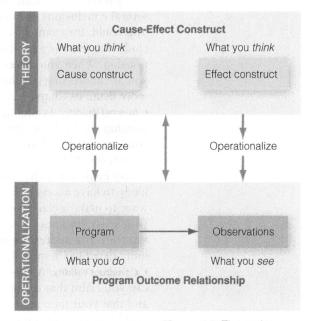

Figure 1.9 The major realms and components of research.

cause construct The abstract idea or theory of what the cause is in a cause-effect relationship you are investigating.

effect construct The abstract idea or theory of what the outcome is in a cause-effect relationship you are investigating.

these two constructs, the cause construct (the website) and the effect construct (understanding), and operationalized them, turning them into realities by constructing the website and a measure of knowledge of the course material. Here are the four validity types and the question each addresses:

- **Conclusion Validity:** In this study, is there a relationship between two variables? For the example at hand, this question might be worded: in this study, is there a relationship between the website and knowledge of course material? There are several conclusions or inferences you might draw to answer such a question. You could, for example, conclude that there is a relationship. You might conclude that there is a positive relationship. You might infer that there is no relationship. When you assess the validity of each of these conclusions or inferences, you are addressing conclusion validity. Conclusion validity will be discussed in more detail in Chapter 11.
- **Internal Validity:** Assuming that there is a relationship in this study, is the relationship a *causal* one? Just because you find that use of the website and knowledge are correlated, you can't necessarily assume that the website use **causes** the knowledge. Both could, for example, be caused by the same factor. For instance, it may be that wealthier students, who have greater resources, would be more likely to have access to a website and would excel on objective tests. When you want to make a claim that your program or treatment caused the outcomes in your study, and not something else, you are assessing the validity of a causal assertion and addressing internal validity. Internal validity will be discussed in more detail in Chapter 8.
- **Construct Validity:** Assuming that there is a causal relationship in this study, can you claim that the program reflected well your *construct* of the program and that your measure reflected well your idea or *construct* of the measure? In simpler terms, did you implement the program you intended to implement, and did you measure the outcome you wanted to measure? In yet other terms, did you operationalize well the ideas of the cause and the effect? Is the website what you intended it would be? Does it look like and work the way you theoretically imagined it would? Does it have the content you thought it should? When your research is over, you would like to be able to conclude that you did a credible job of operationalizing your constructs—that you can provide evidence for the construct validity of such a conclusion. Construct validity will be discussed in more detail in Chapter 5.
- **External Validity:** Assuming that there is a causal relationship in this study between the constructs of the cause and the effect, can you generalize this effect to other persons, places, or times? Would a web-based virtual classroom work with different target groups at different times, on different subject matters? You are likely to make some claims that your research findings have implications for other groups and individuals in other settings and at other times. When you do, you need to address the external validity of these claims. External validity will be discussed in more detail in Chapter 4.

Notice how the question that each validity type addresses presupposes an affirmative answer to the previous one. This is what we mean when we say that the validity types build on one another. **Figure 1.10** shows the idea of the cumulativeness of validity as a staircase, along with the key question for each validity type.

For any inference or conclusion, there are always possible **threats to validity**. Here's the logic. You reach a conclusion about some aspect of your study. There are lots of reasons why you might be wrong in reaching this conclusion. These

conclusion validity The degree to which conclusions you reach about relationships in your data are reasonable.

internal validity The approximate truth about inferences regarding cause-effect or causal relationships.

construct validity The degree to which inferences can legitimately be made from the operationalizations in your study to the theoretical constructs on which those operationalizations are based.

external validity The degree to which the conclusions in your study would hold for other persons in other places and at other times.

threats to validity Reasons your conclusion or inference might be wrong.

The Validity Questions are Cumulative

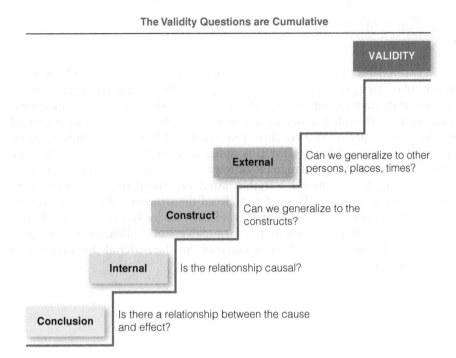

Figure 1.10 The validity staircase, showing the major question for each type of validity.

VALIDITY

External — Can we generalize to other persons, places, times?

Construct — Can we generalize to the constructs?

Internal — Is the relationship causal?

Conclusion — Is there a relationship between the cause and effect?

are the threats to the validity of your conclusion. Some are more reasonable or plausible, others more far-fetched. You essentially want to rule out the plausible threats to validity or "alternative explanations," thereby leaving your explanation as the most reasonable one remaining.

For instance, imagine a study examining whether there is a relationship between the amount of training in a specific technology and subsequent rates of use of that technology. Because the interest is in a relationship, it is considered an issue of conclusion validity. Assume that the study is completed and no significant correlation between amount of training and adoption rates is found. On this basis, it is *concluded* that there is no relationship between the two. How could this conclusion be wrong—that is, what are the threats to conclusion validity? For one, it's possible that there isn't sufficient statistical power to detect a relationship even if it exists. Perhaps the sample size is too small or the measure of amount of training is unreliable. Or maybe assumptions of the correlational test are violated, given the variables used. Perhaps there were random irrelevancies in the study setting or random heterogeneity in the respondents that increased the variability in the data and made it harder to see the relationship of interest (even if there was one there). The inference that there is no relationship will be stronger—have greater conclusion validity—if you can show that these alternative explanations are not credible. The distributions might be examined to see whether they conform to assumptions of the statistical test, or analyses conducted to determine whether there is sufficient statistical power. The concept of statistical power will be explained more thoroughly in Chapter 11.

The theory of validity and the many lists of specific threats provide a useful scheme for assessing the quality of research conclusions. The theory is general in scope and applicability, well-articulated in its philosophical suppositions, and virtually impossible to explain adequately in a few minutes. As a framework for judging the quality of research, it is indispensable and well worth understanding.

SUMMARY

We've covered a lot of territory in this initial chapter, grouped roughly into five broad areas. First, we began by looking at the big picture covering some of the most important emerging concepts in research today, including translational research, the research-practice continuum, research synthesis, evidence-based practice, and a perspective on how research-based knowledge evolves. Next, we briefly considered how research is thought up or conceptualized. Then we turned our attention to the language of research and enhanced your vocabulary by introducing key terms like variable, attribute, causal relationship, hypothesis, and unit of analysis. We next considered the structure of research and the major components of a research project. Finally, we looked at the critical topic of validity in research. It certainly is a formidable and challenging set of topics. But, it provides you with the basic foundation for the material that's coming in subsequent chapters.

Key Terms

alternative hypothesis p. 17
applied research p. 6
attribute p. 19
basic research p. 6
causal p. 13
causal relationship p. 13
causal studies p. 14
cause construct p. 27
conclusion validity p. 28
construct validity p. 28
cross-sectional studies p. 14
deductive p. 22
dependent variable p. 19
descriptive studies p. 14
effect construct p. 27
empirical p. 13
evidence-based practice (EBP) p. 9
evolutionary epistemology p. 9
exhaustive, p. 20
external validity p. 28
guideline p. 8
hierarchical modeling p. 22

hypothesis p. 17
hypothetico-deductive
    model p. 19
impact research p. 7
implementation and
    dissemination research p. 7
independent variable p. 19
inductive p. 23
internal validity p. 28
literature review p. 11
longitudinal studies p. 14
meta-analysis p. 7
mutually exclusive p. 20
negative relationship p. 16
null hypothesis p. 17
one-tailed hypothesis p. 17
operationalization p. 25
peer review p. 11
policy research p. 7
positive relationship p. 16
probabilistic p. 13
quantitative p. 19

quantitative data, p. 20
relational studies p. 14
relationship p. 15
repeated measures p. 15
requests for proposals
    (RFPs) p. 11
research p. 5
research enterprise p. 5
research question p. 24
research synthesis p. 7
research-practice continuum p. 6
systematic review p. 7
theoretical p. 13
third variable or missing variable
    problem p. 15
threats to validity p. 28
time series p. 15
translational research p. 6
two-tailed hypothesis p. 18
unit of analysis p. 21
validity p. 26
variable p. 19

Suggested Websites

The National Library of Medicine PubMed Tutorial.

This web-based tutorial will show you how to search PubMed®, the National Library of Medicine's (NLM™) journal literature search system.

http://www.nlm.nih.gov/bsd/pubmed_tutorial/m1001.html

The Campbell Collaboration.

This international collaboration of social researchers addresses "big picture" questions relevant to public policy by systematically analyzing the results of all available studies in order to generate the best available evidence.

http://www.campbellcollaboration.org

Review Questions

1. When researchers claim that their research is "empirical," they are stating that it is

a. concerned with developing, exploring, or testing theories.
b. based on observations and measurements of reality.
c. concerned with rules that pertain to general cases.
d. concerned with rules that pertain to individual cases.
(Reference: 1.1a)

2. Research is a type of systematic investigation that is _____ in nature and is designed to contribute to _____.

a. theoretical, the Gross National Product
b. theoretical, esoteric books
c. empirical, scientific databases
d. empirical, public knowledge
(Reference: 1.1a)

3. When conducting a survey that examines peoples' opinions about whether the electoral college should continue to be used to select the president of the United States, what type of research is involved?

a. descriptive
b. ecological
c. relational
d. causal
(Reference: 1.3b)

4. Stella Chess and Alexander Thomas have studied temperament for over thirty years, following the same individuals from infancy through adulthood. This is an example of what type of study?

a. cross-sectional study
b. longitudinal study
c. multiple measures study
d. cross-causal study
(Reference: 1.3c)

5. The two main forms of research synthesis are:

a. Case Study and Quasi-Experiment
b. Practice Guidelines and Cross-Case Analysis
c. Meta-analysis and Systematic Review
d. Grounded Theory and Quasi-Synthesis
(Reference: 1.1c)

6. Researchers investigating the effects of aging on cognition have found that, as we get older, our short-term memory deteriorates. This is an example of what kind of pattern of relationship if we are measuring age and short-term memory?

a. a curvilinear relationship
b. a positive relationship
c. a negative relationship
d. no relationship
(Reference: 1.3d)

7. In an experiment, which variable measures the outcome?

a. quantitative variable
b. qualitative variable
c. independent variable
d. dependent variable
(Reference: 1.3f)

8. _____ is a movement designed to encourage or require practitioners to employ practices that have research evidence as reflected through systematically conducted research syntheses or practice guidelines.

a. Researchers Without Borders
b. Evidence-based practice (EBP)
c. The Bayesian Movement
d. Occupy Bethesda
(Reference: 1.1d)

9. If we were conducting an experiment to test a hypothesis that people with religious backgrounds engage in more altruistic behavior than nonreligious people, what would the null hypothesis be?

a. People with less religious backgrounds engage in less altruistic behavior than people with nonreligious histories, or there is no difference between the groups with respect to altruistic behavior.
b. People with religious backgrounds are more altruistic than people with nonreligious backgrounds.
c. People with nonreligious backgrounds are more altruistic than people with religious backgrounds.
d. There would be no null hypothesis for this experiment.
(Reference: 1.3e)

10. If the data are expressed numerically, they are considered

a. qualitative.
b. quantitative.
c. independent.
d. causal.
(Reference: 1.3g)

11. In a research journal, you read an article about final exam performance of self-described "morning" people versus "night" people who take an 8 A.M. class. In this study, the unit of analysis is

a. the class as a whole.
b. the "morning group" versus the "night group."
c. individual students in the class.
d. impossible to quantify accurately.
(Reference: 1.3h)

12. Since research projects generally share the same structure, one resembling an hourglass, the narrowest point of focus occurs when the researcher is engaged in

a. narrowing down the question.
b. operationalizing the question.
c. direct observations or measurements.
d. drawing conclusions.
(Reference: 1.4)

13. In a causal study, the treatment or program variable would be accurately called the

a. independent variable.
b. exhaustive variable.
c. dependent variable.
d. mutually exclusive variable.
(Reference: 1.4a)

14. If two variables are correlated, it does not mean that one causes the other.

a. True
b. False
(Reference: 1.3d)

15. The term "translational research" refers to attempts to translate a measure into various other languages.

a. True
b. False
(Reference: 1.1b)

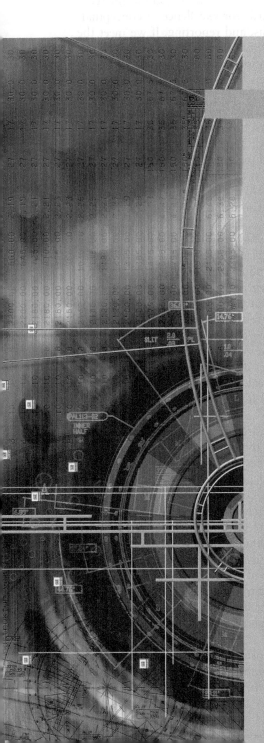

2

Ethics

2.1 Foundations of Ethics in Research

This book has the goal of helping you become a "good" researcher. This means that we hope you will end up with an understanding of what good research is, recognize it when you see it, and ultimately learn to create good research yourself.

If you think about it, there are at least three ways that research can be "good" (Gardner, Csikszentmihalyi, & Damon, 2002). First, it can be good in the technical sense; that is, it meets the current standards for excellence in conceptualization, sampling, measurement, design, analysis, and reporting. If we meet the technical requirements, then we can have more confidence that our research has validity. Since validity is "our closest approximation to truth," then this is clearly one of the main ways that research can aspire to be good. Most of this book is about this particular attribute of good research.

Second, it can be a good experience for you as a person, providing satisfaction from your learning and working with others, creating something you can be proud of, and perhaps becoming your life's work. Quite often researchers become interested in a topic because of their own lived experience. They may have observed interesting or troubling phenomena and want to understand them better. In applied research, the goal may be to find ways of making a difference on such a personally important issue, which can be a great source of meaning and satisfaction for a researcher.

Third, it can be good in the ethical sense, which is the focus of this chapter. Research can be done in a way that respects and cares for the participants, maintains integrity in the process, and results in studies that are reported honestly. From one perspective, ethics in research—as with ethics in any other setting—may be seen as rules for distinguishing between right and wrong, and in that sense, they might appear to be simple common sense. However, if distinguishing between acceptable and unacceptable behavior were simple common sense, there wouldn't be so many moral and ethical conflicts in society today. Because different individuals interpret and apply these rules differently, based on their own values and life experiences, the line between ethical and unethical research is often blurry, and sometimes remains undefined until it is crossed (O'Brien, personal communication, November 27, 2013).

But why should we think about ethics in research in the first place? According to Resnik (2011), there are several reasons why the field of ethics is important in research. First, norms of ethical research promote a variety of other moral and social values, such as human rights, animal welfare, justice, equity, legal compliance, safety, and so on. As you will read in this chapter, throughout history, there have been many instances where ethical lapses in research have caused substantial physical and psychological harm to participants. Therefore, by adhering to ethical principles in research, we also strengthen support to foundational goals of our society.

Second, ethical research also advances key aims of research itself, such as generation of new knowledge, truth, and avoidance of error. For example, guidelines against fabricating, falsifying, or misrepresenting research data promote the truth and avoid error. In addition, ethical norms related to authorship and copyright policies are designed to protect intellectual property interests while also encouraging collaboration among researchers.

Third, many ethical principles such as those relating to **conflict of interest** and research misconduct help ensure that researchers who are funded by public money can be held accountable to the public. In other words, funders can be

conflict of interest
A conflict of interest exists in research when a researcher's primary interest in the integrity of a study is compromised by a secondary interest such as personal gain (e.g., financial profit).

assured that all research they fund adheres to a common set of ethical precepts relating to quality and integrity. Ensuring that all projects abide by these principles also levels the playing field among different researchers. This helps to build broad public support for research in general.

Fundamentally, research ethics are about doing the right thing for all involved, a simple idea with a very important history and considerable complexity in the modern research enterprise. Because research ethics evolved in large part due to a variety of unethical research practices—from the infamous Nazi experiments to the Tuskegee Syphilis Study in Alabama—we begin this chapter by discussing each of these milestone cases individually. We also refer to the various codes of ethical conduct that society developed in response to such tragic cases. Special emphasis is placed on discussing the guidelines contained in the *Belmont Report* and the creation of Institutional Review Boards (IRBs). As you will see for yourself later in this chapter, the *Belmont Report* and IRBs laid the foundation for the modern system of research ethics. We then consider the evolution of ethical debate in clinical research where the emergence of deadly diseases like AIDS and cancer created a tension between participant protection on the one hand, and access to new drugs and therapies on the other. In the last section, we shift our attention from ethical issues in human subject participation to ethics in the production and publication of scholarly work. We conclude the chapter by emphasizing that research ethics should not be an afterthought—instead the researcher should be thinking about each part of the research project (sampling, measurement, design, analysis, and reporting) with potential ethical implications in mind.

2.2 Historical Cases of Unethical Research

Believe it or not, "good" research did not always concern itself with ethics. In fact, the current discourse on research ethics emerged in large part as a response to unethical human experimentation carried out by researchers. In the sections below, we describe some of these milestone cases in detail.

2.2a Nazi Experimentation during WWII and the Nuremberg Code

The history of modern research ethics begins with tragic abuse by Nazi doctors during World War II. When the war ended in 1945, the world learned of the horrors that a large number of prisoners, primarily Jewish, were subjected to in Nazi Germany. In order to improve the survival and treatment of army personnel and to advance the racial and ideological tenets of their worldview, Nazi physicians carried out as many as 30 different types of human experiments on concentration-camp inmates (Tyson, 2000). For instance, to determine the most effective means of treating pilots who had suffered extreme exposure to severe cold, the Nazi doctors conducted various types of freezing experiments. For up to five hours at a time, they placed prisoners into vats of icy water, either in aviator suits or naked; they took others outside in the freezing cold and strapped them down naked. As the victims writhed in pain, foamed at the mouth, and lost consciousness, the doctors measured changes in the patients' heart rate, body temperature, muscle reflexes, and other factors (Tyson,

Figure 2.1 Nazi war criminal Karl Brandt, head of German health service and Hitler's personal physician, on trial in Nuremberg, August 20, 1947.

Nuremberg Code This code was developed following the trial of Nazi doctors after World War II. It includes 10 principles to guide research involving human subjects. The Code has been extremely important as a reference point for all regulations related to the protection of human subjects. Among other things, it established the principles of informed consent, voluntary participation without coercion, clear scientific justification for research, and most important, limits on the risk of harm.

2000). In another example, to learn if a limb or joint from one person could be successfully attached to another one who had lost a limb or a joint, experimenters amputated legs and shoulders from inmates in useless attempts to transplant them onto other victims (Tyson, 2000).

Other experiments included prisoners being burned, poisoned, shot, asphyxiated, infected, injected, and ultimately murdered during the course of the experiments. These studies were, of course, performed without the consent of the victims who suffered indescribable pain, mutilation, permanent disability, or in many cases torturous death as a result. Immediately following the war, a trial was held in Nuremberg, Germany, where 23 German doctors were tried for the atrocities committed. Prosecutors found 15 defendants guilty of war crimes and crimes against humanity; seven were hung (Tyson, 2000).

The modern history of human-subject protections also began at the Nuremberg trial. At the beginning of the trial, the judges had no prior basis in law by which to judge the Nazi physicians. They developed 10 principles in order to do this, and these principles formed the basis of what came to be known as the **Nuremberg Code** for research involving human subjects (**Figure 2.1**). The Code has been extremely important as a reference point for all regulations related to the protection of human subjects. Among other things, it established principles of informed consent, voluntary participation without coercion, clear scientific justification for research, and most important, limits on the risk of harm (Dunn & Chadwick, 2004). We will describe these principles in detail later in the chapter, but for right now, the important takeaway is that, although it did not carry the force of law, the Nuremberg Code was the first international document to state that in any research involving human subjects, participants should give consent and that the benefits of research must outweigh the risks.

2.2b Stanley Milgram's Obedience Studies

After the atrocities of the Holocaust became known, the world wondered what made it possible for ordinary people to commit such horrific acts of genocide. The Nazi physicians on trial at Nuremberg defended their position by pleading that they were only following superiors' orders, or as they put it, "*Befehl ist Befehl*" ("Orders are Orders"). Stanley Milgram, a psychologist from Yale, examined this justification through a famous series of experiments on "obedience" (**Figure 2.2**). In these experiments, he studied the conflict between obedience toward authority and one's personal conscience.

The experiments involved a basic setup. Milgram recruited participants and provided them with a fictitious story that he was trying to explore the effects of punishment on learning behavior. The participants were asked to take on the role of "teachers" and inflict punishment on "learners" for incorrect responses to questions. On the face of it, it seemed to be a simple experiment, but Milgram was after a different question altogether. He and his research assistants played the "authoritative" role (similar to Nazi superiors who gave orders to those on trial at Nuremberg) and asked the "teachers" to

Figure 2.2 The "shocking" reality of Milgram's experiments.

administer electric shocks as a punishment (starting at 15 volts) to the "learner" whenever the "learner" provided an incorrect response to the "teacher's" questions. When more mistakes were made, he instructed the "teacher" to increase the voltage of the shock, despite the "learner's" obvious distress and consequent pleas to stop. At times, the "teacher" questioned Milgram but continued with the punishment even though some participants were obviously very uncomfortable in doing so. The "teacher" was not aware that the "learner" in the study was an actor merely faking discomfort as the "teacher" increased the intensity of electric shocks. This study raised many questions regarding how subjects could bring themselves to administer such high electric shocks under orders from a superior authority.

More central to our interests in this chapter are the many ethical issues inherent in the way Milgram conducted his research: Should deception of subjects be allowed in the pursuit of research? Does the researcher have any right to expose subjects to such stress? Who should decide these issues? The ethical lessons from Milgram's studies may appear obvious now, but at the time were less clear. While the Nazi experiments were related to the issue of physical harm to research subjects, Milgram's work brought about awareness regarding the potential of subjecting research participants to psychological harm.

Despite the fact that Milgram's studies were ethically questionable (given the use of deception), his biographer, Thomas Blass, noted that Milgram should actually be credited with being the first investigator to use the term "debriefing" (2004). The treatment of participants became an ongoing issue within the field and in society in general, as widely read newspapers published negative editorials about Milgram and his procedures. One of the ways that Milgram responded to this criticism was to conduct a survey of participants to identify the consequences of participation. He included a summary of basic findings with the survey questions, both of which are now considered good follow-up steps in studies where some risk of harm is acknowledged (Blass, 2004).

2.2c The Thalidomide Tragedy

In the late 1950s and early 1960s, the drug Thalidomide was approved as a sedative in Europe. Many pregnant women were given this drug to help them sleep and to alleviate morning sickness and nausea. Unfortunately, doctors were unaware that when Thalidomide is given to pregnant women, it grossly interferes with how the fetus develops. The **Thalidomide tragedy** led to severe deformities in the children born to these women. Around the world, some 12,000 babies were born with malformations of their limbs and internal organs (**Figure 2.3**).

Interestingly, while widely used in Europe, the Food and Drug Administration (FDA) did not approve Thalidomide in the United States and was thus able to avert a major domestic tragedy. In 1962, the FDA inspector at the time, Frances Kelsey, prevented the drug's approval within the United States, even though there was intense pressure to approve the drug from Thalidomide's parent pharmaceutical company and from her FDA supervisors. (**Figure 2.4**) Kelsey defended her position by insisting that the application for Thalidomide contained incomplete and insufficient information on the drug's safety and effectiveness.

After the Thalidomide tragedy, the U.S. Senate passed the **Kefauver-Harris Amendments** to the Food, Drug and Cosmetic Act to ensure superior drug safety. As a

Thalidomide tragedy This event involved the occurrence of very serious birth defects in children of pregnant women who had been given Thalidomide as a sedative. The drug side effects should have been known and available to doctors and patients, but were not until much harm had been done.

Kefauver-Harris Amendments After the Thalidomide tragedy, these amendments to the Food, Drug and Cosmetic Act were passed to ensure greater drug safety. For the first time, drug manufacturers were legally required to present evidence on the safety and effectiveness of their products to the FDA before marketing them. It also established the requirement that participants be fully informed about potential risks or harm, and that based on this information, they voluntarily agree to participate in clinical trials.

AP images/Press Association

Figure 2.3 Severe deformations among Thalidomide babies.

Figure: 2.4 Dr. Frances Kelsey receiving the President's Award for Distinguished Federal Civilian Service in 1962 for her role in withholding approval of the drug thalidomide despite pressure from the manufacturer, thereby averting birth defects in infants in the U.S.

result of these amendments, for the first time, drug manufacturers were required to provide evidence on the safety and effectiveness of their products before they could begin to market them to the general public. It also established the requirement that participants be fully informed about potential risks or harm, and that based on this information, they voluntarily agree to participate in clinical trials.

In the international arena, the World Medical Association in 1964 also established recommendations guiding doctors in biomedical research involving human participants. Known as the **Declaration of Helsinki**, some of these guidelines included the principles that "research protocols should be reviewed by an independent committee prior to initiation" and that "research with humans should be based on results from laboratory animals and experimentation."

2.2d The Tuskegee Syphilis Study

The **Tuskegee Syphilis Study** was an infamous clinical study that lasted for 40 years (1932–1972) and was administered by the U.S. Public Health Service in Alabama. Its main goal was to study the health effects of untreated syphilis, which was a deadly sexually transmitted disease at that time. This study led to considerable physical harm, as it involved *not* treating research participants for the disease (leading to their subsequent deaths) even when a cure became available.

All of the participants were poor African American sharecroppers in the area around Tuskegee, Alabama. None of them had a good understanding of the study and were given the impression that they were receiving good medical care when in fact they were subjected to painful procedures that did nothing to address their condition (**Figure 2.5**). Many were not even told that they had the condition. Even worse, they were not given curative treatment—penicillin—when it became available. Thus, a federally sponsored study took advantage of a vulnerable population, was racially motivated, created permanent and serious harm for participants, and left a legacy of ethical misdeeds. As you can imagine, when the media broke the story about Tuskegee, a massive public outrage ensued.

Figure: 2.5 A research technician obtains a blood sample from a participant in the Tuskegee Syphilis Study. From the Records of the Centers for Disease Control and Prevention (http://www.archives.gov /atlanta/exhibit/6.php).

If you think about it, during many of the 40 years of the syphilis studies, the Nuremberg Code and the Declaration of Helsinki were in existence. At the time, each of these international standards delineated ethical principles for informed consent and medical care of research subjects. However, both these documents were disregarded in the syphilis studies. For whatever reasons, the U.S. government continued to sponsor research in humans that was not guided by the Nuremberg and Helsinki principles.

In 1997, U.S. President Bill Clinton issued an apology to the eight surviving members of the study group and their families, and attempted to set the record straight regarding all of the ways that this study created lasting harm to people, science, and society. Nonetheless, the suffering of participants and their families cannot be fully remediated, and the loss of trust in research continues to impact efforts to recruit participants.

For clarification, note that the U.S. government did not deliberately infect Tuskegee participants with syphilis. However, there was another instance when this actually occurred: the work of Wellesley College professor Susan Reverby exposed studies cosponsored by U.S and Guatemalan governments between 1946 and 1948 that involved the deliberate infection of Guatemalan prisoners and mental asylum patients with syphilis. This was done so that researchers could test the effectiveness of penicillin to prevent the spread of the disease. The records indicate that no consent was taken from the participants when they were injected with the infection. On October 1, 2010, U.S. Secretary of State Hillary Clinton and Health and Human Services Secretary Kathleen Sebelius issued a joint statement apologizing for the experiments (McNeil, Jr., 2010).

Declaration of Helsinki
The World Medical Association adopted the Declaration of Helsinki in 1964 in order to provide a set of principles to guide the practice of medical research. The principles include such statements as "research protocols should be reviewed by an independent committee prior to initiation."

2.3 Evolution of a Modern System of Research Ethics

With the emergence of the Nuremberg Code, the Kefauver-Harris Amendments, and the Declaration of Helsinki, a considerable momentum had been developing both internationally as well as domestically on the topic of ethical research. However, until the 1970s, no comprehensive guidelines existed in the United States on the question of what constitutes ethical behavior in research involving human subjects. In addition, except for the Kefauver-Harris Amendments, which were specific to clinical research, the more general principles of ethical research contained in the Nuremberg Code and the Declaration of Helsinki were not legally enforceable.

The first serious attempt at developing a broad ethical framework covering both biomedical and behavioral research (including the creation of a formal system of checks and balances) was triggered by the revelation of the Tuskegee Syphilis Study in 1972. When the press leaked the story, the Tuskegee study in Alabama came to an abrupt end and a massive public outcry ensued over the immorality of knowingly denying treatment to individuals with syphilis. As a result, congressional hearings were initiated and the **National Research Act** was passed in 1974.

In the sections below, we detail the immediate outcomes of this Act (mainly, the *Belmont Report*, related guidelines in ethical research that subsequently emerged, and Institutional Review Boards). Further, in an effort to present a comprehensive picture of a modern system of research ethics, we also discuss recent ethical debates in the field of clinical trials. We end this section by presenting a discussion on ethics in animal research.

Tuskegee Syphilis Study This was a 40-year observational study of the impact of untreated syphilis on men. The participants in the study were low-income African American men who were led to believe they were receiving treatment when in fact they were not. This was one of the major stimuli of the Belmont Conference.

National Research Act An Act passed by the US Congress in 1974. It represents the first serious attempt to build a comprehensive system of research ethics in the U.S. It created a national commission to develop guidelines for human subjects research and to oversee and regulate the use of human experimentation in medicine.

2.3a The Belmont Report

As discussed above, prompted by the Tuskegee study, Congress passed the National Research Act in 1974. This law represents one of the first serious attempts to build a comprehensive system of research ethics in the United States. Perhaps the most important early outcome of the law was the creation of a national commission to examine ethical research practice and to generate a report with recommendations on the topic. This report is known as the **Belmont Report**.

The *Belmont Report* was named for the location of the conference center in Belmont, Maryland, where the commission met to consider the state of research ethics and issue recommendations. The report includes basic standards that should underlie the conduct of any biomedical and behavioral research involving human participants.

Belmont Report The report includes basic standards that should underlie the conduct of any biomedical and behavioral research involving human participants. It emphasizes universal principles that are unbounded by time or technology. The three core principles described in the *Belmont Report* are
1) respect for persons
2) beneficence, and
3) justice.

The report emphasizes universal principles that are unbounded by time or technology. The three core principles described in the report are 1) Respect for persons, 2) Beneficence, and 3) Justice (Dunn & Chadwick, 2004). Each principle has been translated into essential practices that regulate the conduct of research and enable deliberation about proposed studies. Below we describe each of these principles in detail.

Respect for Persons

The principle of **"respect for persons"** means that people are to be treated as independent and autonomous individuals. It further suggests that the well-being of those who are not in a position to be fully autonomous should also be protected. The term **"vulnerable populations"** is used to designate those who may not be fully in control of their decision making. There are several specific groups in this definition including children, prisoners, and people with impaired cognitive capacity (e.g., with mental retardation or dementia). Children under 18 are not of legal age to provide consent. Instead, researchers must seek permission from parents or legal guardians. Children can and should, however, be asked to make a decision about participation. That kind of decision is referred to as **"assent."** Prisoners are vulnerable because incarceration results in a loss of decision-making power as well as the possibility of greater coercion from authorities. Similarly, employees and students who are in a subordinate position to a supervisor or teacher may have their decision-making autonomy compromised by real or perceived risks related to their participation in a study. In addition, pregnant women are considered vulnerable because pregnancy brings additional risks related to health issues for the mother and unborn child.

More broadly, "respect for persons" means that for an individual to make an "informed" decision, she or he must have complete information about the study, including the opportunity to ask any questions. This is commonly known as **informed consent**. Unfortunately, the true meaning of informed consent has, to some degree, been lost in translation. Informed consent is not merely the "consent form." As described in the *Belmont Report*, it is a process that guarantees that a person have three elements met for their consent decision to be well informed: information, comprehension, and voluntariness (Dunn & Chadwick, 2004).

Meeting the "information" criterion means that consent includes a set of specific elements such as purpose, procedures, risks, benefits, alternatives to participation, protection of privacy, contact information regarding rights, and a reminder about the right to withdraw at any time. Comprehension means that the information is presented in a way that enables the participant to fully understand that nature of the study and to ask questions. Researchers must take into account factors such as the age, literacy level, and other characteristics of the sample that could impact comprehension. Voluntariness can only occur if the first two conditions have been met, and it also means that participants are free to decline participation or withdraw at any time. Rewards (e.g., a small monetary reward) for good recall of information in consent forms can be a useful way to assess whether the consent process is working effectively in a specific research context (Festinger, Dugosh, Marlowe, & Clements, 2014).

This means that informed consent is a process that provides every participant all of the information needed in understandable terms so that their ability to volunteer to join a study or withdraw at any point is unimpeded. The principle

respect for persons This principle means that people are to be treated as independent and autonomous individuals. It also means that the well-being of those who are not in a position to be fully autonomous should also be protected. See also "vulnerable populations."

vulnerable populations This is the term used to designate those who may not be fully in control of their decision making. There are several specific groups considered vulnerable, including children, prisoners, and people with impaired cognitive capacity (e.g., with mental retardation or dementia). Children under 18 are not of legal age to provide consent. Instead, researchers must seek permission from parents or legal guardians.

assent Assent means that a child has affirmatively agreed to participate in a study. It is not the mere absence of an objection. The appropriateness of an assent procedure depends on the child's developmental level and is determined by the researcher in consultation with the IRB.

informed consent A policy of informing study participants about the procedures and risks involved in research that ensures that all participants must give their consent to participate.

of **voluntary participation** requires that people not be coerced into participating in research. This is especially relevant where researchers had previously relied on captive audiences for their subjects—prisons, universities, hospitals, and similar settings. Harm can be defined as both physical and psychological. It is also important to note that it is possible in some studies to obtain a waiver of written consent—most often when the risk of harm is minimal. The relative risk of harm in a study is considered under the principle of **beneficence**.

Beneficence

Beneficence represents the expected impact on a person's well-being that may result from participation in research. Researchers should attempt to maximize the benefits of participation and take steps to identify and limit the potential for harm. This is typically done when planning a study by a careful risk/benefit assessment. Risk of harm and probability of benefit is assessed most importantly at the level of the individual participant, but researchers are also obliged to consider the impact on families, communities, and society in general. When assessing risk, researchers should try to think about worst-case scenarios—situations with significant potential for harm even if the probability seems low. The term "no greater than minimal risk" has become an important threshold in risk assessment. It means that the risk of harm is not greater than that experienced in everyday life situations such as typical academic or medical examinations. If an outside authority determines that this threshold is exceeded, then a research proposal will receive greater scrutiny in the review process, and specific safeguards must be implemented in the protocol. If the minimal risk threshhold appears reasonable, then researchers may obtain a waiver of written informed consent. A very common situation in which a signed consent form may not be necessary is online survey research in which consent is implied by participation.

Justice

The *Belmont Report* also established the principle of **justice**. Justice means that participation should be based on fairness and not on circumstances that give researchers access to or control of a population based on status (as in the Tuskegee and Nazi experiments). This idea is very relevant to sampling decisions for two reasons: 1) researchers should not choose samples merely because they are convenient and 2) it would be unjust if studies systematically excluded some groups, thereby denying them the potential benefits of research.

The principle of "justice" also means that recruiting should be done in a way that respects the rights of individuals and does not take advantage of their status as patients, students, prisoners, children, or others not fully able to make independent decisions. Justice is also considered when determining what is a fair trade for participation, whether it is learning, or more practical considerations such as compensation for time or personal expense.

2.3b Related Guidelines on Human Subject Participation

Privacy

In an era of data mining, drones, spying, leaks, cookies, viruses, pop-ups, and social media integrated into nearly every aspect of everyday life, one may expect that comments on privacy will soon only be found in history books. Yet, scientific investigators will always have important scientific, legal, and ethical reasons to

voluntary participation The right of voluntary participation is one of the most basic and important of all research participant rights. It means that fully informed individuals have the right to decide whether or not to participate in a study without coercion. It also means that they can terminate participation at any point without negative consequences.

beneficence The expected impact on a person's well-being that may result from participation in research. Researchers should attempt to maximize the benefits of participation and take steps to identify and limit the potential for harm. This is typically done when planning a study by a careful risk/benefit assessment.

justice Justice means that participation in research should be based on fairness and not on circumstances that give researchers access to or control of a population based on status. The principle of justice also means that recruiting should be done in a way that respects the rights of individuals and does not take advantage of their status as patients, students, prisoners, children, or others not fully able to make independent decisions.

"WE'VE FINALLY COMPUTERIZED YOUR FILES. NOW WE JUST HAVE TO GET THEM OFF FACEBOOK."

Figure: 2.6 Researchers have to take great care to protect privacy in data records.

privacy Privacy in research means protection of personal information about participants. It is typically accomplished through the use of confidentiality procedures that specify who will have access to personally identifying data and the limits of that access. It can also include the use of anonymous data in which no personally identifying information is ever collected.

anonymity The assurance that no one, including the researchers, will be able to link data to a specific individual. This is the strongest form of privacy protection in research because no identifying information is collected on participants.

confidentiality An assurance made to study participants that identifying information about them acquired through the study will not be released to anyone outside of the study.

de-identification The process of removing identifying information from data sets in order to assure the anonymity of individuals.

consider privacy in the context of research studies. **Privacy** is fundamental to the Belmont principles of informed consent, beneficence, and justice (**Figure 2.6**).

Privacy means that information shared by individual participants will not be shared with others. The most certain way of maintaining privacy is through the provision of **anonymity**. Anonymity means that there is no personally identifying information in a data set. That is, there is no way that an individual can be identified from the information stored in a researcher's files. When anonymity is possible, it also has a potential benefit in terms of data quality because respondents may be willing to share information more freely.

However, anonymity may not be possible in some research designs. For example, in a longitudinal study, there has to be a way to link records for each measurement occasion at the individual level, necessitating a coding system based on a unique personal identifier.

Alternatively, **confidentiality** means that the researcher makes a promise that whatever identifying information is shared will be known only by the researcher, unless circumstances dictate exceptions to maintain the well-being of participants. For example, researchers who study sensitive topics such as depression and suicidal thinking may need to maintain a record of the identity of individuals in the event that a "red flag" of significant distress and likelihood of harm appears in the data. In such cases, research protocols spell out the steps that a researcher will take to provide intervention and support to such a person. The informed consent process should always include clear information about what sort of privacy is promised and what exceptions to confidentiality are necessary.

Privacy of participants receives special protection when medical records are part of the study. The Health Insurance Portability and Accountability Act (HIPAA) of 1996 requires that researchers obtain special permission to utilize data kept in medical records. This may include obtaining permission directly from patients. It may also include obtaining a waiver of permission when it is practically impossible to obtain permission, or implementing a process called "**de-identification**" in which all identifying information is removed from the data prior to the researcher being able to access it.

Deception and Debriefing

Deception was famously used by Professor Milgram in the obedience studies. It is now a very carefully controlled aspect of ethical consideration in research proposals. Deception means that either misleading or incomplete information is given to participants in the informed consent process. As you may be thinking, the very act of deceiving a participant compromises informed consent. There are times when deception is regarded as essential to the scientific evaluation of a research question and, in those cases, an assessment of the potential for harm from the deception is conducted, and full **debriefing** is provided following the conclusion of data collection. In addition to having the opportunity to ask any questions during debriefing, participants are given the right to have some or all of their data withdrawn and destroyed (Dunn & Chadwick, 2004). In many studies, participants are given articles to read that are related to the study they have just participated in, or relevant information on community services.

Right to Service

Increasingly, researchers have had to deal with the ethical issue of a person's **right to service**. Good research practice often suggests the use of a **placebo** or a no-treatment, control group—a group of participants who either get something that may look like the real treatment but is inert, or who do *not* get the treatment or program at all. But when that treatment or program may have beneficial effects, persons assigned to the no-treatment control may rightly feel their rights to equal access to services are being curtailed. In designing a comparative study, researchers must carefully weigh the risk of harm if an evidence-based treatment is available. Current guidance on use of placebos includes the following necessary conditions: 1) the placebo is necessary for a valid scientific comparison of the new treatment 2) the placebo does not create a situation with excessive and unnecessary risks and 3) participants are given complete information about the risks associated with the possibility of receiving the placebo and are able to give informed consent (Dunn & Chadwick, 2004).

Research designers may have several options to help resolve the ethical dilemma in use of no-treatment or placebo controls. One is to draw the comparison group from a waiting list and make sure that the treatment is made available to them in a period no longer than the ordinary waiting time would be. If a researcher has a known standard treatment available, then the standard can be compared to the treatment under investigation. Another design option involves the use of a pretest assessment that identifies those in greatest need and then provides the intervention to them. This design, known as the Regression-Discontinuity design, also has some important strengths in internal validity that are discussed in Chapter 10.

2.3c Institutional Review Boards (IRBs)

In addition to the creation of a national commission (which, in 1979, came up with the *Belmont Report*), the National Research Act also mandated the establishment of **Institutional Review Boards (IRBs)** at hundreds of institutions receiving federal funding for research. Administratively, these local IRBs at various institutions come under the Office for Human Subjects Protection (OHRP, http://www.hhs.gov/ohrp/). The OHRP is charged with the mission to provide oversight, guidance, and education for all research conducted with human participants in the United States.

The role of local IRBs is to review all proposed research involving human subjects to ensure that subjects are going to be treated ethically and that their rights and welfare will be adequately protected. Local IRBs are also responsible for monitoring compliance with research protocols. IRBs are typically composed of one or more administrators and a group of qualified peers who are able to review and critique both the scientific and ethical provisions of a proposed study. Since no set of standards can possibly anticipate every ethical circumstance, having a group of knowledgeable peers operating under the supervision of an appointed administrator allows for review of proposals, monitoring implementation of studies, reporting to the federal government, and handling complications such as adverse events during the course of a study. IRBs typically publish guidance on who may submit a protocol for review, when an IRB review is necessary, what the necessary components of a protocol are, and on follow-up monitoring and consultation on approved studies.

The review process is usually well defined with specific requirements for proposed studies posted on the IRB website. IRBs and their activities are defined in

deception The intentional use of false or misleading information in study procedures. Researchers must justify the use of deception on scientific grounds and be certain to provide complete debriefing about the actual nature of the study once it has been completed.

debriefing The process of providing participants with full information about the purpose of a study once a person has completed participation. Participants are typically offered a standard written description of the study and why it was conducted, given the opportunity to ask questions, and given the opportunity to have their data withdrawn if they are not satisfied with what has happened in the study.

right to service The right of study participants to the best available and most appropriate services relevant to a condition that is part of a study.

placebo The use of a treatment that may look like the real treatment, but is inert. The use of placebos in research is allowed if careful assessment of the scientific need and care of participants can be provided.

Institutional Review Boards (IRBs) A panel of people who review research proposals with respect to ethical implications and decide whether additional actions need to be taken to assure the safety and rights of participants.

Title 45, Part 46 of the Code of Federal Regulations (http://www.hhs.gov/ohrp /humansubjects/guidance/45cfr46.html#46.102). Among the regulations governing IRB activity is the very definition of research itself, as follows:

> "Research means a systematic investigation, including research development, testing and evaluation, designed to develop or contribute to generalizable knowledge" (§46.102 accessed from: http://www.hhs.gov/ohrp /humansubjects/guidance/45cfr46.html#46.102).

The latter part of this definition is usually interpreted to mean that the intent of the researcher is to conduct a study that will be presented or published, thereby contributing to generalizable knowledge. This means that some carefully conducted studies, for example a program evaluation conducted for the purpose of program improvement, would not meet the definition of research as defined by the IRB and would therefore not fall under the purview of an IRB or require an IRB review.

The proposal for research is submitted in the form of a **protocol** that includes very specific information about the study background and goals as well as details on all aspects of care of the participants. This includes recruitment materials, informed consent documents, debriefing information, and management of any adverse events that might conceivably occur. In addition, IRBs are responsible for making sure that vulnerable populations receive careful attention and for reporting inclusion of vulnerable populations to federal authorities who monitor research.

In general, IRBs provide guidance on how to properly write a proposal for a study, including all of the specific requirements needed to ensure the well-being of your participants. They make certain that all researchers have met basic educational requirements, most commonly through an online course and test that reviews much of the information presented in this chapter. IRBs also have responsibility for monitoring compliance with ethical requirements. You can think of an IRB-approved research protocol as a contract between the researcher and the institution that guarantees the protection of participants. Many institutions also include a scientific review that may be a prerequisite to the review of ethics. The quality of the proposed science has an ethical dimension, because participants should not be asked to waste their time in a study that is seriously flawed and unlikely to contribute to knowledge.

2.3d Ethics in Clinical Research: Patient Protection versus Access

As in any other field, the discussion on ethics in clinical research has also continued to evolve. After the Thalidomide tragedy in 1962, the FDA's approach to drug testing and clinical research moved to a "go-slow" mentality. There was greater emphasis on risk aversion and patient safety. According to the consensus achieved at that time, it was better to risk denying treatment until there was enough confidence that a therapy's effects were well known, than risk harming innocent people. This approach was characterized in the conservative system of drug trials developed by the FDA, which required evidence from "at least three 'phases' of randomized clinical trials in human subjects performed sequentially" (Epstein, 1996) before a drug could be marketed to the general population. In a **Phase I study,** researchers test a new drug or treatment in a small group of people for the first time to evaluate its safety, determine a safe dosage range, and identify

protocol A detailed document summarizing the purpose and procedures of a study. It provides a reader such as an IRB reviewer with enough information to know how participants are recruited, consented, and treated in all phases of a study. It typically includes details on all measures, the handling and storage of data, consent and assent forms, debriefing materials, and any other item that might have an impact on a participant.

Phase I study A research study designed to test a new drug or treatment in a small group of people for the first time to evaluate its safety, determine a safe dosage range, and identify potential side effects.

potential side effects. In a **Phase II study**, the drug or treatment is given to a larger group of people to see if it is effective and to further evaluate its safety. The system culminates in a **Phase III study**, where the drug or treatment is given to large groups of people to confirm its effectiveness, monitor side effects, compare it to commonly used treatments, and collect information that will allow the drug or treatment to be used safely (NIH, 2008).

While exercising caution is often a good thing, the history of clinical research shows that this may not always be the case. The FDA has been severely criticized for the fact that the implementation of a full three-phase process (often, there is also a preclinical phase devoted to animal testing and a fourth phase that examines long-term effects of the drug) takes an enormous amount of time. By the late 1970s and early 1980s, it could easily take years to move a drug from the beginning of human testing to its dissemination in the market. This criticism of the FDA became even more harsh in the wake of the growing AIDS crisis of the 1980s.

Beginning in the early 1980s, AIDS cases in the United States began doubling and then tripling alarmingly, with the toll of AIDS deaths mounting correspondingly (Messner, 2006). Until a treatment could be found, a diagnosis of AIDS basically translated into a death sentence. However, the FDA's time-intensive procedures (that were also heavily steeped in regulation) meant that it could take a long time for a therapy to get from the lab to the patient. By the mid- to late-1980s, educated, well-organized AIDS activists became aggressive and vocal in demanding access to potentially life-saving experimental drugs and reform of the FDA's drug evaluation and approval practices—demands subsequently embraced by cancer patients and others suffering from life-threatening conditions (Messner, 2006). These activist groups, chief among them ACT UP (AIDS Coalition for Unleashing Power) and Treatment Action Group (TAG), did not see protection from considerable risk in the FDA's cautious approach to drug trials. Instead, they saw insurmountable barriers to accessing drugs that were their only hope for potential survival. Their argument was poignantly and succinctly summarized on an activist poster that read, "Stop Protecting Us to Death" (Killen, Jr., 2008). Activist organizations demanded wider avenues of access by campaigning for trials with minimal enrollment criteria, access to drugs prior to approval, and "fast track" approval of AIDS drugs (Meinert, 2012). These demands were eventually answered by the FDA through "expediting review and approval of new therapies" related to HIV/AIDS, as well as by providing "expanded access" to promising products under investigation. These advocacy organizations were also instrumental in pushing for greater inclusion of women in drug trials—because researchers were worried about fetal harm after the Thalidomide tragedy, women who were pregnant or of child-bearing capacity were excluded from clinical trials. If you are interested in further exploring how AIDS activism mobilized social change and got the FDA to revise its rules, we encourage you to watch David France's incredible documentary, *How to Survive a Plague*.

The realities and desperation of the AIDS epidemic presented the field of research ethics with an entirely new set of issues and concerns regarding access to experimental interventions. It also demanded a fresh look at the interpretation of the principles outlined in the *Belmont Report*. In the wake of Nuremberg, Tuskegee, and Thalidomide, the government developed standards to protect human subjects from potential harm. The AIDS crisis showed that those standards were making it hard for patients who faced a deadly disease to get potentially life-saving new treatments and demonstrated that ethical issues in research were even more complex than they had seemed earlier.

Phase II study A research study designed to test a drug or treatment that is given in a larger group of people than in a Phase I study to see if it is effective and to further evaluate its safety.

Phase III study A research study designed to test a drug or treatment that is given to large groups of people using a highly controlled research design, in order to confirm the intervention's effectiveness, monitor side effects, compare it to commonly used treatments, and collect information that will allow the drug or treatment to be used safely.

2.3e Ethics in Research with Animals

Advocates of the use of animals in research can point to many examples of advances in the quality of life for humans as a direct result of animal-based studies. On the other hand, many individuals and organizations such as People for the Ethical Treatment of Animals (PETA) have taken a strong stand against the use and abuse of animals in research, citing instances that call to mind the methods of the Nazis. In fact, legal regulation of the treatment of animals in research is relatively limited and largely left to the oversight of IRBs and the research ethics codes of professional organizations such as the American Psychological Association. IRBs are required to maintain a committee on the care of animals used in research (Institutional Animal Care and Use Committee, IACUC). The APA has published guidelines for the care of animals that include oversight by the local IACUC and detailed guidance on both scientific justification of such studies and humane care of animals in research (http://www.apa.org/science/leadership/care/guidelines.aspx, accessed 10-13).

2.4 Ethics in the Production and Publication of Scholarly Work

The cases discussed so far in the chapter reflect ethical challenges related to failures to protect research participants from physical and psychological harm. However, a serious ethical compromise also arises when researchers are dishonest in the way they conduct research or report their results. Such offenses make it very difficult for the research community to move forward, because it calls into question whether the research enterprise can be trusted.

The case of William Fals-Stewart provides an example of how research can go wrong when the integrity of the process is violated. William Fals-Stewart was a nationally known researcher at the University of Buffalo who was frequently cited for his work in couples therapy. He had published numerous papers, obtained millions of dollars in federal grants, and received national awards for his work. Things began to unravel when he was accused of falsifying data in 2005. In the course of an investigation that led to a criminal trial, he hired actors to provide testimony regarding their involvement in data collection. The actors, who testified by phone, were told that they were participating in a play. The ruse initially worked, and he was cleared of charges. Not satisfied with this outcome, he sued the state of New York for $4 million. In the course of the next investigation, the attorney general's office discovered the fraud and subsequently arrested Fals-Stewart, charging him with grand larceny, perjury, identity theft, and other charges (http://www.ag.ny.gov/press-release/new-york-state-attorney-general-andrew-m-cuomo-announces-charges-against-former-ub). The week following his arrest, Fals-Stewart was found dead in his home, with no cause of death ever reported.

Another prime example of such an ethical breach is the case of wunderkind and physicist Jan Hendrik Schön, a researcher at Bell Laboratories. Between 2000 and 2002, he published an astounding number of research articles in major journals like *Science* and *Nature*. And, Schön's papers were no ordinary pieces of research. In each of these articles, he announced one unbelievable discovery after another in the field of semiconductors. Even though many of his contributions were initially hailed as "breakthroughs," they were later found to be fraudulent.

After a formal investigation in 2002, Bell Labs dismissed Schön (Kaiser, 2009). The investigating committee concluded that at least 16 of his papers showed clear evidence of scientific misconduct. One of the prime allegations against him included "substitution of data," where Schön recycled graphs representing data from one type of material, changed axes or labels, and passed them off as coming from different materials (Kaiser, 2009). Other charges included the fabrication of entire data sets out of thin air (Kaiser, 2009).

Both of these cases are examples of **research misconduct**. The federal definition of research misconduct is shown below. These cases also raise very important questions about the impact of research misconduct. For example, without a complete audit of every data point in every study, how can the community of scientists and clinicians know how much trust to put in published reports? Further, what is the impact on public opinion about the value of behavioral research when it turns out that studies may not be trustworthy? There is no question that research misconduct compromises the research enterprise.

According to the Office of Research Integrity:

Research misconduct means fabrication, falsification, or plagiarism in proposing, performing, or reviewing research, or in reporting research results.

(a) Fabrication is making up data or results and recording or reporting them.
(b) Falsification is manipulating research materials, equipment, or processes, or changing or omitting data or results such that the research is not accurately represented in the research record.
(c) Plagiarism is the appropriation of another person's ideas, processes, results, or words without giving appropriate credit.
(d) Research misconduct does not include honest error or differences of opinion.
http://ori.dhhs.gov/definition-misconduct

In the sections below, we will consider ethical aspects of the research enterprise in academic and other settings in which research is produced for publication. These issues include several that are specifically addressed in the ethical code of psychologists and other professions, but which can nonetheless become very problematic. These include honesty in reporting, conflict of interest, and fairness in credit for publication.

2.4a Honesty in Reporting

Honesty in reporting includes several specific aspects. First, writers must give credit to authors of any cited work and thereby avoid **plagiarism**. The most blatant form of plagiarism involves verbatim copying of another author's work without providing full details about the source of the material. There are now computer programs that can detect plagiarism in this form, but it not an all-or-none issue. If you think about the world of music, for example, you can readily see that virtually every song is in some way related to other songs, and it would be impossible to provide full credit to all sources. Nonetheless, there have been cases of clear copyright infringement where the similarities between two songs unequivocally indicate that one copied the other. Similarly, the world of research is built on an evolutionary model in which people and ideas co-mingle and influence each other in dynamic ways, increasing the responsibility of authors to take great care in the documentation and crediting of scientific reports. As in music, it is also possible for plagiarism to occur in subtle ways related to the form of research communication. For example, a complete

research misconduct Fabrication, falsification, or plagiarism in proposing, performing, or reviewing research, or in reporting research results, are all prime examples of research misconduct. Fabrication is making up data or results and recording or reporting them. Falsification is manipulating research materials, equipment, or processes, or changing or omitting data or results such that the research is not accurately represented in the research record.

plagiarism The use of another person's work without proper credit.

literature review is supposed to trace the origins and history of thinking and data in an area. It is possible that an important source could be left out of a review accidentally. Further, in attempts to meet journal article word limits, a writer might condense the description of the origins of an idea, creating an incomplete but not directly false impression of a line of research.

There are several aspects of data management and analysis that depend on honesty and integrity as well. These include care in data entry so that only valid and complete data sets are utilized in analysis. In addition, honest reporting requires a full accounting of missing data. Data can be missing for a variety of reasons, including the right of participants to skip questions or to completely withdraw from a study at any time. The researcher's obligation is to report on the extent of missing data and what, if anything, was done to adjust the data or analysis as a result.

Efforts have been made in recent years to eliminate the possibility of bias through inadequate reporting, by requiring the inclusion of flowcharts showing the number of participants at each stage, from initial enrollment to any follow-up. Ethical codes like those of the APA also state that data should be made available to other researchers for follow-up or secondary analysis as a way to ensure integrity of reports. Data-sharing agreements should of course be constructed to protect the rights of the participants, including privacy. This may be easier said than done in some studies—for example, those that utilize video of individuals who may be identifiable.

Errors in data may be unintentional but may also involve deliberate falsification of data in order to enhance the apparent results. We have previously cited the cases of William Fals-Stewart and Jan Hendrik Schön, but there have been many other high-profile cases of data falsification in recent years, including those of prominent psychologists who were found to have literally "made up" data in order to produce results that would support their hypotheses and produce impressive-looking publications and personal benefits, including fame and fortune (Bartlett, 2012; Bhattacharjee, 2013; Carey, 2011). These cases and others fit the formal definition of misconduct held by the Office of Research Integrity of the U.S. Department of Health and Human Services cited previously.

Honest reporting also necessitates that researchers produce reports that are consistent with the original goals of the study. That is, they must complete the study as it was originally intended, including the original analysis. However, in order to do this, researchers must avoid the temptation to "go fishing" for results that they believe enhance publishability. Fishing occurs when someone conducts multiple analyses of the same data without taking this into account statistically, and only reports the results that support their preferred outcome. The very meaning of the term "significance" in statistics therefore has an ethical aspect that is just as important to understand as the statistical and practical significance of a result.

piecemeal and duplicate publication An ethical issue in the dissemination of research referring to the possibility that, in order to maximize personal reward (status, promotion, pay, and such) from publication, a researcher would essentially repackage the results of a single study in multiple articles.

The meaning of "significance" can be corrupted by such practices as well as by **piecemeal and duplicate publication**. The latter refers to the possibility that in order to maximize personal reward (status, promotion, pay, and such) from publishing research, a researcher essentially repackages the results of a single study into multiple publications. There are times when a large and complex data set may produce more results than can be described in a single article, and in such cases authors typically acknowledge the connections between the studies related to the same data set. In addition, secondary analysis of data has become a very important kind of activity, and has been encouraged by the availability of well-documented national data sets in such areas as health and education. But with many people analyzing the same archived data, it is possible for them to inflate the appearance of an effect artificially.

2.4b Conflict of Interest

On June 16, 1966, Dr. Henry K. Beecher published a landmark paper on ethics in clinical research in the *New England Journal of Medicine*. The paper sounded an alarm based on Beecher's finding of ethical problems in many papers published in prominent medical journals. His account provided details on 22 such studies, many focused on inadequate or complete absence of informed consent. In reflecting on the conditions that contributed to the widespread problems, he noted "of transcendent importance is the enormous and continuous increase in available funds" (p. 367), setting the stage for the possibility of personal ambitions to trump scientific integrity. To support his point, he provided a table showing the increase in the total National Institutes of Health (NIH) budget from approximately $700,000 in 1945 to over $430 million in 1965. In case you are curious, NIH Director Francis Collins has submitted a budget request of over $31 billion for fiscal year 2014 (accessed 9-13 from http://www.nih.gov/about/director/budgetrequest /fy2014testimony.htm).

Clearly, Beecher's warning about the potential compromising influence of money on research integrity is no less relevant today, including vast sums of private, profit-oriented research, in addition to the public funds from federal, state, and local agencies. Financial conflicts of interest have in fact been at the center of many important recent cases in research ethics, for example, the tragic case of Jesse Gelsinger.

In 1999, Jesse Gelsinger was an 18-year-old with a rare disease who died during participation in a study. The aftermath of his death included revelations that he should have been excluded based on criteria in the protocol and that informed consent was violated in other ways, including failure to share very serious risks with Jesse and his family. It also turned out that the study investigators were not only clinical scientists, but also investors in a private company that stood to profit from the results of the study, as was the sponsoring institution (the University of Pennsylvania).

This situation represented a conflict of interest because the researchers' judgment about their primary interest (the well-being of participants and the integrity of their study) was likely compromised by a secondary interest (personal gain) (Amdur & Bankert, 2007). This case received major attention in legal, government, and public circles, as it highlighted the way that multiple and conflicting interests can produce harm and compromise research (Dunn & Chadwick, 2004). Among other outcomes from these reviews were guidelines for management of potential conflicts with regards to ethical, scientific, and financial or other personal gain.

The Department of Health and Human Services has placed responsibility for managing conflicts of interest on investigators, IRBs, and institutions. In addition, many journals now require a disclosure of any financial interests that an author may have when a paper is submitted for review. Conflicts of interest have the potential to put people at risk, produce biased results, and undermine the public's trust by reinforcing the notion that participants are "guinea pigs."

2.4c Fairness in Publication Credit

Another very important and sometimes very tricky aspect of honesty in reporting is proper credit in authorship of publications, an issue sometimes described as **fairness in publication credit**. Authorship is supposed to be established on the basis of the quality and quantity of one's contributions to a study, not status, power, or any other factor. However, weighing the relative merit of contributions may vary

fairness in publication credit Authorship credit should be established on the basis of the quality and quantity of one's contributions to a study, rather than on status, power, or any other factor. Many research journals have now adopted guidelines to help authorship groups determine credit.

with one's perspective on their importance in the development of the paper. For example, a study could not be completed without the original idea, but it would also not be completed without the collection and analysis of data, without good writing, and of course without IRB approval. Unless a paper is written by a single author, publication credit must be determined by some set of criteria that takes various kinds of contributions into account. In recent years, some journals have required a listing of the specific contributions made by each person who is an author of a paper. This can be very helpful in establishing authorship order, but in general it is advisable to discuss authorship credit from the outset of a project, and to continue the discussion as the study and resulting forms of communication (presentation, news release, journal article, and so on) are produced.

SUMMARY

This chapter began with an attempt to frame research ethics within a definition of what can make research "good." In conclusion, we propose that ethics are not only one of the key aspects, but an essential and indispensable aspect, of good research. We would also like to encourage you to think about ethics throughout the research design process and not simply as an afterthought. This is relevant because research ethics do not come with a specific formula. Even though all research projects should adhere to the basic principles outlined in the *Belmont Report*, the approach that each researcher takes toward ethics will differ based on the research design adopted by the researcher.

It is important to think about ethical implications from the minute you come up with a research question. You should ask yourself, does this endeavor contribute to beneficence (i.e., doing good)? Is there any way it could cause harm to anyone? Could this research have any unintended consequences? Similarly, when it comes to sampling, try to devote substantial attention to understanding who is included in your research from an ethical perspective. Will your research include human subjects or animal participation? Will it cover vulnerable populations like children? Would you have to submit your proposal to the IRB for a review? If yes, you should plan enough time for an IRB review in your project timeline.

In addition, as described in the section on *justice*, if you use nonprobability sampling techniques (Chapter 4) like convenience sampling (where units are included or excluded for convenience reasons), there might be a danger that some units are excluded needlessly. For instance, suppose a researcher finds that it takes longer than usual to interview individuals for whom English is not their first language, and, for practical reasons, decides to exclude them from his/her sample. When this happens, it raises ethical issues because the findings of research based on such selected samples can be extremely narrow. Ethical considerations may also play an important role when it comes to sample size. If your treatment has potentially negative consequences, and if you use an oversized sample for your study, then the ethical dilemma is whether or not you are unnecessarily exposing a larger number of individuals to these harmful effects.

Ethics are also important when you decide what kinds of measurement instruments to use in your study. For example, informed consent is easier to obtain when you use a structured interview (Chapter 7) to measure response. A structured interview typically uses closed questions, making it easier for a researcher to explain to the respondent beforehand what they will be asking them. If, on the other hand, a researcher employs unobstrusive measures (Chapter 3) or covert observation to collect responses, this may amount to deceiving the respondent. In such cases, it will become important for a researcher to provide strong justifications for why such methods are necessary to employ in the context of their research.

Ethical questions also come to the forefront when you think about the research design process. While randomized experiments (Chapter 8) are sometimes called the "gold standard" in terms of internal validity, they often end up raising important ethical issues. As discussed previously in this chapter, if a given treatment is promising, then aren't we depriving subjects who are randomly assigned to the comparison group (those who get no treatment) of potential benefit? Similarly, if control therapy is known to be efficacious, then aren't we depriving subjects in the treatment group of that known benefit in order to test a new intervention that may entail risk or be ineffective? Other complications arise when a researcher takes on multiple roles in the process of conducting an experiment. For example, in a clinical trial, if a physician is both the experimenter and the doctor, there might arise a conflict where, on the one hand, the physician's sole responsibility is toward the patient, yet in the context of a trial, the physician-as-researcher has a competing obligation to generate high-quality data.

You should be careful about protecting participant privacy and confidentiality in the data reporting stage. For a moment, imagine you are using a quantitative design to examine fairness in the classroom. You survey students within a classroom and then compare responses by nationality. In such an instance, depending upon the school you are looking at, it is possible that there is only one or a small group of children in a particular category (say, children whose home country is Mexico). In this case, presenting results summarized by nationality could enable school administrators or others to identify these research participants. One way to overcome this problem is to present aggregated data. Finally, when you conclude your research and write up your results, it is important that you keep in mind the ethical principles related to citing other people's work, reporting any conflicts of interest, and so on.

In essence, we are trying to emphasize here that without proper thought to ethics throughout the entire research process, it does not matter how well the study was designed or how much the researchers may have found satisfaction in the process. It is important that a researcher build ethics into the research process from the bottom up—that is, the entire research process evolves not only in response to technical considerations, but also to ethical aspects. For this reason, we encourage you to keep ethical implications in mind as you read through the chapters in this book.

By this time, we hope you agree that ethics are often not as simple as black or white. The boundary between ethical and unethical behavior is not always clearly marked in social science research and requires substantial deliberation on the part of the researcher and the institutions that support that research. To conclude, in this chapter, we discussed that the modern research enterprise now encompasses a significant infrastructure of laws and regulations to guide researchers and provide oversight. The policies and procedures that have been developed to protect participants represent an external kind of guidance system. Nonetheless, there will always be a need for internal regulation by each person who conducts research. As Dr. Henry Beecher recommended in summing up the matter of ethical regulation of clinical research: "…there is the more reliable safeguard provided by the presence of an intelligent, informed, conscientious, compassionate, responsible investigator" (1966, p. 1360). Well said, Dr. Beecher.

Key Terms

anonymity p. 42
assent p. 40
Belmont Report p. 39
beneficence p. 41
confidentiality p. 42
conflict of interest p. 34
de-identification p. 42
debriefing p. 42
deception p. 42
Declaration of Helsinki p. 38
fairness in publication credit p. 49
informed consent p. 40

Institutional Review Boards
   (IRBs) p. 43
justice p. 41
Kefauver-Harris Amendments p. 37
National Research Act
   of 1974 p. 39
Nuremberg Code p. 36
Phase I study p. 44
Phase II study p. 45
Phase III study p. 45
piecemeal and duplicate
   publication p. 48

placebo p. 43
plagiarism p. 47
privacy p. 42
protocol p. 44
research misconduct p. 47
respect for persons p. 40
right to service p. 43
Thalidomide tragedy p. 37
Tuskegee Syphilis Study p. 38
voluntary participation p. 41
vulnerable populations p. 40

Suggested Websites

The Office for Human Research Protections (OHRP)

http://www.hhs.gov/ohrp/

The Office for Human Research Protections (OHRP) is a part of the federal Department of Health and Human Services. It provides leadership in the protection of the rights, welfare, and well-being of subjects. The website includes extensive information on all aspects of policy and procedures in regulation of human-participant research ethics.

The National Institutes of Health Human Participant Protections Education for Research Teams

http://phrp.nihtraining.com/users/login.php

This site will allow you to take the online course, complete the self-test, and then provide the documentation required by your local institutional review board to conduct research.

Review Questions

1. What is the Nuremberg Code?

a. This is a law governing the rights of prisoners of all kinds.
b. It is a very important statement of research ethics developed after World War II to govern research with human participants.
c. An evidence-based manual of experimental procedures.
d. A set of guidelines also known as the German Purity Law.
(Reference: 2.2a)

2. Which of the following issues were *not* a major focus in the review of research ethics that followed Stanley Milgram's obedience studies?

a. Deception   c. Debriefing
b. Risk of harm   d. Conflict of interest
(Reference: 2.2b)

3. The *Belmont Report* established these three principles:

a. Deception, Right to Service, and Conflict of Interest
b. Plagiarism, Conflict of Interest, and Standards of Reporting

c. Maleficence, Debriefing, and Justice
d. Respect for Persons, Beneficence, and Justice
(Reference: 2.3a)

4. Information, Comprehension, and Voluntariness are the essential elements of:

a. Right to Service
b. Informed Consent
c. Harm/Benefit assessment
d. Debriefing
(Reference: 2.3a)

5. The strongest form of privacy protection of research participants is:

a. Anonymity
b. Confidentiality
c. Informed Consent
d. Password-protected computers
(Reference: 2.3b)

6. The local entity that oversees the rights and well-being of research participants is called:

a. The Scientific Review Committee (SRC)
b. The National Research Foundation (NRF)
c. The Institutional Review Board (IRB)
d. The Committee for People (CFP)
(Reference: 2.3c)

7. Patient advocacy efforts related to the AIDS crisis:

a. Had an impact on the speed of patient access to new treatments
b. Resulted in patients being cared for in very conservative ways to make sure that treatment side effects were completely controlled
c. Had no impact on research ethics
d. Stimulated the 2nd Belmont Conference
(Reference: 2.3d)

8. Which of the following statements best characterizes the current state of ethical regulation of animal research?

a. Animals have the same rights as humans
b. There is an important distinction in the rights of mammals versus other animals in research ethics regulations
c. There is less of a legal framework in protection of animals than humans, with most of the oversight provided by local IRBs
d. Animals have the same rights as humans except for informed consent
(Reference: 2.3e)

9. _____ means fabrication, falsification, or plagiarism in proposing, performing, or reviewing research, or in reporting research results.

a. Conflict of Interest
b. Research Misconduct
c. The Nuremberg Code
d. The Ivory Tower Syndrome
(Reference: 2.4)

10. The tragic case of Jesse Gelsinger involved _____

a. Conflict of Interest
b. Researchers who "went fishing" for significant results
c. Inadequate Debriefing
d. Right to Service
(Reference: 2.4b)

11. "Piecemeal and Duplicate Publication" is a good thing because it enhances dissemination and therefore speeds up the process of translational research.

a. True
b. False
(Reference: 2.4a)

12. Authorship credit should be established on the basis of the quality and quantity of one's contributions to a study, not status, power, or any other factor.

a. True
b. False
(Reference: 2.4c)

13. In designing the ethical considerations of a study, it is best to guarantee that participants' responses will be both anonymous and confidential.

a. True
b. False
(Reference: 2.3b)

14. The ethical principle of informed consent means that researchers must get a form signed by their participants.

a. True
b. False
(Reference: 2.3a)

15. Research ethics are relevant to every phase of research design.

a. True
b. False
(Reference: Summary)

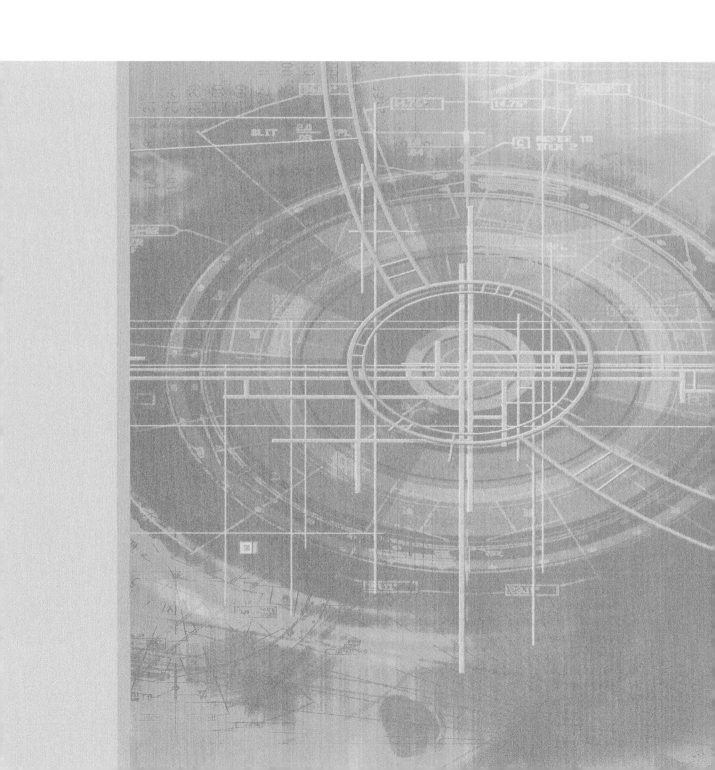

3

Qualitative Approaches to Research

3.1 Foundations of Qualitative Research

Before movies like *The Usual Suspects* and *Memento* came about, there was this remarkable Japanese film, *Rashomon*. If you haven't seen it yet, we highly recommend you watch it! This 1950s classic, directed by Akira Kurosawa, tells the story of four individuals who witness a terrible crime. The witnesses, however, recount the same story from different perspectives, where each account contradicts the other three. At the end of the film, you are left wondering which of the four witnesses was actually telling the truth. In fact, the central theme of the movie revolves around the idea that "truth" or "reality" is subjective and depends on an individual's perspective (**Figure 3.1**).

Similar questions about "truth" are confronted in social science research as well. It is very common to come across a phenomenon that is observed to have various, sometimes even conflicting, interpretations. When trying to understand the different meanings people give to an experience, quantitative methods fall short. One of the best ways to capture and make sense of such complexity is through qualitative research methods. This chapter introduces the qualitative traditions, the different types of qualitative methods, the link between qualitative and quantitative data, and, finally, the standards for judging qualitative work. The chapter includes discussion of design and analysis as well as measurement, because we think presenting the qualitative approach holistically fits the method better. However, you should be aware that entire volumes have been written about each of the approaches we discuss.

Qualitative methods have recently become more prominent in social science and education, but they actually have a relatively long history in many fields. Some fifty years ago a sociologist by the name of William Bruce Cameron commented on the necessity of qualitative methods. He said: "It would be nice if all of the data which sociologists require could be enumerated because then we could

Figure 3.1 The classic film *Rashomon* is an example of how "truth" is subjective.

DAIEI FILMS/Album/Newscom

run them through IBM machines and draw charts as the economists do. However, not everything that can be counted counts, and not everything that counts can be counted" (1963, p. 13). Cameron's point reminds us of the limits of quantitative measurement as well as the inevitability of encountering phenomena that are difficult or impossible to quantify. We need all sorts of methods if our goal is to know the truth of human experience. This is where qualitative approaches come into focus.

Most qualitative methods prescribe different roles for the researcher in how they engage with study participants. For example, in some cases the researcher might wish to observe the population of interest from a distance, and in other cases it might be more useful to interact extensively with participants. These variations strongly contrast with the role of a quantitative experimenter who follows the exact same set of procedures with every participant in a controlled setting.

3.2 The Context for Qualitative Research

Qualitative research is a vast and complex area of methodology that concerns itself with analyzing how people interpret their experience and the world in which they live. The term **qualitative** refers to the nonnumerical representation of some object. It is often contrasted with the term quantitative, which is applicable only when an object is represented in numerical form. **Qualitative measures** are any measures where the data are recorded in a nonnumerical format. These include short written responses on open-ended surveys; interviews; detailed contextual descriptions such as those in anthropological field research; video and audio data recording; and many others, all of which are collected in the form of words and/or pictures rather than numbers.

Qualitative research is typically the approach of choice in situations where you want to:

- Generate new constructs, theories, or hypotheses
- Develop detailed stories to describe a phenomenon
- Achieve a deeper understanding of issues
- Improve the quality of quantitative measures

These goals are addressed in detail below.

3.2a Generating New Theories or Hypotheses

One of the major reasons for doing qualitative research is to understand a phenomenon well enough to be able to form some initial theories, hypotheses, or hunches about how it works. Very often in applied social research, students jump from doing a quick review of the literature on a particular topic to writing an entire research proposal, based just on existing research and their own thinking. What they miss is the direct experience of the phenomenon. Think about the difference in "knowing" that comes from reading about something versus that which comes from experiencing it—either firsthand, or by asking those who have experienced it. If you have the benefit of such an experience, you are likely to approach the existing literature on the topic with a fresh perspective. You might also have your own insights about what causes what to happen. This is where many interesting and valuable new theories and hypotheses originate.

qualitative The descriptive nonnumerical characteristic of some object. A qualitative variable is a descriptive nonnumerical observation.

qualitative measures Data not recorded in numerical form.

Qualitative research can play a major role in the initial stages of theory development because it employs an inductive approach that focuses on asking open-ended questions. In fact, if you examine the history of any well-known theory (for example, Piaget's theory about the nature and development of human intelligence), you'll likely find that it began with qualitative work. Often these methods are extremely useful when you construct new theories because they make it possible to integrate concepts from multiple different disciplines. Finally, these methods also lend themselves to refining preliminary hypotheses through initial testing.

3.2b Developing Detailed Stories to Describe a Phenomenon

Qualitative research excels at generating detailed information to tell stories. We can understand why that is important when we look at how social research is used in policy development and decision making. There's an informal saying among social researchers that goes something like, "one good personal story trumps pages of quantitative results." In legislative hearings and organizational boardrooms, a well-researched case study is often what compels decision makers. We are not suggesting that all we need to do in social research is produce stories—evidence based on quantitative assessment is often quite persuasive. However, if that is all we present, the impersonal numbers may not translate well for decision makers and may not connect to their experience. Illustrating the implications of quantitative data through well-researched qualitative anecdotes and stories is often essential to persuading decision makers.

Because of its complexity, some of the best qualitative research is published in book form, often in a style that almost approaches a narrative story. One of our favorite writers (and, we daresay, one of the finest qualitative researchers) was Studs Terkel. He wrote intriguing accounts of the Great Depression (*Hard Times*), World War II (*The Good War*), and socioeconomic divisions in America (*The Great Divide*), among others. In each book, he followed a similar qualitative methodology: identifying informants who directly experienced the phenomenon in question; interviewing them at length; and then editing the interviews so that the collection tells a rich and multilayered story that addresses the question of interest in a way that no one story alone would convey (**Figure 3.2**).

So, qualitative research, and the stories it can generate, enables you to describe the phenomena of interest with great detail. In intervention studies, qualitative methods provide the benefit of a richer understanding of the key "independent variable," that is, of the treatment or the program that is being evaluated. Too often, researchers treat this variable like a dichotomous one—whether the study participant did or did not receive the treatment. Such a specification may fail to take account of the degree of construct validity of a treatment—did the participants who received the program receive exactly what was conceived/intended by the research team in the planning stages, or did the process of implementation lead to the

Dennis Van Tine/Landov

Figure 3.2 Writer Studs Terkel's work is an excellent example of qualitative research.

provision of a different kind of treatment (see Chapter 5 for a discussion on construct validity)?

For example, is it possible that your study participants were simultaneously involved in other programs similar to the one you administered? If yes, then it may be that the effects you see (or don't see!) are a result of a combination of programs and not just attributable to the one program you administered. Also, in some multisite studies, there may be considerable variation in the way a program is implemented at different places, based on site-specific characteristics. Here again, a 0-1 specification of receiving or not receiving a treatment may be misleading. In cases like these, it is difficult to know exactly what intervention you are evaluating. The rich, descriptive capacity of qualitative methods gets around limitations like these by providing a much more comprehensive articulation of the program, as well as that of the context in which it was implemented. As a result, this ultimately leads to a better interpretation of the study results.

3.2c Achieving Deeper Understanding of the Phenomenon

As discussed above, qualitative research enables us to get at the rich complexity of the phenomenon, to deepen our understanding of how things work. While quantitative research can describe a phenomenon generally across a group of respondents, it is very difficult to understand from a quantitative study how the phenomenon is understood and experienced by multiple respondents, and how it interacts with other issues and factors that affect their lives. In addition, in social research there are many complex and sensitive issues that almost defy simple quantitative summarization. For example, if you are interested in how people view topics like spirituality, human sexuality, the death penalty, gun control, and so on, our guess is that you would be hard-pressed to develop a quantitative methodology that would do anything more than summarize a few key positions on these issues. While this does have its place (and it's done all the time), if you really want to try to achieve a deep understanding of how people think about these topics, some type of in-depth interviewing or observation is almost certainly required.

One of the great advantages of qualitative methods is that they work to enhance a researcher's "peripheral vision" (Sofaer, 1999). Not only are they beneficial in providing a rich description of events, but they also help to understand how and why the "same" events are seen in different light by different stakeholders. In other words, just like in the movie *Rashomon*, qualitative methods lend themselves very well to analyzing "relative" truths. Often, qualitative researchers use these methods to bring to the fore perspectives of those who are rarely heard, for example illegal immigrants, many of whom are alienated from the society's formal institutional processes. Giving weight to their voices contributes significantly to the process of policy making, policy implementation, and policy evaluation. It provides for a more nuanced, comprehensive, and meaningful explanation of the phenomenon under study.

3.2d Improving the Quality of Quantitative Measures

Qualitative methods play an important role in the development and improvement of the quality of quantitative methods such as survey instruments. In the initial stage, qualitative data help focus the research on particular constructs and terms. Then, after the development of a survey instrument, researchers often realize that the items on the survey may mean different things to different people. This is problematic because if questions are understood and perceived in different ways by different people (say, by people of different ethnic groups), the subsequent analysis would be an apples-to-oranges comparison. Qualitative research is therefore imperative to improve the quality of quantitative measures during the development and pilot testing process.

A range of qualitative procedures may be used in strengthening quantitative measures. These methods include focus groups, interviews, input from experts, and other approaches. For example, focus groups may be used to determine how different groups think about and approach the key construct in question. If some individuals do not feel comfortable about discussing their perspectives in a focus group, in-depth interviews may be conducted to gain additional information on sensitive issues. Thereafter, insights from the focus group sessions and interviews can be combined to develop a preliminary survey questionnaire. Then the questionnaire may be provided to a panel of experts for comments. Input from experts can be used to make any further changes to the survey instrument. In such ways, qualitative methods can be used to improve the reliability and validity of quantitative measures. Here again, because qualitative methods have the potential to draw out and synthesize the full variety of experiences from multiple different participants as well as experts, they play a useful role in developing high-quality quantitative measures.

3.3 Qualitative Traditions

A qualitative tradition is a general way of thinking about conducting qualitative research. It describes, either explicitly or implicitly, the purpose of the qualitative research, the role of the researcher(s), the stages of research, and the method of data analysis. Here, four of the major qualitative traditions are introduced: ethnography, phenomenology, field research, and grounded theory.

3.3a Ethnography

The ethnographic approach to qualitative research comes largely from the field of anthropology. The emphasis in **ethnography** is on studying a phenomenon in the context of its culture. Originally, the idea of a culture was tied to the notion of ethnicity and geographic location, but it has been broadened to include virtually any group or organization (so, it's reasonable to talk of an "organizational culture").

Ethnography is an extremely broad area with a great variety of practitioners and methods. However, the most common ethnographic approach is **participant observation**, which is conducted as a part of field research. The ethnographer becomes immersed in the culture as an active participant and records extensive field notes. As in grounded theory (below), there is no preset limiting of what will be observed and no obvious ending point in an ethnographic study.

An example of this approach is Myra Bluebond-Langner's series of studies of critically ill children and their families. Her original study took place in a cancer hospital in which she interviewed children and their families in order to better understand children's knowledge of their situation as well as the socialization processes that involved everyone connected to the ill child.

ethnography Study of a culture using qualitative field research.

participant observation A method of qualitative observation where the researcher becomes a participant in the culture or context being observed.

Her careful observations included not only what children and others said, but also what they did and how they interacted. Among many other findings, she discovered that children knew much more about their condition than most adults realized (Bluebond-Langner, 1978). Bluebond-Langner's work led to reconsideration of communication strategies and to a new emphasis on honest and open communication with children.

At the end of her book, she included a chapter with a personal description of what it was like to do the fieldwork as a participant in the setting. For example, she described the centrality of building trust with the young patients and how she went about doing that. The comment below provides an example of the importance of the relationship in obtaining genuine and qualitatively rich data. After you've read this, think of the difference in her research process versus an alternate strategy of using structured interviews, written surveys, psychometric scales, unobtrusive measures, or other methods with the children in the hospital.

> "As one eight-year-old boy said to me after I watched TV for one solid hour without a word, "All right. You're OK. What do you want to know?" In general, I moved slowly and cautiously. There were times when I wanted to pursue a question, but refrained from doing so for fear of breaking off communication." (p. 247)

The outcomes of this approach included stimulation of theories of adaptation, quantitative studies of communication and coping, as well as additional research using similar methods to examine siblings and parents.

3.3b Phenomenology

The phenomenology tradition emphasizes the study of how the phenomenon is experienced by respondents or research participants. Originating with German philosopher Edmund Husserl, it has a long history in several social research disciplines including psychology, sociology, and social work. **Phenomenology** is a school of thought that focuses on people's subjective experiences and interpretations of the world. That is, the phenomenologist wants to understand how the world is experienced by others from their perspective.

Robinson, Giorgi, and Ekman (2012) conducted a phenomenological study to understand the process of adaptation to early Alzheimer's disease, with annual interviews over four years with a woman who had signs of early dementia. The researchers followed a structured protocol to collect and analyze the data. The analysis consisted of four steps, beginning with reading the entire transcript several times to "get a sense of the whole experience" (p. 221), and concluding with a description of the **meaning units** discovered in the analysis. For example, the researchers discussed the experience of being confronted by others who were critical of increasing forgetfulness, thus generating distress over something the woman was not aware of and could not understand. The study concluded by tying the detailed account of one person's experience to more general issues related to the process of adaptation and support of people with early dementia.

3.3c Field Research

Field research can also be considered either a broad tradition of qualitative research or a method of gathering qualitative data. The essential idea is that the researcher

phenomenology A philosophical perspective as well as an approach to qualitative methodology that focuses on people's subjective experiences and interpretations of the world.

meaning units In qualitative data analysis, a small segment of a transcript or other text that captures a concept that the analyst considers to be important.

field research A research method in which the researcher goes into the field to observe the phenomenon in its natural state.

goes into the field to observe the phenomenon in its natural state or "in situ" (on site). As such, it is probably most related to the method of participant observation. The field researcher typically takes extensive field notes that are subsequently coded and analyzed for major themes.

Within this general set of methods, participatory action research (PAR) perhaps goes the furthest in varying the researcher–participant–data relationship by turning researchers into participants, participants into researchers, and data into action. An example of this is a series of studies of prison-based college programs for women by Professor Michelle Fine and her colleagues. The studies are multi-faceted, richly documented and impossible to summarize briefly. However, one distinguishing aspect of the study design was including prisoners as researchers. Among the many notable consequences of this approach was the insight provided by those who lived in the setting as prisoners, as well as their critical perspective. As Fine et al. (2003) suggested, a study of prison college programs conducted exclusively by researchers from the outside might have turned into a "sugar-coated greeting card of praise for the program" (p. 195), instead of a critical view of the less positive outcomes such as dissatisfaction and dropout. In addition, the research process became a collaborative process of knowledge development that generated enduring change within the research–participant team as well as within the program itself.

3.3d Grounded Theory

Grounded theory is a qualitative research tradition that was originally developed by Glaser and Strauss in the 1960s (Glaser & Strauss, 1967). The purpose of grounded theory is to develop theory about the phenomena of interest; but they are not talking about abstract theorizing. Instead the theory needs to be grounded or rooted in observations; hence the term.

Grounded theory is a complex dynamic iterative process in which the development of a theory and the collection of data related to that theory build on each other. The research begins with the raising of generative questions that help guide the research but are not intended to be either static or confining. As the researcher begins to gather data, core theoretical concept(s) are identified. Tentative linkages are developed between the theoretical core concepts and the data. This early phase of the research tends to be open, and it can take months. Later on, the researcher is more engaged in verification and summary. The effort tends to evolve toward one core category that is central. Eventually you approach a conceptually dense theory as each new observation leads to new linkages that lead to revisions in the theory and more data collection. The core concept or category is identified and fleshed out in detail.

What do you have when you're finished? Presumably you have an extremely well-considered explanation for some phenomenon of interest—the grounded theory. This theory can be explained in words and is usually presented along with much of the contextually relevant detail.

This approach was used in a study of self-management of HIV infections in a sample of ethnically diverse men and women by Wilson, Hutchinson, and Holzemer (2002). The goal of the study was to understand how people with HIV might choose to comply with treatment or find alternative methods of managing their condition. Through the constant comparative method of analysis, they developed a theory of "Reconciling Incompatibilities" that modeled complex factors influencing a person's "State of Mind," resulting in a range of choices.

grounded theory A theory rooted in observation about phenomena of interest. Also, a method for achieving such a theory.

3.4 Qualitative Methods

Now that we have a general understanding of qualitative research, let's discuss how to go about *doing* such research practically. A variety of methods are common in qualitative measurement. In fact, the methods are limited primarily by the imagination of the researcher. Here, we discuss a few of the more widely used methods.

3.4a Participant Observation

One of the most common methods for qualitative data collection, participant observation, is also one of the most demanding. It requires that the researcher become a participant in the culture or context being observed. The literature on participant observation discusses how to enter the context, the role of the researcher as a participant, the collection and storage of field notes, and the analysis of field data. Participant observation often requires months or years of intensive work, because the researcher needs to become accepted as a natural part of the culture to ensure that the observations are of the natural phenomenon.

3.4b Direct Observation

Direct observation is distinguished from participant observation in a number of ways. First, a direct observer doesn't typically try to become a participant in the context. Instead, the direct observer strives to maintains some distance so as not to bias the observations by their presence. Second, direct observation suggests a more detached perspective. The researcher is watching rather than both watching and taking part. Consequently, technology can be a useful part of direct observation. For instance, you can videotape the phenomenon or observe from behind one-way mirrors. Third, direct observation tends to be more structured than participant observation. The researcher is observing certain sampled situations or people rather than trying to become immersed in the entire context. Finally, direct observation tends not to take as long as participant observation. For instance, one might observe child–mother interactions under specific circumstances in a laboratory setting, looking especially for the nonverbal cues being used.

direct observation The process of observing a phenomenon to gather information about it. This process is distinguished from participant observation, in that a direct observer doesn't typically try to become a participant in the context and does strive to be as unobtrusive as possible so as not to bias the observations.

3.4c Unstructured Interviewing

Unstructured interviewing involves direct interaction between the researcher and a respondent or group. It differs from traditional structured interviewing in several important ways. First, although the researcher may have some initial guiding questions or core concepts to ask about, there is no formal structured instrument or protocol. Second, the interviewer is free to move the conversation in any direction of interest that may come up. Consequently, unstructured interviewing is particularly useful for exploring a topic broadly. However, there is a price for this lack of structure. Because each interview tends to be unique with no predetermined set of questions asked of all respondents, it is usually more difficult to analyze unstructured interview data, especially when synthesizing across respondents.

Unstructured interviewing may very well be the most common form of data collection of all. You could say it is the method being used whenever anyone asks someone else a question! It is especially useful when conducting site visits or casual focus groups designed to explore a context or situation.

unstructured interviewing An interviewing method that uses no predetermined interview protocol or survey and where the interview questions emerge and evolve as the interview proceeds.

3.4d Case Studies

A **case study** is an intensive study of a specific individual, event, organization, or specific context. For instance, Freud developed case studies of several individuals as the basis for the theory of psychoanalysis, and Piaget did case studies of children to study developmental phases (**Figure 3.3**). Case studies are extensively used in business, law, and policy analysis, with the level of analysis varying from a particular individual to the history of an organization or an event. There is no single way to conduct a case study, and a combination of methods (such as unstructured interviewing and direct observation) is often used. We include case studies in our discussion of qualitative research strategies, but quantitative approaches to studying cases are quite possible and becoming more common with new technology. For example, sometimes researchers provide participants with electronic data collection devices (sometimes called *ambulatory data loggers*) to capture a stream of lived events in the natural context. This kind of data can be examined using many kinds of graphic and time series analyses.

case study An intensive study of a specific individual or specific context.

Sometimes qualitative case studies can become a type of intervention as well as a way of observing or measuring. An interesting recent example of this is the Most Significant Change (MSC) technique (Dart & Davies, 2003). The MSC approach generates stories directly from program participants by asking them to describe the most significant change they have experienced or observed in a given period as a result of the program. This form of case study is well suited to understanding change processes as they unfold, but as Dart and Davies pointed out, it can also be used to summarize change at the conclusion of a program and may include both quantitative and qualitative indicators.

3.4e Focus Groups

Focus groups have become extremely popular in marketing and other kinds of social research, because they enable researchers to obtain detailed information about attitudes, opinions, and preferences of selected groups of participants. These methods can be used to generate as many ideas on a topic as possible and to achieve consensus in a group. Sometimes a focus group can be effectively used as a first stage in development of a survey through the identification of potential items relevant to a topic or population. Careful planning of a focus group includes the following considerations:

Bill Anderson/Science Source

Figure 3.3 Piaget wrote significant case studies of children while studying developmental phases.

- *What will the specific focus be?* It is wise to keep the number of focus questions limited to about five to seven.
- *Who will participate?* Generally speaking, seven to twelve participants per group will be optimal, but the number of groups you conduct will depend on how much diversity you want to include in your sample.
- *How will you record the observations?* Audiotaping and videotaping, transcripts, and detailed note taking can be used solely or in combination.
- *How will you analyze the data?* There are several approaches to focus group analysis, but perhaps the main consideration is to have a written plan prior to conducting your groups.

It is also very important to think carefully about the ethics of inviting people to discuss topics in a focus group format, especially if the topic is a sensitive one and if your participants are in some way considered vulnerable or have ongoing relationships with one another.

3.4f Unobtrusive Methods in Qualitative Research

In all of the methods we've presented to this point, researchers have some interaction with respondents in the course of conducting studies. For example, direct observation and participant observation require the researcher to be physically present. This can lead the respondents to alter their behavior so that they look good in the eyes of the researcher or to conform to what they think the researcher would like to see. A questionnaire is an interruption in the natural stream of behavior. Respondents may tire of filling out a survey or become resentful of the questions asked.

Unobtrusive measures are methods of collecting data that do not interfere in the lives of the respondents. In most cases the respondent is not even aware that they are being collected. Unobtrusive measurement presumably reduces the biases that result from the intrusion of the researcher or measurement instrument. However, unobtrusive measures depend on the context and, in many situations, are simply not available or feasible. For some constructs, there may not be any sensible way to develop unobtrusive measures. In addition, a researcher may also miss out on relevant information as a result of not engaging with the study audience.

It is important to note that unobtrusive methods may be used in both quantitative and qualitative research. For example, secondary analysis of data is an unobtrusive method that is almost exclusively used in quantitative research. In this section, we will discuss two approaches to unobtrusive measurement in qualitative research: indirect measures and content analysis.

Indirect Measures

An **indirect measure** is an unobtrusive measure that occurs naturally in a research context. The researcher is able to collect data without the respondent being aware of it.

The types of indirect measures that may be available are limited only by the researcher's imagination and inventiveness. For instance, let's say you would like to measure the popularity of various exhibits in a museum. It may be possible to set up some type of mechanical measurement system that is invisible to the museum patrons. In one study, the system was simple. The museum installed new floor tiles in front of each exhibit it wanted a measurement on, and, after a period of time, researchers measured the wear-and-tear on the tiles as an indirect measure of patron traffic and interest. You might be able to improve on this approach considerably, using more contemporary electronic instruments. For instance, you might construct an electrical device that senses movement in front of an exhibit, or place hidden cameras and subsequently code patron interest based on videotaped evidence. Similarly, if you want to study which sections of a book (say, the Yellow Pages) are used the most; you could look for dirty page edges, dirt smudges, and underlining on pages.

One of our favorite indirect measures occurred in a study of radio station listening preferences. Rather than conducting an obtrusive, costly, and time-consuming survey or interviewing people about their favorite radio stations, the researchers went to local auto dealers and garages and checked all cars that were

unobtrusive measures Methods used to collect data without interfering in the lives of the respondents.

indirect measure An unobtrusive measure that occurs naturally in a research context.

being serviced to see what station the radios were tuned to when the cars were brought in for servicing. Of course, we need to be careful about how we interpret indirect measures. Just checking radio station settings of cars brought in for servicing can be deceptive. We can't automatically conclude that the driver of the car was the one who actually preferred that station (wait till you have kids!), or when it was being listened to, or how often it was tuned in. Another interesting use of unobtrusive methods involved studying art objects to gain an understanding of how infants and small children are held. Researchers noted that from early paintings of *Madonnas* to contemporary sculptures and photographs, women universally tend to hold children against their left side. Portrayals of men do not exhibit this bias. This finding was confirmed in an observational study of live adults holding children (Finger, 1975).

There is an explosion of new approaches to indirect measurement based on new technologies, including tracking and counting capacities that are built into the smart phone that is likely to be in your pocket right now. Other examples include "smart garbage cans" that allow researchers to identify specific materials recycled or disposed as garbage, and "smart pill dispensers" that can track the time and amount of medication removed by a patient. In addition, cameras and GPS devices are producing mountains of data that could be coded. The number and kind of such applications is beyond inventorying at any moment because of the rate of innovation.

These examples illustrate one of the most important points about indirect measures; you have to be careful about ethics when using this type of measurement. In an indirect measure you are, by definition, collecting information without the respondents' knowledge. In doing so, you may be violating their right to privacy and you are certainly not using informed consent. Of course, some types of information may be public and therefore do not involve an invasion of privacy, but you should be especially careful to review the ethical implications of the use of indirect measures.

Content Analysis

Content analysis is the systematic analysis of text. The analysis can be quantitative, qualitative, or both. Typically, the major purpose of content analysis is to identify patterns in text. Content analysis is an extremely broad area of research. For qualitative research, it includes the following types of analysis:

- *Thematic analysis of text:* The identification of themes or major ideas in a document or set of documents. The documents can be any kind of text including field notes, newspaper articles, technical papers, or organizational memos.
- *Indexing:* A variety of automated methods for rapidly indexing text documents exists. For instance, Key Words in Context (KWIC) analysis is a computer analysis of text data. A computer program scans the text and indexes all key words. A key word is any term in the text that is not included in an exception dictionary. Typically an **exception dictionary** would exclude nonessential words like "is," "and," and "of." All remaining key words in the text are alphabetized and listed with the text that precedes and follows it, so the researcher can see the word in the context in which it occurred in the text. In an analysis of interview text, for instance, you could easily identify all uses of the term "abuse" and the context in which they were used.

In some types of content analysis, you might want to find out which words or phrases were used most frequently in the text. Again, this type of analysis is most often done directly with computer programs. At this point, content and qualitative

content analysis The analysis of text documents. The analysis can be quantitative, qualitative, or both. Typically, the major purpose of content analysis is to identify patterns in text.

exception dictionary A dictionary that includes all nonessential words like "is," "and," and "of," in a content analysis study.

analysis can start to blend into or merge with quantitative. If we are counting words, the objects (the words) are qualitative but the results (counts) are numbers.

Content analysis typically includes several important steps or phases. First, when there are many texts to analyze (e.g., newspaper stories, organizational reports) the researcher often has to begin by sampling from the population of potential texts to select the ones that will be used. Second, the researcher usually needs to identify and apply the rules that are used to divide each text into segments or "chunks" that will each be treated as a separate unit of analysis in the study, a process referred to as **unitizing**. For instance, you might extract each identifiable assertion from a longer interview transcript. Third, the content analyst constructs and applies one or more codes to each unitized text segment, a process called **coding**. The development of a coding scheme is based on the themes that you are searching for or uncover as you classify the text. Finally, you analyze the coded data, very often both quantitatively and qualitatively to determine which themes occur most frequently, in what contexts, and how they might be correlated.

Content analysis has several potential limitations you should keep in mind. First, you are limited to the types of information available in text form. If you are studying the way a news story is being handled by the news media, you probably would have a ready population of news stories from which you could sample. However, if you are interested in studying people's views on capital punishment, you are less likely to find an archive of text documents that would by itself be appropriate. Second, you have to be especially careful with sampling to avoid bias. For instance, a study of current research on methods of treatment for cancer might use the published research literature as the population. This would leave out both the writing on cancer that did not get published for one reason or another (including potential publication bias) as well as the most recent work that has not yet been published. Finally, you have to be careful about interpreting results of automated context analyses. A computer program cannot determine what someone meant by a term or phrase. It is relatively easy in a large analysis to misinterpret a result because you did not take into account the subtleties or context of meaning.

However, content analysis has the advantage of being unobtrusive and, depending on whether automated methods exist, can be a relatively rapid method for analyzing large amounts of text.

3.5 Qualitative Data

Qualitative data are extremely varied in nature. They include virtually any information that can be captured in nonnumerical form. Qualitative data, say, from interviews or field research, can be recorded in numerous ways including stenography, audio recording, video recording, and written notes. Data can also be image-based; for example, a courtroom drawing of witnesses is a type of qualitative data collected through direct observation methods. Further, qualitative data can also be in the form of written documents. Usually these are existing documents such as newspapers, magazines, books, websites, memos, transcripts of conversations, and annual reports, but they could also be transcripts of interviews you conduct. Usually written documents are analyzed with some form of content analysis. One current example is the analysis of *Tweets*. At the moment of this writing (November 2013), a Google search of the term "Twitter analysis" produced over 1.5 billion results, and it is common to hear reports of Twitter content analysis as an indicator of public opinion.

unitizing In content analysis, the process of dividing a continuous text into smaller units that can then be analyzed.

coding The process of categorizing qualitative data.

3.5a How Different Are Quantitative and Qualitative Data?

While the assumptions and philosophical underpinnings of quantitative and qualitative approaches can be easily contrasted, many researchers argue that the dichotomy between quantitative and qualitative *data* is false. There does seem to be some truth in this argument, for the following reasons:

- **All qualitative data can be coded quantitatively.**
- **All quantitative data are based on qualitative judgment.**

We'll consider each of these reasons in turn.

All Qualitative Data Can Be Coded Quantitatively

What we mean here is simple. Anything that is qualitative can be assigned meaningful numerical values. These values can then be manipulated numerically or quantitatively to help you achieve greater insight into the meaning of the data so you can examine specific hypotheses. Consider an example.

Many surveys have one or more short, open-ended questions that ask the respondent to supply text responses. The most familiar instance is probably the sentence that is often tacked onto a short survey, "Please add any additional comments." The immediate responses are text-based and qualitative; but you can always (and usually will) perform some type of simple classification of the text responses. You might sort the responses into simple categories, for example. Often, you'll give each category a short label that represents the theme in the response. What you don't often recognize is that even the simple act of categorizing can be viewed as a quantitative one. For instance, let's say that you develop five themes that the respondents express in their open-ended responses. Assume that you have ten respondents. You could easily set up a simple coding table like the one in Table 3.1. In this example, Person 1 had a comment that touched on Themes 1 and 3. The coder put an "X" mark in each of those columns for Person 1.

This is a simple qualitative thematic coding analysis. But, you can represent exactly the same information quantitatively, as shown in Table 3.2.

Table 3.1 Coding of qualitative data into five themes for ten respondents.

Person	Theme 1	Theme 2	Theme 3	Theme 4	Theme 5
1	X		X		
2	X	X		X	
3	X	X			
4			X	X	
5		X			X
6	X		X		
7					
8	X	X	X	X	X
9		X		X	X
10		X			

Table 3.2 Quantitative coding of the data in Table 3.1.

Person	Total	Theme 1	Theme 2	Theme 3	Theme 4	Theme 5
1	2	1	0	1	0	0
2	3	1	1	0	1	0
3	2	1	1	0	0	0
4	2	0	0	1	1	0
5	2	0	1	0	0	1
6	2	1	0	1	0	0
7	0	0	0	0	0	0
8	5	1	1	1	1	1
9	3	0	1	0	1	1
10	1	0	1	0	0	0
Total		5	6	4	4	3

Notice that this is exactly the same raw data, except that we added up the number of "Xs" per person and per theme. The first table (Table 3.1) would probably be called a qualitative coding, while the second (Table 3.2) is clearly quantitative. The quantitative coding gives you additional useful information and makes it possible to do analyses that you couldn't do with the qualitative coding. For instance, simply by adding down the columns, you can say that Theme 2 was the most frequently mentioned and, by adding across the rows, you can say that most of the respondents touched on two or three of the five themes.

The point is that the line between qualitative and quantitative data is less distinct than we sometimes imagine. All qualitative data can be quantitatively coded in an almost infinite variety of ways. This doesn't detract from the qualitative information. You can still do any judgmental syntheses or analyses you want; but recognizing the similarities between qualitative and quantitative information opens up new possibilities for interpretation that might otherwise go unutilized.

All Quantitative Data Are Based on Qualitative Judgment

Numbers in and of themselves can't be interpreted without understanding the assumptions that underlie them. Take, for example, a simple 1-to-5 rating variable, shown in **Figure 3.4**.

Here, the respondent answered 2=Disagree. What does this mean? How do you interpret the value 2 here? You can't really understand this quantitative value unless you dig into some of the judgments and assumptions that underlie it, such as:

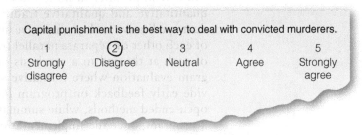

Figure 3.4 A rating illustrates that quantitative data are based on qualitative judgments.

- **Did the respondent understand the term capital punishment?**
- **Did the respondent understand that "2" means that he or she is disagreeing with the statement?**

- Does the respondent have any idea about alternatives to capital punishment (otherwise, how can he or she judge what's best)?
- Did the respondent read carefully enough to determine that the statement was limited only to convicted murderers (for instance, rapists were not included)?
- Does the respondent care, or was he or she just arbitrarily circling responses?
- How was this question presented in the context of the survey (for example, did the questions immediately before this question bias the response in any way)?
- Was the respondent mentally alert (especially if this is late in a long survey or the respondent had other things going on earlier in the day)?
- What was the setting for the survey (lighting, noise, and other distractions)?
- Was the survey anonymous? Was it confidential?
- In the respondent's mind, is the difference between a "1" and a "2" the same as between a "2" and a "3" (meaning, is this an interval scale)?

We could go on and on, but our point should be clear. All numerical information involves numerous judgments about what the number means. Quantitative and qualitative data are, at some level, virtually inseparable. Neither exist in a vacuum; neither can be considered totally apart from the other. To ask which are better or more valid or have greater verisimilitude ignores the intimate connection between them. To do the highest-quality research, you need to incorporate both the qualitative and the quantitative approaches.

Mixed Methods Research

One of the most important areas in applied social research these days is called **mixed methods research**. In mixed methods research (Greene & Caracelli, 1997) we simultaneously or sequentially conduct both qualitative and quantitative research, to achieve the advantages of each and mitigate their weaknesses. The proponents of mixed methods research argue that the combination of qualitative and quantitative methods provides for a better understanding of the research problem than any one approach alone. An example of mixed methods research was provided in the previous section, where the use of qualitative methods to improve quantitative measures such as survey instruments was discussed.

There are several different ways to accomplish the mixing of methods. These tend to differ in how, and at what stage of the research, you bring the quantitative and qualitative traditions together. For instance, you can conduct qualitative and quantitative substudies as though they are independent of each other on separate parallel tracks, where you bring together the results of each at the end in a synthesis or summary. This is very common in program evaluation where formative studies—those that are conducted to provide early feedback on program implementation—often employ qualitative, open-ended methods, while summative evaluations for studying outcomes of a program, once the implementation has stabilized, usually use quantitative methods. In this case, the overall evaluation methodology uses a mix of both methods. Another example of this type of sequential mixing is when a researcher uses a survey instrument to obtain quantitative data, but follows up with interviews to get in-depth responses and details on contextual elements from the participants.

You might even mix quantitative and qualitative data collection methods throughout, analyzing the results together and examining similarities and

mixed methods research Research that uses a combination of qualitative and quantitative methods.

contrasts. Or, you can integrate the qualitative and quantitative approaches into a new synthetic method, such as when we combine qualitative brainstorming and quantitative rating approaches into a single synthesized approach. Or, you can integrate the paradigmatic perspectives of qualitative and quantitative traditions at all stages of a research project, repeatedly and dynamically using each to question and improve the results of the other.

Quantitative research excels at summarizing large amounts of data and reaching generalizations based on statistical estimations. Qualitative research excels at telling the story from the participant's viewpoint, providing the rich descriptive detail that sets quantitative results into its human context. We are only beginning to learn about how we can best integrate these great traditions of qualitative and quantitative research, and many of today's social research students will spend much of their careers exploring this idea.

3.6 Assessing Qualitative Research

Qualitative research is often criticized as biased, small-scale, anecdotal, and/or lacking rigor; however, when it is carried out properly it is unbiased, indepth, valid, reliable, credible, and rigorous (Campbell, 1975). In qualitative research, there needs to be a way of assessing whether claims are adequately backed up by convincing evidence. Some qualitative researchers reject the framework of validity that is commonly accepted in more quantitative research in the social sciences (Lincoln & Guba, 1985). They reject the idea that there is a single reality that exists separate from our perceptions. In their view, each of us sees a different reality because we see it from a different perspective and through different experiences. They don't think research can be judged using the criteria of validity. Research is less about getting at the "truth" than it is about reaching meaningful conclusions, deeper understanding, and useful results. These qualitative researchers argue for different standards of judging the quality of qualitative research.

For instance, Lincoln and Guba (1985) proposed four criteria for judging the soundness of qualitative research and explicitly offered these as an alternative to four common criteria often used in the quantitative tradition. They felt that their four criteria better reflected the underlying assumptions involved in much of qualitative research. Their proposed criteria and the analogous quantitative criteria are listed in Table 3.3.

Table 3.3 Criteria for judging research quality from a more qualitative perspective.

Traditional Criteria for Judging Quantitative Research	Alternative Criteria for Judging Qualitative Research
Internal validity	Credibility
External validity	Transferability
Reliability	Dependability
Objectivity	Confirmability

3.6a Credibility

The credibility criterion involves establishing that the results of qualitative research are believable from the perspective of the participant in the research. In other words, to what extent do they make sense? Because the purpose of qualitative research is to describe or understand the phenomena of interest from the participant's eyes, in this view the participants are the only ones who can legitimately judge the credibility of the results.

3.6b Transferability

Transferability refers to the degree to which the results of qualitative research can be generalized or transferred to other contexts or settings. This is similar to the external validity idea. From a qualitative perspective, transferability is primarily the responsibility of the one doing the generalizing. The qualitative researcher can enhance transferability by doing a thorough job of describing the research context and the assumptions that were central to the research. The person who wishes to transfer the results to a different context is then responsible for making the judgment of how sensible the transfer is. This is often achieved in the case study context through the choice of cases that are included in the research.

3.6c Dependability

The traditional quantitative view of reliability is based on the assumption of replicability or repeatability (see reliability in Chapter 5, "Introduction to Measurement"). Essentially, it is concerned with whether you would obtain the same results if you could observe the same thing twice. However, you can't actually measure the same thing twice; by definition, if you are measuring twice, you are measuring two different things. This thinking goes back at least to the ancient Greek Democritus who argued that we can never step into the same river twice because the river is constantly changing. To estimate reliability, quantitative researchers construct various hypothetical notions (for example, true score theory as described in Chapter 5) to try to get around this fact.

The idea of dependability, on the other hand, emphasizes the need for the researcher to account for the ever-changing context within which research occurs. The researcher is responsible for describing the changes that occur in the setting and how these changes might affect the conclusions that are reached. Reliability emphasizes the researcher's responsibility to develop measures that, in the absence of any real change, would yield consistent results. Dependability emphasizes the researcher's responsibility to describe the ever-changing research context.

3.6d Confirmability

Qualitative research tends to assume that each researcher brings a unique perspective to the study. Confirmability refers to the degree to which the results could be confirmed or corroborated by others. There are a number of strategies for enhancing confirmability. The researcher can actively search for and describe negative instances that contradict prior observations. After the study, a researcher can conduct a **data audit** that examines the data collection and analysis procedures and makes judgments about the potential for bias or distortion. Additionally, triangulation of data can be a useful tool to demonstrate confirmability.

data audit The process of systematically assessing the quality of data in a qualitative study.

SUMMARY

One way to sharpen the focus on qualitative methods is to integrate quantitative methods into the discussion. In **Figure 3.5**, we have attempted to provide a framework for thinking about these two broad approaches by organizing them on two dimensions: 1) Qualitative versus Quantitative and 2) Primary versus Secondary.

The latter dimension refers to whether the data collection occurred in the course of the study, making it primary, or whether the data were previously collected, making it a secondary analysis. **Secondary analysis** uses data that was collected for some other purpose or study (e.g., census bureau data, crime records, standardized testing data, economic data, consumer data) either to check the analyses done previously or to explore entirely new research hypotheses. Crossing the two dimensions allows us to place many of the methods that were discussed in this chapter in this larger context. Qualitative methods like participant observation, direct observation, focus groups, case studies, and interviews involve primary data collection. In addition, the researcher closely engages with study participants in all of these methods.

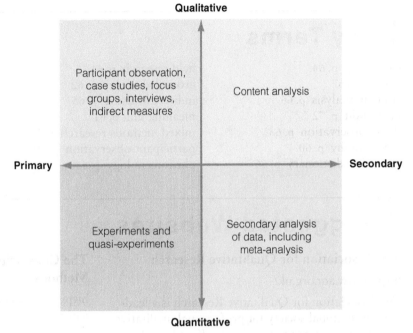

Figure 3.5 A two-dimensional model of research methods.

In contrast, certain unobtrusive qualitative methods, like indirect measures, are used to collect primary data where those being studied are unaware that they are being studied. Content analysis involves the systematic assessment of existing texts (or secondary data) and, because it does not require original data collection, is also typically considered unobtrusive.

Given the different strengths of qualitative and quantitative methods, it is important to choose the most appropriate methods based on your particular research inquiry. Neither qualitative nor quantitative research is appropriate to answer every research question, and therefore, as a researcher, you need to think carefully about your objectives. Is your aim to study a particular phenomenon in-depth (for example, students' perceptions of effective instruction)? Or are you more interested in making standardized comparisons and accounting for variance (for example, analyzing differences in examination grades upon the introduction of after-school tutoring)? Clearly, a quantitative approach would be more appropriate in the latter example. As with any research project, a clear research objective has to be identified to know which methods should be applied. If you choose to employ qualitative research methods, then the standards for judging the quality of your work might include credibility, transferability, dependability, and confirmability.

In conclusion, qualitative measurement comes from a long tradition of field research, which originated in anthropology and then subsequently developed in psychology, sociology, and the other social sciences. This systematic tradition of

secondary analysis
Quantitative analysis of existing data that is done either to verify or extend previously accomplished analyses or to explore new research questions. Using existing data can be an efficient alternative to collecting original primary data and can extend the validity, quality and scope of what can be accomplished in a single study.

inquiry is extremely diverse, and there's probably as much variation and dispute within the tradition as there is across qualitative and quantitative traditions in general. Even the simple notion that qualitative means nonquantitative has begun to break down as we recognize the interconnectedness between the two streams. That said, one theme that underlines much of qualitative measurement is the idea of naturalistic inquiry—that is, it focuses on how people behave when they are absorbed in real-life experiences in natural settings.

Key Terms

case study p. 64
coding p. 67
content analysis p. 66
data audit p. 72
direct observation p. 63
ethnography p. 60
exception dictionary p. 66

field research p. 61
grounded theory p. 62
indirect measure p. 65
meaning unit p. 61
mixed methods research p. 70
participant observation p. 60
phenomenology p. 61

qualitative p. 57
qualitative measures p. 57
unitizing p. 67
unobtrusive measures p. 65
unstructured interviewing p. 63

Suggested Websites

The Association for Qualitative Research

http://www.aqr.org.uk/

The Association for Qualitative Research is a leading international society for professional qualitative researchers. Founded in the United Kingdom about forty years ago, the organizations's website provides you with a sense of the activities and interests of qualitative researchers, including commercially oriented researchers, along with some sense of professional standards of writing and publishing qualitative work.

The Consortium on Qualitative Research Methods

http://www.maxwell.syr.edu/moynihan/programs/cqrm/

The consortium is based at the Maxwell School at Syracuse University. This group works to promote the use of qualitative methods through a variety of courses, an annual institute, and a newsletter.

Review Questions

1. Which of the following statements is correct?

a. Qualitative data can only be coded qualitatively.
b. Quantitative data are based on qualitative judgment.
c. Quantitative data are often collected in text format for content analysis.
d. Qualitative data are the primary type of unobtrusive measure.
(Reference: 3.5a)

2. Which of the following is not an approach to qualitative research?

a. field research
b. ethnography
c. factor analysis
d. grounded theory
(Reference: 3.3)

3. What type of qualitative approach to data collection begins with a set of generative questions, then identifies core concepts as data are gathered, with

linkages developed between the core concepts and the data?

a. grounded theory
b. ethnography
c. focus groups
d. field research
(Reference: 3.3d)

4. In general, which of the following methods is the most demanding qualitative method?

a. direct observation
b. analyzing existing documents
c. participant observation
d. case studies
(Reference: 3.4a)

5. For which of the following purposes is qualitative research *not* well suited?

a. generalizing themes across a population
b. generating new theories or hypotheses
c. achieving a deep understanding of complex and sensitive issues
d. generating information that is very detailed
(Reference: 3.2)

6. According to Lincoln and Guba, the alternative criterion for judging the "internal validity" of qualitative research is

a. credibility.
b. confirmability.
c. transferability.
d. dependability.
(Reference: 3.6)

7. According to Lincoln and Guba, the alternative criterion for judging the "objectivity" of qualitative research is

a. credibility.
b. confirmability.
c. transferability.
d. dependability.
(Reference: 3.6d)

8. When assessing the confirmability of qualitative research results, a researcher can

a. actively search for and describe negative instances that contradict prior observations.
b. do a data audit after the fact, to make judgments about potential bias.

c. have another researcher play "devil's advocate" with respect to the results, documenting the process.
d. All of the above will increase the confirmability of qualitative research results.
(Reference: 3.6d)

9. Which of the following is *not* an unobtrusive measure?

a. indirect measure
b. content analysis
c. participant observation
d. secondary analysis
(Reference: 3.4f)

10. Which unobtrusive measure involves the systematic analysis of text?

a. indirect measure
b. content analysis
c. participant observation
d. secondary analysis
(Reference: 3.4f)

11. Qualitative measures are any measures where the data are not recorded in numerical form.

a. True   b. False
(Reference: 3.5)

12. Quantitative research is better than qualitative research for generating new theories.

a. True   b. False
(Reference: 3.2a)

13. All qualitative data can be coded quantitatively.

a. True   b. False
(Reference: 3.5a)

14. One of the best ways to reduce the impact of measurement procedures or the presence of researchers as observers is known as unobtrusive measurement.

a. True   b. False
(Reference: 3.4f)

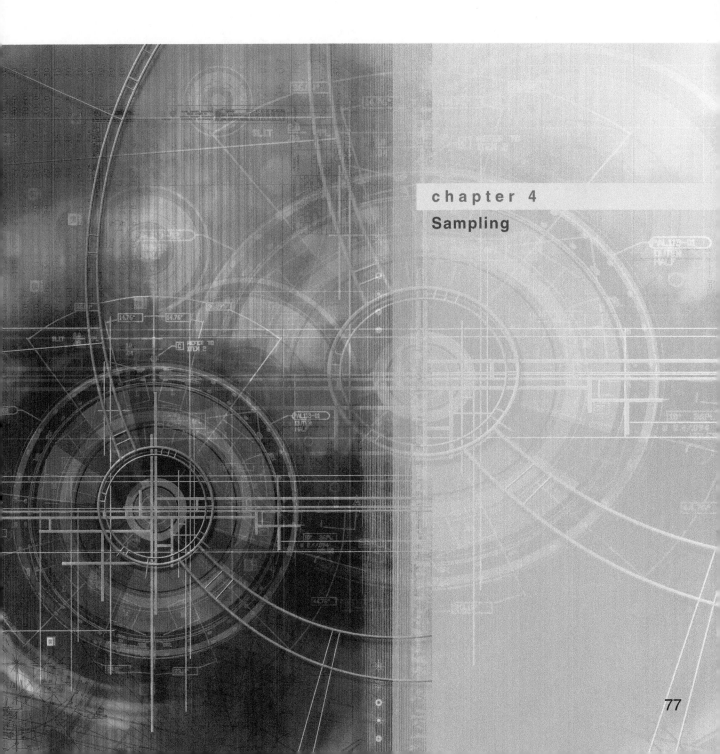

2 Sampling

chapter 4

Sampling

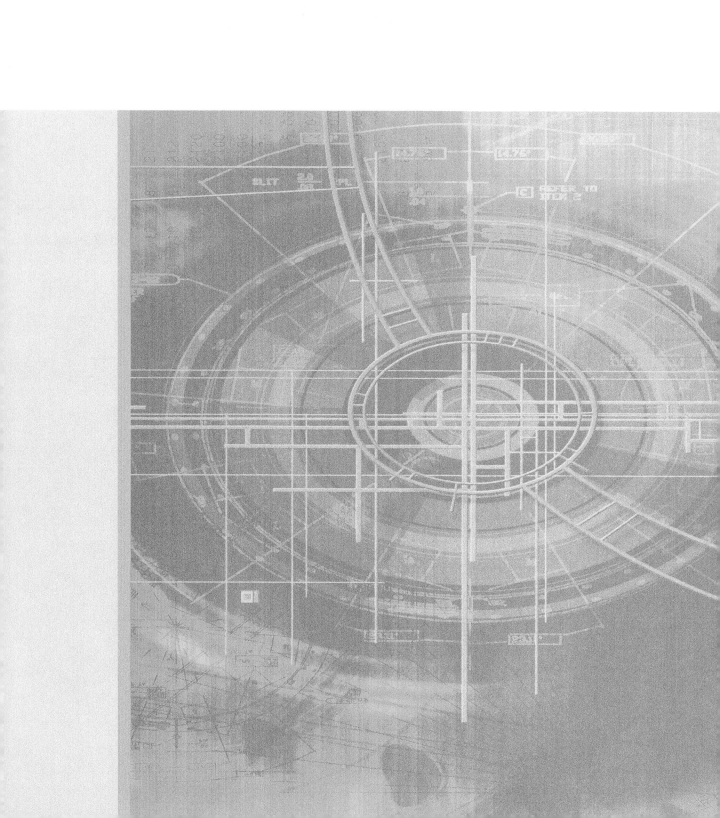

4

Sampling

4.1 Foundations of Sampling

We bet that you and many of the other students in your class have already been in at least one research study. In fact, most of what we know about human behavior has probably come from studies of captive audiences of undergraduates! Okay, maybe that is a bit of an exaggeration, but it's close enough to the truth that it has been a criticism of research in basic and applied social sciences over the years (**Figure 4.1**). And it has everything to do with the subject matter of this chapter: Who participates in studies and why—a process that researchers call "sampling."

Sampling is the process of selecting units (most often people, groups, and organizations, but sometimes texts like diaries, Internet discussion boards and blogs, or even graphic images) from a population of interest, so that by studying the sample you can generalize your results to the population from which the units were chosen.

As will become clearer later in this chapter, the nature of your research question has important implications for the type of sampling design that you chose. More descriptive and exploratory studies generally rely on a small number of cases that can be studied intensively (i.e., up-close and personal). In contrast, some studies aim for results that are very general and that represent populations (like large surveys or studies that are used to test drugs for deadly diseases), so they include larger samples of carefully chosen participants. And you have probably heard of a census, like the one that the U.S. government conducts every ten years. A census study doesn't pick a sample from a population; it attempts to include the whole population (**Figure 4.2**).

There may be times you won't be concerned about generalizing. Maybe you're evaluating a program in a local agency and don't care whether the program would work with other people in other places and at other times. In that case, sampling and generalizing might not be of interest. In fact, in this case your study would not even be considered research by the federal government

sampling The process of selecting units (e.g., participants) from the population of interest.

Figure 4.1 Would you say this "sample" is representative of college students?

UNIVERSAL/THE KOBAL COLLECTION

or your local IRB, precisely because you are *not* looking for generalizable knowledge. Or, you might be conducting a case study in which you're primarily interested in a detailed look at one specific case, be it a person, program, organization, or event. In addition, many, if not most, qualitative studies are focused on capturing the essence of a particular phenomenon in a particular context rather than on comparisons to other people or situations.

In this chapter, you will see that there are two major approaches to identifying a sample for a study. One is called probabilistic (because it is—surprise—based on probability theory) or random sampling. It is probably the best sampling approach when you are able to accomplish it. It can provide a credible and rigorous way to extend your conclusions beyond just the group you studied to the broader population you are interested in. However, as you'll see, it's not always possible to do a probability sample. There are a number of alternatives that are called nonprobability sampling approaches, for situations when using a probability-based sampling is not feasible. We'll review examples of both kinds of sampling strategy.

We'll also cover many of the key terms that researchers need to know to in order to critique or design a study, like population and sampling frame. And, along the way we'll discuss how researchers can make judgments about the quality of the sample using statistical concepts such as sampling distribution and sampling error.

Rhoda Sidney/PhotoEdit

Figure 4.2 The U.S. Census attempts to measure the entire population.

4.2 Sampling Terminology

As with anything else in life you have to learn the language of an area if you're going to ever hope to navigate successfully. Here, we introduce several different terms for the major groups that are involved in a sampling process and discuss the role that each group plays in the logic of sampling.

The group you wish to generalize to is called the **population** in your study (see **Figure 4.3**). This is the group you would like to sample from because this is the group you ultimately want your results to be about. Let's imagine that you want to generalize to urban homeless males between the ages of thirty and fifty in the United States. If that is the population of interest, you are likely to have a hard time developing a reasonable sampling plan. You are probably not going to find an accurate listing of this population, and even if you did, you would almost certainly not be able to mount a national sample across hundreds of urban areas. So we can make a distinction between the population you would like to generalize to, and the population that is accessible to you. We'll call the former the **theoretical population** and the latter the **accessible population**. In this example, the accessible population might be homeless males between the ages of thirty and fifty in six selected urban areas across the United States.

population The group you want to generalize to and the group you sample from in a study.

theoretical population A group which, ideally, you would like to sample from and generalize to. This is usually contrasted with the accessible population.

accessible population A group that reflects the theoretical population of interest and that you can get access to when sampling. This is usually contrasted with the theoretical population.

Figure 4.3 The different groups in the sampling model.

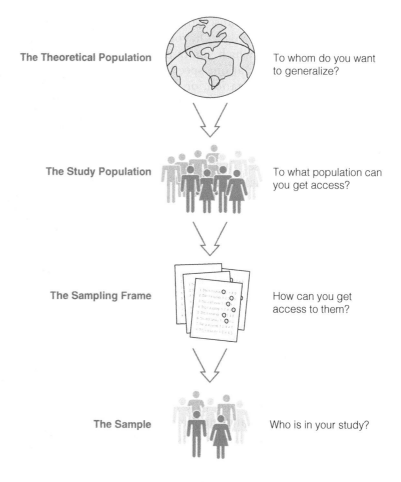

The Theoretical Population — To whom do you want to generalize?

The Study Population — To what population can you get access?

The Sampling Frame — How can you get access to them?

The Sample — Who is in your study?

After you identify the theoretical and accessible populations, you have to do one more thing before you can actually draw a sample: get a list of the members of the accessible population. Or, if you cannot get a complete listing of the accessible population, you have to spell out in detail the procedures you will follow to contact them, to assure representativeness. For example, when doing quota sampling of passers-by at the local mall for a short interview, you will not have a list of people in advance. In this case the sampling frame is the population of people who pass by within the time frame of your study and the rule(s) you use to decide whom to select. In either case, the listing of the accessible population from which you'll draw your sample is called the **sampling frame**. If you were doing a survey and selecting names from a directory of email addresses, the email address list would be your sampling frame. By the way, that wouldn't be a great way to sample because significant subportions of the population either don't have an email address or have changed it since the last directory was posted.

The **sample** is the group of people you select to be in your study. Notice that we didn't say that the sample was the group of people who are actually *in* your study. You may not be able to contact or recruit all of the people you actually sample, or some could drop out over the course of the study. The group that actually completes your study is a subsample of the sample; it doesn't include nonrespondents or dropouts. (The problem of nonresponse and its effects on a study will be addressed in Chapter 8, "Design," when discussing mortality threats to internal validity.)

sampling frame The list from which you draw your sample. In some cases, there is no list; you draw your sample based upon an explicit rule.

sample The actual units you select to participate in your study.

At this point, you should appreciate that sampling is a difficult multistep process and that you can go wrong in many places. In fact, as you move from each step to the next in identifying a sample, there is the possibility of introducing systematic error or **bias**. For instance, even if you are able to identify perfectly the population of interest, you may not have access to all of it. Even if you do, you may not have a complete and accurate sampling frame from which to select. And if that's not bad enough, you may not draw the sample correctly or accurately. And, even if you do, your participants may not all come and they may not all stay. Depressed yet? Sampling is a difficult business indeed. At times like this we are reminded of what Donald Campbell (a famous social scientist) used to say (we'll paraphrase here): "Cousins to the amoeba, it's amazing that we know anything at all!" Well, one difference between the amoeba and humans is that we have statistics and they don't (let's hope that's not the primary difference). Once we get over our envy of our cousins' stat-free single-cell life, we can put statistics to work in making our research better.

4.3 External Validity

When we conduct research we are often interested in reaching conclusions not just about our sample in the time and place where we conducted our study. We're often interested in making some conclusions that are broader than that, by concluding what is likely to happen with other people at other times and in other places than just those represented by our sample. When we try to reach conclusions that extend beyond the sample in our study, we say that we are generalizing.

Generalizing and **generalizability** are possible when a study has good evidence of external validity. This means you can confidently say that what you found in the study sample holds for the broader population (that is, it appears to be *generally* true). In other words, external validity is the degree to which the conclusions in your study would hold for other persons in other places and at other times. Our best chance of obtaining this kind of evidence of external validity is if we have a representative sample. The best chance of having a representative sample is if it is randomly selected so that every member of the population has an equal chance of participating. And as you might imagine, a truly random sample depends on being able to obtain a list of every member of the population. In reality, the available sampling frame (the list of accessible members of a population) may or may not include every member. Ultimately, the idea of external validity should guide our selection of sampling.

4.3a Two Major Approaches to External Validity in Sampling

Research uses two major approaches to gather evidence to support generalizations. We'll call the first approach the **sampling model**. In the sampling model, you start by identifying the population you would like to generalize to (see **Figure 4.4**). Then, you draw a representative sample from that population and conduct your research with the sample. Finally, because the sample is representative of the population, you can automatically generalize your results back to the population. This sounds simple enough, but there are several ways it can become difficult. First, at the time of your study, you might not know what part of the population you will ultimately want to generalize to. Second, you may not be able to

bias A systematic error in an estimate. A bias can be the result of any factor that leads to an incorrect estimate. When bias exists, the values that are measured do not accurately reflect the true value.

generalizing, generalizability The process of making an inference that the results observed in a sample would hold in the population of interest. If such an inference or conclusion is valid we can say that it has generalizability.

sampling model A model for generalizing in which you identify your population, draw a fair sample, conduct your research, and finally generalize your results from the sample to the population.

Population

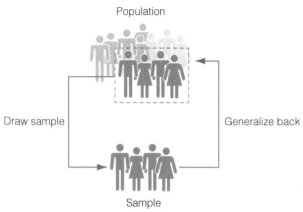

Draw sample

Generalize back

Sample

Figure 4.4 The sampling model for external validity. The researcher draws a sample for a study from a defined population to generalize the results to the population.

proximal similarity model A model for generalizing from your study to other contexts based upon the degree to which the other context is similar to your study context.

gradient of similarity The dimensions along which your study context can be related to other potential contexts to which you might wish to generalize. Contexts that are closer to yours along the gradient of similarity of place, time, people, and so on can be generalized to with more confidence than ones that are further away.

draw a fair or representative sample easily. Third, it's impossible to sample across all times that you might like to generalize to (such as next year). Researchers who study real people in real-life settings deal with these issues frequently. Sometimes these problems are controllable, other times they are issues that must be worked around by making careful choices in design, measurement, and analysis.

The second approach is called the **proximal similarity model** (Campbell, 1986) (see **Figure 4.5**). Proximal means nearby and similarity means . . . well, it means similarity. In the proximal similarity approach, you begin by thinking about different contexts you might want to generalize to, and which are more similar to your study and which are less so. For instance, you might imagine several settings, some having people who are more similar to the people in your sample, others having people who are less similar. This process also holds for times and places. Now picture this in terms of gradations of closeness to your study. Some combinations of people, places, and times are "closer" to those in your study, while others are more "distant" or less similar. The technical term for this idea is the **gradient of similarity**. There are some groups, for instance, who are more similar to your sample and would be closer along this gradient. Others would be less similar and further away along this imaginary gradient.

For example, imagine that the sample in your study consisted of females between the ages of 18–22 in a small midwestern college town. Who would likely be more similar to your sample, a group of females aged 18–22 from another midwestern college town, or a group of females aged 18–22 from Paris? Most likely you would argue that the group from another small midwestern college town are "closer" to your sample than the one from Paris. That is, you would be saying they are closer along a gradient of similarity.

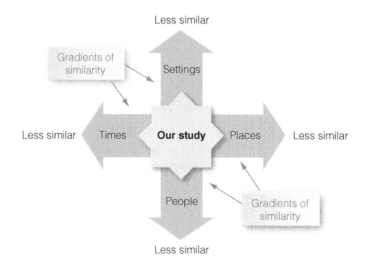

Figure 4.5 The Proximal Similarity Model for external validity.

So, how do you use the idea of the gradient of similarity from the proximal similarity framework to generalize? Simple. You generalize the results of your study with greater confidence to other persons, places, or times that are more like (that is, more proximally similar to) your study. Notice that here, you never generalize with certainty; such generalizations are always a question of more or less similar, always a question of degree.

Which approach is right? Which is better? There's no simple way to answer these questions. The sampling model is the more traditional and widely accepted of the two. If you are able to draw the sample well (a real challenge in many situations), and if you don't have too many subjects who don't respond (nonresponse is a growing problem in social research), then the sampling model makes sense. But the proximal similarity approach applies in any study as a guideline for thinking about external validity and generalizability. Even in situations where you aren't able to sample well from a population or when you have many subjects who don't respond or who drop out, you can always ask yourself what groups, places, and times are more or less similar to those in your study. Therefore, when it comes to external validity, you should *always* think in terms of proximal similarity but use the sampling model to the extent it is practical.

4.4 Sampling Methods

Sampling methods can be classified into two broad categories: probability and nonprobability sampling. While probability samples are selected in such a way that every member of the population has a known probability of being included in the sample, nonprobability samples are selected on the basis of researcher judgment (Henry, 1990). As discussed previously, smaller descriptive studies (interviews, focus groups, case studies) tend to utilize nonprobability sampling methods, whereas larger surveys or controlled studies generally employ probability sampling techniques. That being said, it is important to keep an open mind when matching your research question with a particular sampling method.

For instance, nonprobability sampling methods tend to be used during the early stages of testing a new program, intervention, or treatment. In medical research, we conduct "clinical trials" in which a medication or other treatment is tested in multiple phases through a sequence of studies. The early phases are typically smaller studies looking for evidence that the treatment works without doing harm. Later phases include more tightly controlled studies that enable researchers to eliminate the possibility of a treatment result being due to something other than the treatment. Much like the phases in clinical trials, during the earlier stages of your research, you may be interested in an in-depth exploration of a social phenomenon (let's say, global warming) and in examining whether two variables—the potential "cause" and "effect" (for example, carbon emissions and surface temperatures) are at all related. Here, nonprobability sampling methods are likely to be more suitable compared to probability sampling ones. Both types of methods, as well as their advantages and disadvantages, are discussed in detail below.

4.5 Nonprobability Sampling

The difference between nonprobability and **probability sampling** is that **nonprobability sampling** does not involve **random selection** and probability sampling does. Does that mean that nonprobability samples aren't representative of the population? Not necessarily; but it does mean nonprobability samples cannot depend upon the rationale of probability theory. At least, with a probabilistic sample, you know the odds or probability that you have represented the population well. You can estimate **confidence intervals (CIs)** for the statistic (there is an example of a confidence interval calculation in Section 4.6c below). With nonprobability samples, you may or may not represent the population well, and it will often be hard for you to know how well you've done so.

In general, researchers prefer probabilistic or random sampling methods over nonprobabilistic ones and consider them to be more accurate and rigorous. In nonprobability sampling, the sample is selected in such a way that the chance of each unit being selected is unknown. The selection of the subjects may be based on the researcher's experience, judgment, and access to potential participants. The trade-off in this type of sampling is in external validity; we won't have the same confidence about generalizing as we do when we can use probability sampling. In the following paragraphs, we will discuss a variety of nonprobabilistic sampling techniques.

Nonprobability sampling methods are divided into two broad types: *accidental* or *purposive*. Most sampling methods are purposive in nature because the sampling problem is usually approached with a specific plan in mind. The most important distinctions among nonprobability sampling methods are between the different types of purposive sampling approaches.

4.5a Accidental, Haphazard, or Convenience Sampling

One of the most common methods of sampling goes under the various titles listed here: accidental, haphazard, or convenience. We would include in this category the traditional person-on-the-street interviews conducted frequently by television news programs to get a quick (nonrepresentative) reading of public opinion. We would also argue that the typical use of college students in much psychological research is primarily a matter of convenience. (You don't really believe that psychologists use college students because they think they're representative of the population at large, do you?) In clinical practice, you might use clients available to you as your sample. In many research contexts, you sample by asking for volunteers. In a rather unusual example of convenience sampling, MIT neuroscientist Pawan Sinha—who studies how the brain interprets what our eyes see—strapped a camera on his newborn son to record everything the baby looked at. The ethical debate notwithstanding, it is commonplace for many scientists, especially those who study child development, to use their own children as "convenience" research subjects (Belluck, 2009)!

Clearly, the problem with all these types of samples is that you have no evidence that they are representative of the populations you're interested in generalizing to, and in many cases, you would suspect that they are not. It is for this reason that convenience sampling is often used in pilot or feasibility studies, because it allows the researcher to obtain basic data that can be used subsequently to provide a justification for a larger study.

probability sampling Method of sampling that utilizes some form of *random selection.*

nonprobability sampling Sampling that does not involve *random* selection.

random selection Process or procedure that assures that the different units in your population are selected by chance.

confidence intervals (CIs) A confidence interval is used to indicate the precision of an estimate of a statistic. The CI provides the lower and upper limits of the statistical estimate at a specific probability level. For instance, a 95% confidence interval for an estimate of a mean or average (the statistic) is the range of values within which there is a 95% chance that the true mean is likely to fall.

4.5b Purposive Sampling

In purposive sampling, you sample with a *purpose* related to the kind of participant you're looking for. Usually you would be seeking one or more specific kinds of people or groups. For instance, have you ever run into people in a mall or on the street carrying clipboards and stopping various people and asking to interview them? Most likely, they are trying to get a purposive sample (and most likely, they are engaged in market or political research). They might be looking for Caucasian females between 30–40 years old. They size up the people passing by and stop people who look to be in that category and ask whether they will participate. One of the first things they're likely to do is verify that the respondent does in fact meet the criteria for being in the sample (**Figure 4.6**). Purposive sampling can be useful in situations where you need to reach a targeted sample quickly and where sampling for proportionality is not the primary concern. With a purposive sample, you are likely to get the opinions of your target population, but you are also likely to overrepresent subgroups in your population that are more readily accessible.

All of the methods that follow can be considered subcategories of purposive sampling methods. You might sample for specific groups or types of people as in modal instance, expert, or quota sampling. You might sample for diversity as in heterogeneity sampling; or you might capitalize on informal social networks to identify specific respondents who are hard to locate otherwise, as in snowball sampling. In all of these methods, you know what you want—you are sampling with a purpose.

4.5c Modal Instance Sampling

In statistics, the *mode* is the most frequently occurring value in a distribution. In sampling, when you do a **modal instance sample**, you are sampling the most frequent case, or the typical case. Many informal public opinion polls, for instance, interview what they call a "typical voter." For instance, you see this approach used

modal instance sample Sampling for the most typical case.

Figure 4.6 Purposive sampling in a street survey. This researcher has found the kind of person she needs in her sample. We'll learn all about the techniques involved in conducting a good interview in Chapter 7 on Survey Research.

Janine Wiedel Photolibrary/Alamy

Figure 4.7 A "typical" couple that meets the study's definition in a modal instance sample.

© iStockphoto.com/bonnie jacobs

all the time on evening news programs. The newscaster goes out and interviews a "typical" voter to get an idea of what voters are thinking about (**Figure 4.7**). Or they interview a few people from a "typical" town (an example of modal sampling of towns rather than individuals). This sampling approach has a number of problems: First, how do you know what the typical or modal case is? You could say that the modal voter is a person of average age, educational level, and income in the population; but, it's not clear that using the averages of these is the fairest (consider the skewed distribution of income, for instance). Additionally, how do you know that those three variables—age, education, income—are the ones most relevant for classifying the typical voter? What if religion or ethnicity is an important determinant of voting decisions? Clearly, modal instance sampling is only sensible for informal sampling contexts.

This method is also sometimes used in implementation research for planning new programs. For example, Aitken et al. (2004) utilized modal instance sampling to investigate knowledge, practices, and beliefs of All-Terrain Vehicle (ATV) users in order to develop effective educational strategies to promote safer ATV use. The study targeted "typical" adolescent ATV users who were males attending a rural community school and who were also enrolled in an agricultural or a hunter education program.

4.5d Expert Sampling

Expert sampling involves assembling a sample of persons with known or demonstrable experience and expertise in some area. A common example of this type of sample is a panel of experts (**Figure 4.8**). There are actually multiple reasons why you might do expert sampling. First, it is the best way to elicit the views of persons who have specific expertise. In such a case, it's clear why we consider expert sampling a subtype of purposive sampling. When someone says they used expert sampling in a study, all they are saying is that the specific purpose for their sampling was to identify experts. Second, you might use expert sampling to provide evidence for the validity of another sampling approach you've chosen. For instance, let's say you do modal instance sampling and are concerned that the criteria you used for defining the modal instance are subject to criticism. You might

expert sampling A sample of people with known or demonstrable experience and expertise in some area.

Figure 4.8 Experts are an important source of sampling in many studies. Zen Master Shunryu Suzuki said: "In the beginner's mind, there are many possibilities but in the expert's, there are few." Do you agree? Is it a good or bad thing for researchers using expert sampling?

assemble an expert panel consisting of persons with acknowledged experience and insight into that field or topic and ask them to examine your modal definitions and comment on their appropriateness and validity. The advantage of doing this is that you aren't trying to defend your decisions on your own; you have some acknowledged experts to back you.

Expert sampling is also used when there is insufficient data on a particular topic. In such cases, expert sampling procedures serve as a means to synthesize the (limited) available knowledge, which can then be used to inform policy development before conclusive scientific evidence becomes available. This method is routinely employed in many environmental impact studies. For instance, Knol et al. (2009) assembled a panel of clinicians, toxicologists, and epidemiologists to investigate the causal relationship between exposure to certain airborne pollutants and health outcomes (such as mortality and cardiovascular and respiratory hospital admissions). Despite the method's many advantages, it is important to keep in mind that even experts can be, and often are, biased and wrong.

4.5e Quota Sampling

In **quota sampling,** you select people nonrandomly according to some fixed quota until you achieve a specific number of sampled units for each subgroup of a population. There are several different types of quota sampling. In **proportional quota sampling,** you want to represent the major characteristics of the population by sampling a proportional amount of each. For instance, if you know the population is composed of 77 percent whites, 13 percent blacks, and 10 percent individuals from other races, and that you want a total sample size of 100, you should keep sampling until you get those percentages (by race) and then stop. Once you meet the quota of 77 white individuals, you wouldn't include any more whites even if legitimate respondents came along. You would do the same thing

quota sampling Any sampling method where you sample until you achieve a specific number of sampled units for each subgroup of a population.

proportional quota sampling A sampling method where you sample until you achieve a specific number of sampled units for each subgroup of a population, where the proportions in each group are the same.

Jorge Núñez/EFE/Newscom

for blacks and other races—you would continue to sample until you reach your quota of 13 and 10 individuals, respectively. Of course, the hard part is figuring out the specific characteristics on which the quota will be based. Will you decide to use gender, age, education, race, religion, or countless other possible variables?

Nonproportional quota sampling is less restrictive. In this method, you specify the minimum number of sampled units you want in each category. Here, you're not concerned with having numbers that match the proportions in the population. Instead, you simply want to have enough to assure that you will be able to talk about even small groups in the population. This method is similar to stratified random sampling (Section 4.7c below) in that it is typically used to assure that smaller groups are adequately represented in your sample.

An issue with quota sampling, as with all nonprobability sampling, is the potential overrepresentation of those individuals who are more convenient to reach. This can lead to biased results, because those who are accessible are also more likely to have socioeconomic characteristics that make them less representative of the entire population. This shortcoming was the reason why 1948 was the last presidential election in which the Gallup presidential poll prediction used the quota sampling method (Fishkin, 2006). In this case, even though quota sampling was designed to obtain opinion from the nation's "demographic mirror," it let to faulty predictions because interviewers were free to survey whomever they wanted. In 1948, as now, Republicans were, on average, richer than Democrats; hence, they were more likely to have phones, live on good streets, and were thus more convenient to interview. An overrepresentation of Republicans in the Gallup poll led researchers to predict Thomas E. Dewey as the winner, when Harry Truman actually won the election and the presidency.

4.5f Heterogeneity Sampling

You sample for heterogeneity when you want to include all opinions or views, and you aren't concerned about representing these views proportionately. Another term for this is *sampling for diversity*. In many brainstorming or nominal group processes, you would use some form of heterogeneity sampling because your primary interest is in getting a broad spectrum of ideas, not identifying the average typical ones. In effect, what you would like to be sampling is not people, but ideas. You imagine that there is a "population" of all possible ideas relevant to some topic and that you want to sample this population, rather than the population of people who have the ideas. Clearly, to get all of the ideas, and especially the unusual ones, you have to include a broad and diverse range of participants (**Figure 4.9**). **Heterogeneity sampling** is, in this sense, almost the opposite of modal instance sampling.

For example, Magin, Adams, Heading, Pond and Smith (2006) used heterogeneity sampling to study patient perceptions of acne causes. Researchers aimed to increase the degree of representativeness by recruiting a wide variety of participants based on different individual-specific characteristics. These characteristics included: severity of acne, care by general practitioner or specialist dermatologist, age, and gender. In effect, by introducing the desired level of variation, researchers were able to identify common themes across a broad and diverse range of acne beliefs.

4.5g Snowball Sampling

In **snowball sampling**, you begin by identifying people who meet the criteria for inclusion in your study. You then ask them to recommend others they know who

nonproportional quota sampling A sampling method where you sample until you achieve a specific number of sampled units for each subgroup of a population, where the proportions in each group are not the same.

heterogeneity sampling Sampling for diversity or variety.

snowball sampling A sampling method in which you sample participants based upon referral from prior participants.

Figure 4.9 Sampling for heterogeneity is designed to get at the variability in a population.

also meet the criteria (**Figure 4.10**). Although this method would hardly lead to representative samples, at times it may be the best method available. Snowball sampling is especially useful when you are trying to reach populations that are inaccessible or hard to find. For instance, if you are studying the homeless, you are not likely to be able to find good lists of homeless people within a specific geographical area. However, if you go to that area and identify one or two, you may find that they know who the other homeless people in their vicinity are and how you can find them.

Researchers routinely use the snowball sampling method to recruit members of stigmatized populations such as sex workers and illicit drug users. For example, Ding et al. (2005) used this method to estimate rates of HIV infection and sexually transmitted diseases (STDs) among female commercial sex workers (FSWs) in China. The recruited FSWs were responsible for recruiting additional FSWs to participate in the study.

Figure 4.10 In snowball sampling, the original people sampled recommend others, who recommend still others, and so on.

The advantage of snowball sampling is that it can achieve broad coverage of a population, because respondents, including those who do not attend public venues, are reached through their social networks. However, the biggest disadvantage of this method is that because respondents are not randomly selected, and are dependent on the subjective choices of the first respondents, snowball samples are more likely to be biased.

A recent development in sampling methodology, **respondent-driven sampling** (RDS), was designed to overcome some of these limitations by providing breadth of coverage with statistical validity. RDS combines a modified form of chain-referral, or snowball sampling, with a mathematical system for weighting the sample to compensate for its not having been drawn as a simple random sample (Heckathorn, 1997). The method is based on mathematical modeling that is possible when certain information about the respondents and their social network can be collected. The development of RDS means that hard-to-reach populations (e.g., homeless drug users) can now be studied with confidence that approaches that of probabilistic methods, a development that has the potential to revolutionize sampling and have a large impact on social science.

4.5h Summary of Nonprobability Methods

We have seen that there are a variety of sampling methods that can be used that are not based on probability theory. Sometimes these methods are a good fit for the kind of research question that is being asked, especially in the early phases of research in an area where little is known. At other times these methods are the only available options for practical reasons (e.g., the sampling frame is impossible to define). The major limit we encounter when using these kinds of methods is in the degree of external validity that is possible.

4.6 Probability Sampling: Theory

In the remaining portion of this chapter, you will see how the use of probability-based sampling methods provides stronger external validity and lowers the chances that results are biased or merely "locally valid." The great benefit of using procedures based on probability is that we take much of the guessing out of the question of what our sample, and our study results, actually represent. In fact, this is not just a theoretical benefit; we can actually quantify how close our sample estimate is to the answer we'd get if we could study everyone in the population.

4.6a The Sampling Distribution

respondent-driven sampling A nonprobability sampling method that combines chain-referral or snowball sampling, with a statistical weighting system that helps compensate for the fact that the sample was not drawn randomly.

response A specific measurement value that a sampling unit supplies.

statistic A value that is estimated from data.

Let's begin by discussing some simple statistical terms that are relevant to probability sampling. If you've ever wondered what's behind those "margin of error" comments in the news, pay attention. When you sample, the units that you sample—usually people—supply you with one or more responses. In this sense, a **response** is a specific measurement value. In **Figure 4.11**, the person responding to a survey instrument gives a response of "4". When you summarize the numerical responses from a group of people, you use a **statistic**. There are a wide variety of statistics you can use: mean, median, mode, and so on. In this example, the mean or average for the sample is 3.72; but the reason you sample is to get an estimate for the population from which you sampled. If you could, you would probably prefer to measure the entire population. If you measure the entire population and calculate a value like a mean or average, this is not referred to as a statistic; it is a **population parameter**.

So how do you get from a sample statistic to an estimate of the population parameter? A crucial concept you need to understand is the sampling distribution. To understand it, let's do a thought experiment. Imagine that instead of just taking a single sample like you do in a typical study, you took three independent samples of the same population. Furthermore, imagine that for each of your three samples, you collected a single response (like a response to a single question) and computed a single statistic, say, the mean for each sample. This is depicted in the top part of **Figure 4.12**. Even though all three samples came from the same population, you wouldn't expect to get exactly the same statistic from each. They would differ slightly due to the random luck of the draw or to the natural fluctuations of drawing a sample. However, you would expect all three samples to yield a similar statistical estimate because they were drawn from the same population.

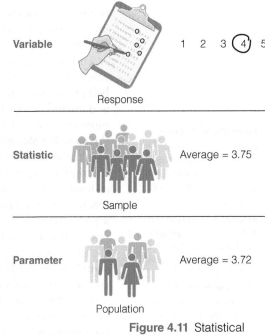

Figure 4.11 Statistical terms in sampling.

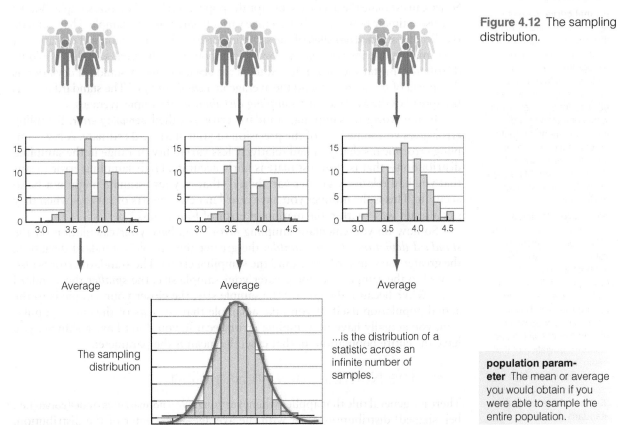

Figure 4.12 The sampling distribution.

The sampling distribution

...is the distribution of a statistic across an infinite number of samples.

population parameter The mean or average you would obtain if you were able to sample the entire population.

**sampling distribu-
tion** The theoretical
distribution of an infinite
number of samples of the
population of interest in
your study.

standard deviation An
indicator of the variability of a set of scores in a
sample around the mean
of that sample.

standard error The
spread of the averages
around the average of
averages in a sampling
distribution.

sampling error Error in
measurement associated
with sampling.

normal curve A common
type of distribution where
the values of a vari-
able have a bell-shaped
histogram or frequency
distribution. In a normal
distribution, approximately
68 percent of cases occur
within one standard deviation of the mean or center,
95 percent of the cases
fall within two standard
deviations, and 99 percent
are within three standard
deviations.

bell curve Also known
as a normal curve. A type
of distribution where the
values of a variable have
a smoothed histogram
or frequency distribution
that is shaped like a bell.
In a normal distribution,
approximately 68 percent
of cases occur within one
standard deviation of the
mean or center, 95 percent
of the cases fall within two
standard deviations, and
99 percent are within three
standard deviations.

Now, for the leap of imagination! Imagine that you took an *infinite* number of samples from the same population and computed the average for each
one. If you plotted the averages on a histogram or bar graph, you should find
that most of them converge on the same central value, and that you get fewer and fewer samples that have averages farther above or below that central
value. In other words, the bar graph would often be well described by the
bell-curve shape that is an indication of a normal distribution in statistics. This
is depicted in the bottom part of Figure 4.12. The distribution of an infinite
number of samples of the same size as the sample in your study is known as
the **sampling distribution**.

You don't ever actually construct a sampling distribution. Why not? Because
to construct one, you would have to take an *infinite* number of samples and,
at least the last time we checked, on this planet infinite is not a number we can
reach. So why do researchers even talk about a sampling distribution? Now that's
a good question! Because you need to realize that your sample is just one of a
potentially infinite number of samples that you could have taken. When you keep
the sampling distribution in mind, you realize that while the statistic from your
sample is probably near the center of the sampling distribution (because most of
the samples would be there), you could have gotten one of the extreme samples
just by chance. If you take the average of the sampling distribution—the average
of the averages of an infinite number of samples—you would be much closer to
the true population average—the parameter of interest.

4.6b Sampling Error

So we can estimate the average score for the population based on our sample. But we
can also estimate how much that average varies from sample to sample. That estimate
is called the **standard deviation** of the sampling distribution. The standard deviation of
the sampling distribution tells us something about how different samples would be
distributed. In statistics it is referred to as the **standard error**. A standard deviation is
the spread of the scores around the average in a *single sample*. The standard error is
the spread of the averages in a *sampling distribution* of sample averages.

In the world of sampling, standard error is called **sampling error**. Sampling
error gives you some idea of the precision of your statistical estimate. A low sampling error means that you had relatively less variability or range in the sampling
distribution—that is, your estimate is fairly accurate. High sampling error means
that you had relatively greater variability—that is, your estimate might not be so
accurate. This is the number you have heard in news reports of surveys when they
say a result was "within the margin of error of four points."

So how do you calculate sampling error? You base your calculation *on the
standard deviation of your sample*: the greater the sample's standard deviation,
the greater the standard error (and the sampling error). The standard error is also
related to the sample size: the greater your sample size, the *smaller* the standard
error. Why? Because the greater the sample size, the closer your sample is to the
actual population itself. If you take a sample that consists of the entire population, you actually have no sampling error because you don't have a sample; you
have the entire population. In that case, the mean *is* the parameter.

4.6c The Normal Curve in Sampling

There is a general rule that applies whenever you have a **normal curve** or **bell curve** (i.e.,
bell-shaped) distribution. Start with the average—the center of the distribution.

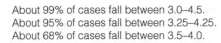

About 99% of cases fall between 3.0–4.5.
About 95% of cases fall between 3.25–4.25.
About 68% of cases fall between 3.5–4.0.

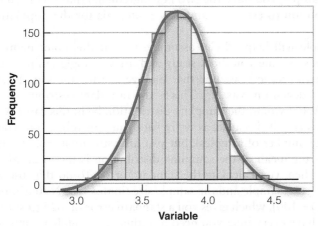

Figure 4.13 The 68, 95, 99 Percent Rule.

The distribution has a mean of 3.75 and a standard deviation of .25.

If you go up and down (that is, left and right) from the center one standard unit, you will include approximately 68 percent of the cases in the distribution (68 percent of the area under the curve). If you go up and down two standard units, you will include approximately 95 percent of the cases. If you go plus or minus three standard units, you will include 99 percent of the cases.

Notice that we didn't specify in the previous few sentences whether we were talking about standard *deviation* units or standard *error* units. That's because the same rule holds for both types of distributions (the raw data and sampling distributions). For instance, in **Figure 4.13**, the mean of the distribution is 3.75 and the standard unit is .25. (If this were a distribution of raw data, we would be talking in standard-deviation units. If it were a sampling distribution, we'd be talking in standard-error units.) If you go up and down one standard unit from the mean, you would be going up and down .25 from the mean of 3.75. Within this range—3.5 to 4.0—you would expect to see approximately 68 percent of the cases. We leave it to you to figure out the other ranges.

What does this all mean, you ask? If you are dealing with raw data, and if you can assume a normal distribution, and if you know the mean and standard deviation of a sample, you can *predict* the intervals within which 68, 95, and 99 percent of your cases would be expected to fall. We call these intervals the— you guessed it—68, 95, and 99 percent confidence intervals.

Now, here's where everything should come together in one great "aha!" experience if you've been following along. If you have a *sampling distribution,* you should be able to predict the 68, 95, and 99 percent confidence intervals for where the population parameter should be, which is why we can refer to the "68, 95, and 99 Percent Rule" as a *rule.* Now, isn't that why you sampled in the first place? So that you could predict where the population is on that variable? There's only one hitch. You don't actually have the sampling distribution. However, you do have the distribution for the sample itself; and from that distribution, you can estimate the standard error (the sampling error) because it is based on the standard deviation, and you have that. Of course, you don't actually know the population parameter value; you're trying to find that out, but you can use your best estimate for that—the sample statistic. For example, if you have an estimate of the mean of the sampling distribution (that is, the mean from your sample), and you have an estimate of the standard error, which you calculate from your sample, you have the two key ingredients that you need for your sampling distribution to estimate confidence intervals for the population parameter.

Perhaps an example will help. (We're assuming that at this point many of you are confused!) Let's assume you did a study and drew a single sample from the population. Furthermore, let's assume that the average for the sample was 3.75 and the standard deviation was .25. This is the raw data distribution depicted in **Figure 4.14**. What would the sampling distribution be in this case? Well, you don't actually construct it (because, as we're sure you remember, you would need to take an infinite number of samples), but you *can* estimate it. For starters, you must assume that the mean of the sampling distribution is the mean of the sample, that is, 3.75. Then, you calculate the standard error. To do this, use the standard deviation for your sample and the sample size (in this case N (the abbreviation for sample size) = 100), which gives you a standard error of .025 (just trust us on this). Now you have everything you need to estimate a confidence interval for the population parameter. You would estimate that the probability is 68 percent that the true parameter value falls between 3.725 and 3.775 (3.75 plus or minus .025); that the 95 percent confidence interval is 3.700 to 3.800; and that you can say with 99 percent confidence that the population value is between 3.675 and 3.825. Using your sample, you have just estimated the average for your population (that is, the mean of the sample which is 3.75), and you have given odds that the actual population mean falls within certain ranges.

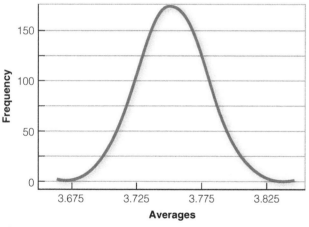

About 99% of cases fall between 3.675–3.825.
About 95% of cases fall between 3.70–3.80.
About 68% of cases fall between 3.725–3.775.

The sampling distribution has a mean of 3.75 and a standard error of .025.

Figure 4.14 Estimating the population using a sampling distribution.

4.7 Probability Sampling: Procedures

A probability sampling procedure is any method of sampling that uses some form of random selection. To have a random or probability selection method, you must set up some process or procedure that assures that the different units in your population have equal probabilities (chances) of being chosen. Humans have long practiced various forms of random selection, such as picking a name out of a hat, or choosing the short straw. These days, we tend to use computers as the mechanism for generating random numbers that are used for random selection.

At the beginning of the chapter we noted that our ultimate goal in sampling is to do so in a way that gives us the best evidence of external validity. The best way we know how to do this is by using methods based on probability. By selecting potential participants so that every member of a population has an equal chance of being selected, you can see that, even without looking at a single formula, you are doing things in a way that is both fair and representative. As we noted in the previous section, it isn't always possible, but if you are able to use the ideas and procedures discussed in the rest of this chapter, you will be in a much stronger position with regard to the external validity of your results.

4.7a Initial Definitions

Before we can explain the various probability methods, we must define the following basic terms:

- N is the number of cases in the sampling frame.
- n is the number of cases in the sample.
- $_NC_n$ is the number of combinations (subsets) of n from N.
- $f = n/N$ is the sampling fraction.

That's it. Now that you understand those terms, we can define the different probability sampling methods.

4.7b Simple Random Sampling

The simplest form of random sampling is called **simple random sampling**. Pretty tricky, huh? Here's the quick description of simple random sampling:

- **Objective: To select n units out of N such that each $_NC_n$ has an equal chance of being selected.**
- **Procedure: Use a table of random numbers, a computer random-number generator, or a mechanical device to select the sample.**

Let's see if we can make this somewhat general description a little more real. How do you select a simple random sample? Let's assume that you are doing some research with a small service agency to assess clients' views of quality of service over the past year. First, you have to get the sampling frame organized. To accomplish this, you go through agency records to identify every client over the past twelve months. If you're lucky, the agency has accurate computerized records and can quickly produce such a list (see **Figure 4.15**). Then, you have to draw the sample and decide on the number of clients you would like to have in the final sample.

For the sake of the example, let's say you want to select 100 clients to survey and that there were 1,000 clients over the past twelve months. Then, the sampling fraction is $f = n/N = 100/1000 = .10$ or 10 percent. To draw the sample, you have several options. You could print the list of 1,000 clients, tear it into separate strips with one per name, put the strips in a hat, mix them up, close your eyes, and pull out the first 100. This mechanical procedure would be tedious, and the quality of the sample would depend on how thoroughly you mixed up the paper strips and how randomly you reached into the hat. Perhaps a better procedure would be to use the kind of ball machine that is popular with many of the state lotteries. You would need three sets of balls numbered 0 to 9, one set for each of the digits from 000 to 999. (If you select 000 you call that 1,000.) Number the list of names from 1 to 1,000 and then use the ball machine to select the three digits that select each person. The obvious disadvantage here is that you need to get the ball machines. (Where do they make those things, anyway? Is there a ball machine industry?)

simple random sampling A method of sampling that involves drawing a sample from a population so that every possible sample has an equal probability of being selected.

Kari Thams
Anette Christianson
Peter Wright
Inger Curwin
Vanessa Dupont
Tyron Johnson
Beatrice Klassen
Fernando Dewitt
Chantal Dubeh
Arash Dabbo
Steve Roegner
Anne Burt

List of clients

Random subsample

Figure 4.15 Simple random sampling.

Neither of these mechanical procedures is typically feasible, and there is a much easier way. Here's a simple procedure that's especially useful if you have the names of the clients already on the computer. Many computer programs can generate a series of random numbers. Let's assume you copy and paste the list of client names into a column in an Excel spreadsheet (let's call it column A). Then, in the column right next to it (column B), paste next to each name the function = RAND(), which is Excel's way of putting a random number between 0 and 1 in the cells. However, it is not possible to sort these numbers because they are based on a formula. So, select and copy all the cells containing random numbers that have just been generated. Now, on the *Home* tab, from the *Paste* pull-down menu, select *Paste Values* and insert the values in column C. Remember to delete Column B because it is no longer needed. Column C now contains random numerical values that can be sorted. Then, sort both column A and column C—the list of names and the random number—by the column with the random numbers. This rearranges the name list in random order from the lowest to the highest random number. Then, all you have to do is take the first 100 names in this sorted list. Pretty simple. You could probably accomplish the whole thing in under a minute.

Simple random sampling is easy to accomplish and explain to others. Because simple random sampling is a fair way to select a sample, it is reasonable to generalize the results from the sample back to the population. Simple random sampling is not the most statistically efficient method of sampling and you might not—just because of the luck of the draw—get a good representation of subgroups in a population. To deal with these issues, you have to turn to other sampling methods.

4.7c Stratified Random Sampling

Stratified random sampling, also sometimes called *proportional* or *quota* random sampling, involves dividing your population into homogeneous subgroups and then taking a simple random sample in each subgroup. A homogeneous subgroup is a group in which all of the members are relatively similar. The following restates this in more formal terms:

OBJECTIVE Divide the population into nonoverlapping groups (*strata*) N_1, N_2, N_3, ... N_j, such that $N_1 + N_2 + N_3 + \cdots + N_j = N$. Then do a simple random sample of $f = n/N$ in each strata.

You might prefer stratified sampling to simple random sampling for several reasons. First, it assures that you will be able to represent not only the overall population, but also key subgroups of the population, especially small minority groups. If you want to be able to talk about subgroups, this may be the only way to ensure effectively that you'll be able to do so. If the subgroup is extremely small, you can use different sampling fractions (f) within the different strata to randomly oversample the small group. (Although you'll then have to weight the within-group estimates using the sampling fraction whenever you want overall population estimates. What does this mean? In a weighted analysis, you typically multiply the estimate by the weight, in this case, the sampling fraction.) When you use the same sampling fraction within strata, you are conducting *proportionate* stratified random sampling. Using different sampling fractions in the strata is called *disproportionate* stratified random sampling. Second, stratified random sampling has more statistical precision than simple random sampling if the strata or groups are homogeneous. If they are, you would expect the variability within groups to be lower than the variability for the population as a whole. Stratified sampling capitalizes on that fact.

stratified random sampling A method of sampling that involves dividing your population into homogeneous subgroups and then taking a simple random sample in each subgroup.

For example, let's say that the population of clients for your agency can be divided as shown in **Figure 4.16** into three groups: Caucasian, African American, and Hispanic American. Furthermore, let's assume that both the African Americans and Hispanic Americans are relatively small minorities of the clientele (10 percent and 5 percent, respectively). If you just did a simple random sample of $n = 100$ with a sampling fraction of 10 percent, you would expect by chance alone to get 10 and 5 persons from each of the two smaller groups. And, by chance, you could get even fewer than that! If you stratify, you can do better. First, you would determine how many people you want to have in each group. Let's say you still want to take a sample of 100 from the population of 1,000 clients over the past year; but suppose you think that to say anything about subgroups, you will need at least 25 cases in each group. So, you sample 50 Caucasians, 25 African

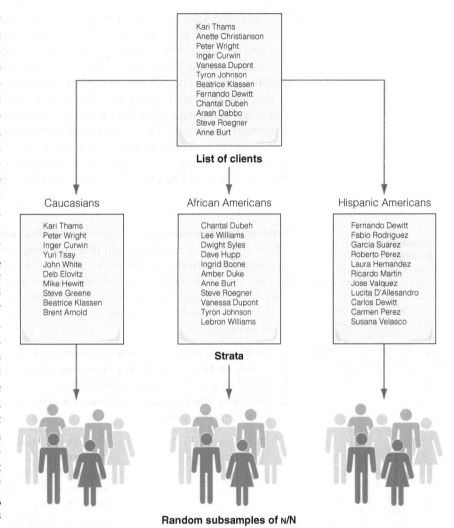

Figure 4.16 Stratified random sampling.

Americans, and 25 Hispanic Americans. You know that 10 percent of the population, or 100 clients, are African American. If you randomly sample 25 of these, you have a within-stratum sampling fraction of $25/100 = 25$ percent. Similarly, you know that 5 percent, or 50 clients, are Hispanic American. So your within-stratum sampling fraction will be $25/50 = 50$ percent. Finally, by subtraction you know there are 850 Caucasian clients. Your within-stratum sampling fraction for them is $50/850 = $ about 5.88 percent. Because the groups are more homogeneous within group than across the population as a whole, you can expect greater statistical precision (less variance), and, because you stratified, you know you will have enough cases from each group to make meaningful subgroup inferences.

Given these advantages, stratified random sampling is widely used in social and medical research. A study analyzing the association between obesity and depression in middle-aged women used this technique to oversample women with a higher body mass index (BMI). The population was stratified into three groups: women who last reported a BMI of 30 or more were sampled at 100 percent, women who last reported a BMI less than 30 were sampled at 12 percent, and women who did not complete a screening questionnaire in the last five years were sampled at 25 percent.

This stratified sampling procedure was intended to increase the efficiency of the survey and to permit correction for differences in response rates (Simon et al., 2008).

Based on the above discussion, you might find yourself thinking, how does stratified sampling differ from quota sampling? Both sampling methods involve dividing the target population into subgroups and then selecting a certain number of elements from each subgroup. However, the key difference between stratified and quota sampling is that the former utilizes simple random sampling once the subgroups are created; but quota sampling utilizes availability sampling. In other words, quota sampling is simply a nonprobability form of stratified sampling. Because stratified sampling is a probability sampling method, we can estimate statistics like sampling error. This is not possible with quota samples.

4.7d Systematic Random Sampling

systematic random sampling A sampling method where you determine randomly where you want to start selecting in the sampling frame and then follow a rule to select every xth element in the sampling frame list (where the ordering of the list is assumed to be random).

Systematic random sampling is a sampling method where you determine randomly where you want to start selecting in a sampling frame list and then follow a rule to select every xth element in the list (where the ordering of the list is assumed to be random—it should be ensured that there is no cyclical or periodic element to the ordering of elements in the sampling frame). This method is also sometimes called interval random sampling. To achieve a systematic random sample, follow these steps:

1. **Number the units in the population from 1 to N.**
2. **Decide on the n (sample size) that you want or need.**
3. **Calculate $k = N/n =$ the interval size.**
4. **Randomly select an integer between 1 and k.**
5. **Take every kth unit.**

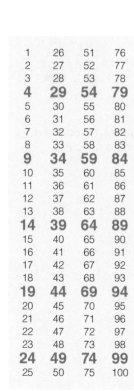

1 Determine the number of units. $N = 100$

2 Determine the sample size (n). Want $n = 20$

3 The interval size is $K = N/n$. $100/20 = 5$ $K = 5$

4 Select a random integer from 1 to K; choose 4.

5 Select every Kth unit.

Figure 4.17 Systematic random sampling.

All of this will be much clearer with an example. Let's assume, as shown in **Figure 4.17**, that you have a population that only has $N = 100$ people in it and that you want to take a sample of $n = 20$. To use systematic sampling, the population must be listed in a random order. The sampling fraction would be $f = 20/100 = 20$ percent. In this case, the interval size, k, is equal to $N/n = 100/20 = 5$. Now, select a random integer from 1 to 5. In this example, imagine that you chose 4. Now, to select the sample, start with the 4th unit in the list and take every kth unit (every 5th, because $k = 5$). You would be sampling units 4, 9, 14, 19, and so on to 100, and you would wind up with 20 units in your sample.

For this to work, it is essential that the units in the population are randomly ordered, at least with respect to the characteristics you are measuring. Why would you ever want to use systematic random sampling? For one thing, it is fairly easy to do. You only have to select a single random number to start things off. It may also be more precise than simple random sampling. Finally, in some situations there is simply no easier way to do random sampling.

Walsh, Jones, Cross & Lippert (2010) used systematic sampling to study the impact of evidence type on prosecution rates in child sexual abuse cases in Dallas, Texas. Data for this study (in the form of reported child sexual abuse cases with adult offenders) were obtained mainly from the Dallas Children's Advocacy Center. Systematic sampling was employed by taking every third case from the listing of all cases that was provided by the Children's Advocacy Center and from comparison community agencies (e.g., Child Protective Services, police). If there were multiple victims in the same family or multiple perpetrators per case, data collection focused on one randomly selected victim or perpetrator. The initial sample for this analysis included 360 cases.

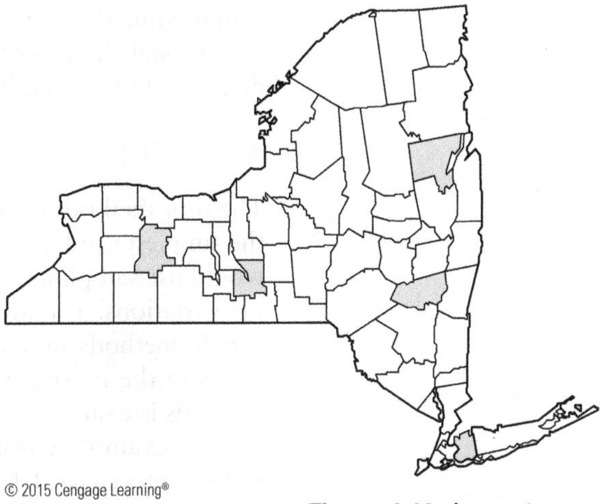

© 2015 Cengage Learning®

Figure 4.18 A county-level map of New York state used for cluster (area) random sampling.

4.7e Cluster (Area) Random Sampling

The problem with random sampling methods, when you have to sample a population that's spread across a wide geographic region, is that you will have to cover a lot of ground geographically to get to each of the units you randomly sample. Imagine taking a simple random sample of all the residents of New York state to conduct personal interviews. By the luck of the draw, you will wind up with respondents who come from all over the state. Your interviewers are going to have a lot of traveling to do. It is precisely to address this problem that **cluster random sampling** (sometimes also called **area random sampling**) was invented.

In cluster sampling, you follow these steps:

1. **Divide population into clusters (usually along geographic boundaries).**
2. **Randomly sample clusters.**
3. **Measure *all* units within sampled clusters.**

For instance, **Figure 4.18** shows a map of the counties in New York. Let's say that you have to do a survey of town governments that requires you to go to the towns personally to interview key town officials. If you do a simple random sample of towns statewide, your sample is likely to come from all over the state, and you will have to be prepared to cover the entire state geographically. Instead, you can do a cluster sampling of counties, let's say five counties in this example (shaded in the figure). Once these are selected, you go to *every* town government in the five county areas. Clearly this strategy will help you economize on mileage. Instead of having to travel all over the state, you can concentrate exclusively within the counties you selected. Cluster or area sampling is useful in situations like this, and is done primarily for efficiency of administration.

You might be wondering why all this sounds similar to stratified sampling. Well, that's because it is similar, but it is not exactly the same. In stratified sampling, once the categories (strata) are created, a random sample is drawn from each stratum. On the other hand, in cluster sampling, elements are not selected from each cluster (at least, not in the first stage). In single-stage cluster sampling, after a random sample of clusters is drawn (the five counties in the example above), *all* elements in the selected cluster are included in the sample. (As described in the next section, in two-stage cluster sampling and multistage cluster sampling, a random sample of clusters is drawn and then elements are randomly selected from the selected clusters.) Another difference between the two methods

cluster random sampling or area random sampling A sampling method that involves dividing the population into groups called clusters, randomly selecting clusters, and then sampling each element in the selected clusters. This method is useful when sampling a population that is spread across a wide area geographically.

is that while the main purpose of cluster sampling is to reduce costs and increase operational efficiency (that's why you usually cluster by geographical unit), stratified sampling is mainly used to improve statistical precision and representation.

4.7f Multistage Sampling

The four methods covered so far—simple, stratified, systematic, and cluster—are the simplest random sampling strategies. In most real applied social research, you would use sampling methods that are considerably more complex than these simple variations. The most important principle here is that you can combine these simple methods in a variety of useful ways to help you address your sampling needs in the most efficient and effective manner possible. Combining sampling methods is called **multistage sampling**.

For example, consider the idea of sampling New York state residents for face-to-face interviews. Clearly you would want to do some type of cluster sampling as the first stage of the process. You might sample townships or census tracts throughout the state. In cluster sampling you would then measure everyone in the clusters you selected. But, even if you are sampling census tracts, you may not be able to measure *everyone* who is in the census tract. So, you might set up a systematic random sampling process within the clusters. In this case, you would have a two-stage sampling process with systematic samples within cluster samples.

Alternatively, consider the problem of sampling students in grade schools. You might begin with a national sample of school districts stratified by economics and educational level. Within selected districts, you might do a simple random sample of schools; within schools, you might do a simple random sample of classes or grades; and, within classes, you might even do a simple random sample of students. In this case, you have three or four stages in the sampling process, and you use both stratified and simple random sampling. By combining different sampling methods, you can achieve a rich variety of probabilistic sampling methods to fit a wide range of social research contexts.

Multistage cluster sampling is a technique that is routinely used in public health studies. For instance, the World Health Organization has been using the "30 by 7 cluster survey" to estimate the prevalence of immunized children within +/− 10 percentage points (Henderson & Sundaresan, 1982). That is, if the true prevalence was 50 percent, one would expect an estimate between 40 percent and 60 percent when using the 30 by 7 method. This survey is a two-stage cluster sample. In the first stage, you randomly select 30 census blocks (or clusters) from a list of all census blocks in the county and then randomly select seven interview sites (households) per block. In most surveys like this, census blocks in the first stage are chosen through a method known as "Probability Proportionate to Size," which means that a census block with more households is more likely to be included than one with fewer households.

The National Home and Hospice Care Survey (NHHCS)—a continuing series of surveys of home and hospice care agencies in the United States—is another example of a complex, multistage sampling design. The first stage (carried out by the Centers for Disease Control and Prevention's National Center for Health Statistics) includes the selection of home health and hospice agencies from the sampling frame of over 15,000 agencies, representing the universe of agencies providing home health care and hospice services in the United States. The primary sampling strata of agencies are defined by agency type and metropolitan statistical area (MSA) status. Within these sampling strata, agencies are sorted by census region, ownership, certification status, state, county, ZIP code, and size (number of employees). For the 2007 NHHCS, 1,545 agencies were systematically and randomly

multistage sampling The combining of several sampling techniques to create a more efficient or effective sample than the use of any one sampling type can achieve on its own.

sampled with probability proportional to size. The second stage of sample selection was completed by the interviewers during the agency interviews. The current home health patients and hospice discharges were randomly selected by a computer algorithm, based on a census list provided by each agency director. Up to ten current home health patients were randomly selected per home health agency, and up to ten hospice discharges were randomly selected per hospice agency (CDC, 2007).

4.7g How Big Should the Sample Be?

Estimating how big a sample you need in your study should be a key consideration. It is challenging to obtain the "right" estimate of sample size for both statistical and practical reasons. The ideal sample size can depend on many factors—an important one being resource constraints. Bigger samples are more expensive and harder to obtain and manage than smaller ones. A study's sample size also depends on the level of precision you desire—in other words, how sure do you want to be of your conclusions? Larger sample sizes generally have smaller sampling errors, thus increasing the precision of your results. Similarly, if your target population has a large degree of variability, you will probably need a bigger sample size. You will also need some idea of the kind of result that will be meaningful (e.g., how big a difference or correlation will be important to detect?). Sample size estimation is intimately related to the idea of statistical power (the bigger the sample, the more "power" there will be in your result), which will be discussed in Chapter 11.

Additionally, the size of the sample also depends on the sampling strategy you choose. For instance, if you plan to carry out a multistage sampling survey, like the NHHCS, you will need a large sample to get an appropriate level of precision. However, if you are using a nonprobability sampling method like expert sampling, a relatively smaller sample size may be more desirable. Finally, the ideal sample size also depends upon the purpose of your research study—for example, sample sizes can be vastly different depending on whether you are conducting an exploratory study or a more confirmatory one, and on the degree of precision you want.

4.7h Summary of Probabilistic Sampling

Sampling procedures based on probability theory have clear advantages over nonprobability methods. Their main advantage is in building the best possible case for fair, externally valid, generalizable, and unbiased results. In this sense we are able to achieve both scientific and ethical ideals, a notable feat in any field. In addition, we can utilize probability theory to obtain estimates of the precision of our results, and at the same time quantify our confidence about the likelihood that we have obtained an accurate result.

4.8 Threats to External Validity

Threats to external validity are explanations of how you might be wrong in making a generalization. For instance, imagine that you conclude that the results of your study (which was done in a specific place, with certain types of people, and at a specific time) can be generalized to another context (for instance, another place, with slightly different people, at a slightly later time). In such a case, three major threats to external validity exist because there are three ways you could be wrong: people, places, or times. Your critics could argue that the results of your study

threats to external validity
Any factors that can lead you to make an incorrect generalization from the results of your study to other persons, places, times, or settings.

were due to the unusual type of people who were in the study. Or they could claim that your results were obtained only because of the unusual place in which you performed the study. Perhaps you did your educational study in a college town with lots of high-achieving, educationally oriented kids. They might suggest that you did your study at a peculiar time. For instance, if you did your smoking-cessation study the week after the U.S. Surgeon General issued the well-publicized results of the latest smoking and cancer studies, you might get different results than if you had done it the week before.

4.9 Improving External Validity

How can you improve external validity? One way, based on the sampling model, suggests that you do a good job of drawing a sample from a population. From the perspective of the sampling model, you should use random selection, if possible, rather than a nonrandom procedure. Additionally, once selected, you should try to assure that the respondents participate in your study and that you keep your dropout rates low. A second approach would be to use the theory of proximal similarity more effectively. How? You could do a better job of describing the ways your contexts differ from others by providing data about the degree of similarity between various groups of people, places, and even times. Perhaps the best approach to criticisms about generalizations is simply to show critics that they're wrong, by replicating your study in a variety of places, with different people, and at different times. That is, your external validity (ability to generalize) will be stronger the more you **replicate** your study in different contexts.

replicate, replication A study that is repeated in a different place, time, or setting.

SUMMARY

So, those are the basics of sampling methods. Quite a few options, aren't there? How about a table to summarize the choices and give you some idea of when they might be appropriate? Table 4.1 shows each sampling method, when it might best be used, and the major advantages and disadvantages of each.

Sampling is a critical component in virtually all social research. When a research report is written, it should include many aspects of the discussion in this chapter: the population, the sampling method, the actual sample, details about nonrespondents, and the accuracy of your results (the confidence interval). While we've presented a wide variety of sampling methods in this chapter, it's important that you keep them in perspective. The key is not which sampling method you use. The key is external validity—how valid the inferences from your sample are. You can have the best sampling method in the world but it won't guarantee that your generalizations are valid (although it does help!). Alternatively, you can use a relatively weak nonprobability sampling method and find that it is perfectly useful for your context. Ultimately, whether generalizations from your study to other persons, places, or times are valid is a judgment. Your critics, readers, friends, supporters, funders, and so on, will judge the quality of your generalizations, and they may not even agree with each other in their judgment. What might be convincing to one person or group may fail with another. Your job as a social researcher is to create a sampling strategy that is appropriate to the context and that will assure that your generalizations are as convincing as possible to as many audiences as is feasible.

Table 4.1 Summary of sampling methods showing when to use each and their advantages and disadvantages.

Sampling Method	Use	Advantages	Disadvantages
Accidental, haphazard, or convenience nonprobability sampling	Anytime.	Very easy to do; almost like not sampling at all.	Very weak external validity; likely to be biased.
Modal instance nonprobability sampling (purposive)	When you only want to measure a typical respondent.	Easily understood by nontechnical audiences.	Results only limited to the modal case; little external validity.
Expert nonprobability sampling (purposive)	As an adjunct to other sampling strategies.	Experts can provide opinions to support research conclusions.	Likely to be biased; limited external validity.
Quota nonprobability sampling (purposive)	When you want to represent subgroups.	Allows for oversampling smaller subgroups.	Likely to be more biased than stratified random sampling; often depends on who comes along when.
Heterogeneity nonprobability sampling (purposive)	When you want to sample for diversity or variety.	Easy to implement and explain; useful when you're interested in sampling for variety rather than representativeness.	Won't represent population views proportionately.
Snowball nonprobability sampling (purposive)	With hard-to-reach populations.	Can be used when there is no sampling frame.	Low external validity.
Simple random sampling	Anytime.	Simple to implement; easy to explain to nontechnical audiences.	Requires a sample list (sampling frame) to select from.
Stratified random sampling	When concerned about underrepresenting smaller subgroups.	Allows you to oversample minority groups to assure enough for subgroup analyses.	Requires a sample list (sampling frame) from which to select, and variable(s) to stratify by.
Systematic random sampling	When you want to sample every kth element in an ordered set.	You don't have to count through all of the elements in the list to find the ones randomly selected.	If the order of elements is nonrandom, there could be systematic bias.
Cluster (area) random sampling	When organizing geographically makes sense.	More efficient than other methods when sampling across a geographically dispersed area.	Usually not used alone; coupled with other methods in a multistage approach.
Multistage random sampling	Anytime.	Combines sophistication with efficiency.	Can be complex and difficult to explain to nontechnical audiences.

Key Terms

accessible population p. 81
bell curve p. 94
bias p. 83
cluster random sampling p. 101
confidence interval p. 86
expert sampling p. 88
generalizing,
    generalizability p. 83
gradient of similarity p. 84
heterogeneity sampling p. 90
modal instance sampling p. 87
multistage sampling p. 102
nonprobability sampling p. 86
nonproportional quota
    sampling p. 90

normal curve p. 94
population p. 81
population parameter p. 93
probability sampling p. 86
proportional quota
    sampling p. 89
proximal similarity model p. 84
quota sampling p. 89
random selection p. 86
replicate, replication p. 104
respondent-driven sampling p. 92
response p. 92
sample p. 82
sampling p. 80
sampling distribution p. 94

sampling error p. 94
sampling frame p. 82
sampling model p. 83
simple random sampling p. 97
snowball sampling p. 90
standard deviation p. 94
standard error p. 94
statistic p. 92
stratified random
    sampling p. 98
systematic random sample p. 100
theoretical population p. 81
threats to external validity p. 103

Suggested Websites

The National Opinion Research Center.

http://www.norc.org/Pages/default.aspx

The National Opinion Research Center is one of the largest and most respected research organizations in the country. The scope of research activities is very wide, and if you visit the site you'll be able to see excellent examples of large-scale studies and their accompanying sampling plans.

U.S. Health Resources and Services Administration

http://bphc.hrsa.gov/policiesregulations/performancemeasures/patientsurvey/calculating.html

This web page provides a brief review of some of the sampling methods discussed in this chapter. In addition, you can find formulas to calculate sample size, a web-based calculator, and some good examples of sampling plans in various kinds of studies.

Review Questions

1. One way to generalize the results in a study to other persons, places, and times is to develop an idea of the degree to which the other contexts are similar to the study context. What is the phrase that describes this theoretical similarity framework?

a. the sampling model
b. similarity matrix
c. gradient of similarity
d. threats to similarity
(Reference: 4.3a)

2. What is the sampling term used to describe the population that a researcher has access to—the population from which he or she will draw a sample?

a. the theoretical population
b. the accessible population
c. the drawn population
d. the random population
(Reference: 4.2)

3. If we looked at the average college entrance-exam score for all first-year college students in the United States, we would be studying a

a. statistic of that population.
b. response of that population.
c. sample distribution of that population.
d. parameter of that population.
(Reference: 4.6a)

4. The standard error is

a. the spread of scores around the average of a single sample.
b. the spread of scores around the average of averages in a sampling distribution.
c. the spread of scores around the standard deviation of a single sample.
d. the spread of scores around the standard deviation in a sampling distribution.
(Reference: 4.6b)

5. Which of the following statements about sampling error is most accurate?

a. The smaller the size of a sample in a research project, the smaller the standard error.
b. The larger the size of a sample in a research project, the smaller the standard error.
c. The standard error is completely independent of the size of a sample.
d. The standard error is synonymous with a sampling error.
(Reference: 4.6b)

6. Picking a name out of a hat is a simple form of

a. random selection.
b. stratified selection.
c. systematic selection.
d. cluster selection.
(Reference: 4.7)

7. Three hundred adolescents sign up for a research project that only 100 can complete. If each adolescent has a 33.3 percent chance of being selected for the project, then the sample is considered a

a. simple random sampling.
b. stratified random sampling.
c. systematic random sampling.
d. cluster random sampling.
(Reference: 4.7b)

8. If 1,000 college sophomores sign up for a research project designed to assess gender and ethnic differences in dating, what must researchers do to ensure that they have a proportionate stratified random sample?

a. Divide the groups into subgroups according to gender; then randomly draw a sample of equal males and females from the pool.
b. Divide the groups into subgroups according to ethnic identity and gender; then randomly draw a sample population that would include an equal number of members from each subgroup.
c. Divide the groups into homogeneous subgroups according to ethnic identity and gender; then randomly draw a sample population with the same percentage in each subgroup.
d. Divide the groups into heterogeneous subgroups according to ethnic identity and gender; then randomly draw a sample population that would reflect the percentage of subgroup membership reflected in the general population.
(Reference: 4.7c)

9. Which sampling process begins with the selection of a random number in a list of elements in your sampling frame, and assumes that the characteristics being measured are randomly distributed in the population?

a. simple random sampling
b. stratified random sampling
c. systematic random sampling
d. cluster random sampling
(Reference: 4.7d)

10. What is the sampling technique that is best used when there is a large geographical area to cover?

a. simple random sampling
b. stratified random sampling
c. systematic random sampling
d. cluster (area) random sampling
(Reference: 4.7e)

11. The normal distribution is often referred to as the "bell curve" because it provides the "ring of truth."

a. True
b. False
(Reference: 4.6c)

12. Surveys reported in the media almost always mention that the numbers presented are accurate within a few percentage points. The statistic used to determine the accuracy of such results is called the standard error.

a. True
b. False
(Reference: 4.6b)

13. A study based on a nonprobability sampling method can never be considered representative of the population.

a. True
b. False
(Reference: 4.5)

14. A researcher was trying to study a hard-to-research population (for example, homeless adolescents, migrant workers, cocaine dealers, etc.). The researcher decided to try sampling by tapping the social network of the local population, beginning with the first person she could find and then asking that person to help identify others, who would then be asked to further identify possible participants. This researcher is using a sampling technique known as "avalanche sampling" because pretty soon she could expect to have a very large number of participants.

a. True
b. False
(Reference: 4.5g)

15. The main advantage of multistage sampling is that it combines sophistication with efficiency, while the main disadvantage is that it can be complex and difficult to explain to nontechnical audiences.

a. True
b. False
(Reference: 4.7f)

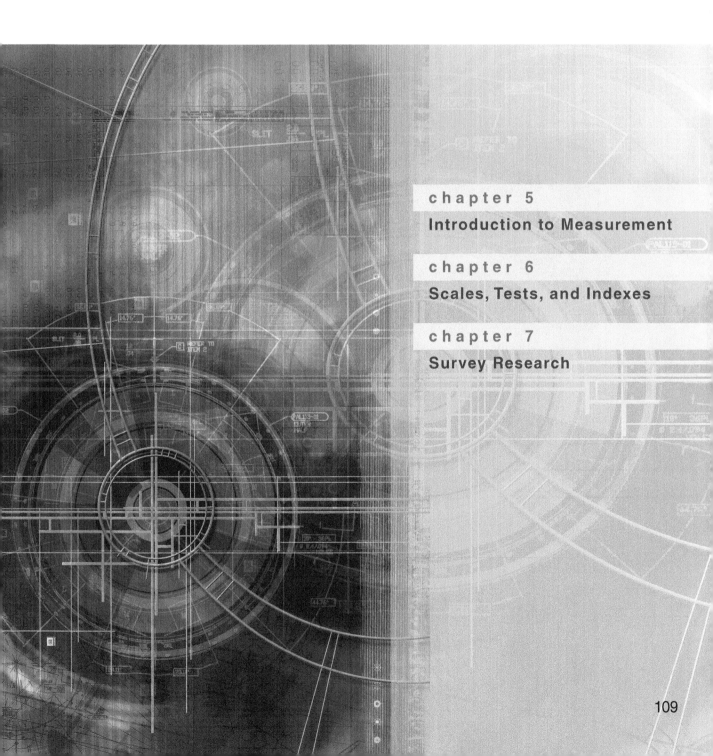

3 Measurement

Introduction to Measurement

Scales, Tests, and Indexes

Survey Research

5

Introduction to Measurement

5.1 Foundations of Measurement

In the last chapter, we tried to give you a clearer idea about how to sample participants. The next step is to think about what information you're going to collect from or about these participants. In applied social research, many of the outcomes of interest (for example, *quality of life*) are abstract or theoretical in nature. Measurement involves thinking about how to translate these abstract concepts so that we can observe them consistently and accurately. It also covers the development and application of instruments for collecting data on these variables. Simply put, measurement is the process of observing and recording the observations that are collected as part of a research effort.

In this chapter, we focus on how we think about and assess the quality of our observations. We begin by describing the four major levels of measurement: nominal, ordinal, interval, and ratio. Then, we discuss the two main indicators of the quality of measurement: reliability and validity. In the section on reliability, we consider the consistency or dependability of our observations, including the idea of true score theory and several ways to estimate reliability. In the section on validity, we focus on construct validity and discuss the importance of accurately translating theoretical concepts into what we observe. Finally, the chapter concludes by showing the relationship between reliability and validity. Having covered the measurement landscape broadly, our discussion then moves to the very specific, as we describe some of the most common methods of measuring things with scales, tests, indexes, and surveys, in Chapters 6 and 7.

Okay, we know this sounds like a really exciting lineup of topics. So, here's the deal: It turns out that if you can understand the ideas in this chapter you're going to be in a much better position to master what comes later. We're not saying you will, mind you, but you'll be better prepared to do so. Our job is to try to make the stuff in this chapter as enjoyable as we can. Since we actually do like this stuff (really, there are a few of us out there who do) we think we can make this at least mildly entertaining while getting across the basic ideas. So, here goes.

The first time anyone ever measured you, you probably were not even born yet. Yes, those fetal measures of your heart rate, weight, height and so on were the first "data points" in the ongoing assessment of life (**Figure 5.1**). Maybe your

Figure 5.1 Fetal heartbeat in an ultrasound.

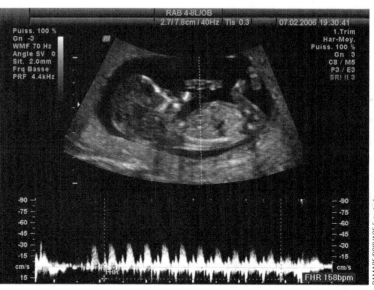

RAMARE/BSIP/AGE Fotostock

parents still have the ultrasound photos! Just think about how many ways you have been measured since then, and think about some of the consequences of those measurements, from the trivial to the most important. From your first shoes to your most recent exam, there have been thousands of scores, scales, indexes and other measures collected about you. OK, that's kind of a depressing thought—unless you're a researcher!

Let's begin with a fundamental measurement truth—no measure is perfect. No matter how precisely we attempt to measure something, it would always be possible to improve on the accuracy. Take for example an apparently simple thing to measure: your weight. Any of you who've ever been on a diet will immediately recognize how challenging it is to measure weight precisely. It varies based on the time of day, whether you just gorged out on another pizza, and the precision of the scale you use. And no matter how precise the scale, it is still possible to imagine a more precise one. Try weighing yourself on your favorite bathroom scale. Then get off the scale and step back on. Does it give

"No, that doesn't make any difference either, Miss Jones."

Kes/CartoonStock

Figure 5.2 Measurement is tricky, even on a scale measuring weight!

exactly the same number every time? Okay, maybe stepping on and off is the most exercise you've had all week and the weight is dropping off you like magic. But even so, there is a certain amount of error in even the simplest measures (**Figure 5.2**).

This isn't cause for despair. (At least not yet. Despair comes later in the chapter.) Even though no measure is perfect, we can estimate how great the potential error is. This chapter will help you understand how social scientists approach the job of estimating how consistent (reliable) and accurate (valid) our measurement observations are.

5.1a Levels of Measurement

Let's start with a simple example. Suppose we are trying to measure the type of political thinking a person has—the construct of *political ideology*. We might use which U.S. political party the person is affiliated with as a very rough approximation of their ideology. In this case our observation might have three attributes—Democrat, Republican, or Independent. Further, for purposes of analyzing the results of this variable, we also often assign numerical values to represent word categories. For instance, we could arbitrarily assign the values 1, 2, and 3 to these three party affiliations. The **level of measurement** describes the relationship among these three numerical values. In this case, the relationship between the three values does not represent any particular rank order. Higher values don't mean more of something and lower numbers don't signify less. The numbers simply indicate the *party affiliation* for each study participant.

level of measurement The relationship between numerical values on a measure. There are different types of levels of measurement (nominal, ordinal, interval, ratio) that determine how you can treat the measure when analyzing it. For instance, it makes sense to compute an average of an interval or ratio variable but does not for a nominal or ordinal one.

Figure 5.3 Relationship between attributes and values in a measure.

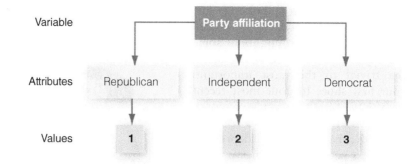

On the other hand, suppose respondents were asked to identify their *political beliefs* as one of the following four categories on a survey: "very liberal," "liberal," "conservative," and "very conservative," where each category is assigned the value 1, 2, 3, and 4, respectively. Now, in this case it is possible to think of this as a ranking or ordering of the responses in a particular dimension, say, degree of conservativeness. A value of 1 describes the lowest level of conservativeness, while a value of 4 reflects the highest. This is a different level of measurement as compared to the previous case on *party affiliation*. In essence, the level of measurement is based on the relationship between the values assigned to the attributes of a variable (**Figure 5.3**).

There are four levels of measurement that are most commonly discussed (Stevens, 1946) (see **Figure 5.4**): Nominal, Ordinal, Interval, and Ratio.

- **Nominal Level of Measurement:** Nominal means name, so this level of measurement is similar to the name of something. For example, the case of *party affiliation* is a nominal measurement because the numerical values simply name the attribute uniquely. No ordering of the cases is implied. Or, jersey numbers in basketball are measures at the nominal level. A player with number 30 is not more of anything than a player with number 15, and is certainly not twice whatever the player with number 15 is.

- **Ordinal Level of Measurement:** In ordinal measurement the attributes can be rank-ordered. Ranking of *political beliefs* described above is one example of ordinal measurement. However, here, distances between attributes do not have any meaning. For example, on a survey you might code Educational Attainment as 0 = less than high school; 1 = some high school; 2 = high school degree; 3 = some college; 4 = college degree; 5 = post college. In this measure, higher numbers mean *more* education. But, is the distance from 0 to 1 the same as 3 to 4? Of course not. The interval between values is not interpretable in an ordinal measure.

- **Interval Level of Measurement:** In interval measurement the distance between attributes is interpretable. For example, when we measure temperature, the difference between 30 degrees Fahrenheit and 40 degrees Fahrenheit is the same as the difference between 70 degrees Fahrenheit and 80 degrees Fahrenheit. Simply put, when each interval represents the same increment of the thing being measured, the measure is called an interval measure. This is very important in analysis because it makes sense to compute an average of an interval variable, while it doesn't make sense to do so for nominal scales. In other words, it makes sense to discuss the average temperature. But it doesn't make sense to talk about the average basketball jersey number. So, this is a hint: the level of measurement is important to know because it has an effect on the type of analysis you can do on the data.

nominal level of measurement Measuring a variable by assigning a number arbitrarily in order to name it numerically so that it might be distinguished from other objects. The jersey numbers in most sports are measured at a nominal level.

ordinal level of measurement Measuring a variable using rankings. Class rank is a variable measured at an ordinal level.

interval level of measurement Measuring a variable on a scale where the distance between numbers is interpretable. For instance, temperature in Fahrenheit or Celsius is measured on an interval level.

- **Ratio Level of Measurement:** In interval measurement, ratios don't make any sense; 80 degrees is not twice as hot as 40 degrees (although the numeric value we assign is twice as large). In ratio measurement there is always a meaningful absolute zero. This means that you can construct a meaningful fraction (or ratio) with a ratio variable. Weight is a ratio variable. We can say that a 100-lb bag weighs twice as much as a 50-lb one. Similarly, age is also a ratio variable. In applied social research, most count variables are ratio, for example, the number of clients in the past six months. Why? Because you can have zero clients and because it is meaningful to say, "We had twice as many clients in the past six months as we did in the previous six months."

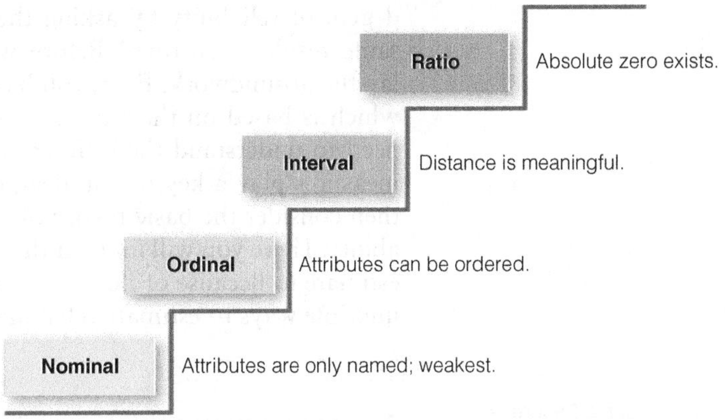

Figure 5.4 The hierarchy of levels of measurement.

It's important to recognize that there is a hierarchy implied in the level of measurement idea. At lower levels of measurement, assumptions tend to be less restrictive and data analyses tend to be less sensitive. At each level up the hierarchy, the current level includes all of the qualities of the one below it and adds something new. In general, it is desirable to have a higher level of measurement (such as interval or ratio) rather than a lower one (such as nominal or ordinal).

ratio level of measurement Measuring a variable on a scale where the distance between numbers is interpretable and there is an absolute zero value. For example, weight is a ratio measurement.

Why Is Level of Measurement Important?

First, knowing the level of measurement helps you decide how to interpret the data from that variable. When you know that a measure is nominal, you know that the numerical values are simply placeholders for the text names. Second, knowing the level of measurement helps you decide what statistical analysis is appropriate on the values that were assigned. If you know that a measure is nominal, then you would automatically know that you don't average the data values (except in certain circumstances like the use of "dummy" variables described in Chapter 12). Why? Because it makes no sense to add "names" and then divide them by the number of names, which is how an average is calculated (see Chapter 12). And, it also means that all statistical analyses that depend on the average or use it as part of their calculation (e.g., the *t*-test as described in Chapter 12), would not be appropriate.

5.2 Quality of Measurement

The two key criteria for evaluating the quality of measurement are reliability and validity. Reliability refers to *consistency* of measurement, while validity describes the *accuracy* with which a theoretical construct is translated into an actual measure. Both of these criteria are discussed in detail below.

5.2a Reliability

Reliability is an important aspect of measurement quality. In its everyday sense, reliability is the consistency or stability of an observation. You can infer the degree of reliability by asking the question—does the observation provide the same results each time? Before we can define reliability precisely, we have to lay the groundwork. First, you have to learn about the foundation of reliability, which is based on the true score theory of measurement. Along with that, you need to understand the different types of measurement error, because errors in measures play a key role in degrading reliability. With this foundation, you can then consider the basic theory of reliability, including a precise definition of reliability. There you will find out that you cannot calculate reliability—you can only estimate it. Because of this, there are a variety of different types of reliability and multiple ways to estimate reliability for each type.

True Score Theory

True score theory is one of the foundational theories in measurement. Like all theories, you need to recognize that it is not proven; it is postulated as a model of how the world operates. Also, similar to many powerful models, true score theory is a simple one. Essentially, true score theory maintains that every observable score is the sum of two components: true ability (or the true level) of the respondent on that measure; and random error. The true score is essentially the score that a person would have received if the score were perfectly accurate. This fundamental equation in true score theory is provided in **Figure 5.5**. You make an observation: get a score on a test or the scale value for a person's weight. It's important to keep in mind that you observe the **X** score (what is on the left side of the equation in the figure); you never actually see the true (T) or error (e) scores. For instance, a student may get a score of 85 on a math achievement test. That's the score you observe, an X of 85. However, the reality might be that the student is actually better at math than that score indicates. Let's say the student's true math ability is 89 (T = 89). That means that the error for that student is −4. What does this mean? Well, while the student's true math ability may be 89, he/she may have had a bad day, may not have had breakfast, may have had an argument with someone, or may have been distracted while taking the test. Factors like these can contribute to errors in measurement that make the students' observed abilities appear lower or higher than their true or actual abilities.

Why is true score theory important? For one thing, it is a simple yet powerful model for measurement. It is a reminder that most measurement will inevitably have an error component. Second, true score theory is the foundation of reliability theory, which will be discussed later in this chapter. A measure that has no random error (is all true score) is perfectly reliable; a measure that has no true score (is nothing but random error) has zero reliability. Minimizing measurement error is the key aim of developing measures that are more reliable. Third, true score theory can be used in computer simulations as the basis for generating observed scores with certain known properties.

You should know that the true score model is not the only measurement model available. Measurement theorists continue to come up with more and more complex models that they think represent reality even better (e.g., Item Response Theory, the Rasch model, Generalizability theory). However, these models are complicated enough that they lie outside the boundaries of this book. In any event, true score theory should give you an idea of why measurement models are important at all and how they can be used as the basis for defining key research ideas.

Measurement Error

While true score theory is a classic measurement theory, it may not always be an accurate reflection of reality. In particular, it assumes that any observation is

Observed score =
True ability + Random error

$$X = T + e$$

Figure 5.5 The basic equation of true score theory.

true score theory A theory that maintains that an observed score is the sum of two components: true ability (or the true level) of the respondent; and random error.

composed of the true value plus some random error value. But is that reasonable? What if all error is not random? Isn't it possible that some errors are systematic, by which we mean errors that are not introduced by chance and tend to fall consistently in a particular direction? One way to deal with this notion is to revise the simple true score model by dividing the error component into two subcomponents, random error and systematic error. **Figure 5.6** shows these two components of measurement error.

$$X = T + e$$
$$X = T + e_r + e_s$$

Figure 5.6 Random and systematic errors in measurement.

What Is Random Error?

Random error is caused by any factors that randomly affect measurement of the variable across the sample. For instance, people's moods can inflate or deflate their performance on any occasion. In a particular testing, some children may be in a good mood and others may be depressed. If mood affects the children's performance on the measure, it might artificially inflate the observed scores for some children and artificially deflate them for others.

The important thing about random error is that it does not have any consistent effects across the entire sample. Instead, it pushes observed scores up or down randomly. This means that if you could see all the random errors in a distribution they would have to sum to 0—random errors tend to balance out on average. There would be as many negative errors as positive ones. (Of course you can't see the random errors because all you see is the observed score X). The important property of random error is that it adds variability to the data but does not affect average performance for the group (see **Figure 5.7**). Because of this, random error is sometimes considered *noise*.

What Is Systematic Error?

Systematic error is caused by any factors that consistently affect measurement of the variable across the sample. For instance, if there is loud traffic going by just outside of a classroom where students are taking a test, this noise is liable to affect all of the children's scores—in this case, systematically lowering them. Another source of systematic error may stem from nonresponse to a survey instrument. If nonrespondents differ systematically from respondents in relation to the

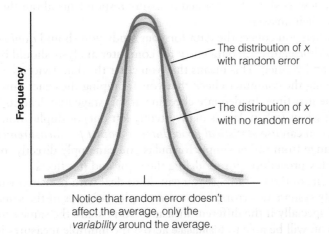

Notice that random error doesn't affect the average, only the *variability* around the average.

Figure 5.7 Random error adds variability to a distribution but does not affect central tendency (the average).

random error A component or part of the value of a measure that varies entirely by chance. Random error adds noise to a measure and obscures the true value.

systematic error A component of an observed score that consistently affects the responses in the distribution.

Figure 5.8 Systematic error affects the central tendency of a distribution.

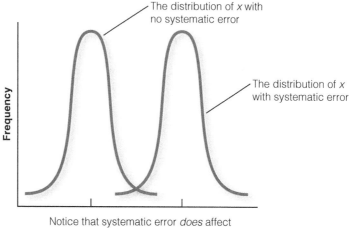

The distribution of *x* with no systematic error

The distribution of *x* with systematic error

Frequency

Notice that systematic error *does* affect the average— we call this a *bias*.

attributes that your instrument is trying to measure, then survey results may be biased in one direction. Unlike random error, systematic errors tend to be either positive or negative consistently; because of this, systematic error is sometimes considered to be bias in measurement (see **Figure 5.8**).

Reducing Measurement Error

So, how can you reduce measurement errors, random or systematic? Here is a list of things you can do.

1. One thing you can do is to *pilot test* your instruments to get feedback from your respondents regarding how easy or hard the measure was, and information about how the testing environment affected their performance. In the pilot run, you should also check if there was sufficient time to complete the measure.
2. Second, if you are gathering measures using people to collect the data (as interviewers or observers), you should make sure you *train* them thoroughly so that they aren't inadvertently introducing error. To improve response rates, the interviewers should be trained to assure respondents about the confidentiality of their answers.
3. Third, when you collect the data for your study you should *double-check the data* thoroughly. All data entry for computer analysis should be double-entered and verified. This means that you enter the data twice, the second time having the computer check that you are typing the exact same data you typed the first time. Inadequate checking at this stage may lead to loss of data/omissions (when data is inadvertently left out) or duplication.
4. Fourth, you can *use statistical procedures to adjust for measurement error*. These range from rather simple formulas you can apply directly to your data to complex procedures for modeling the error and its effects.
5. Finally, one of the best things you can do to deal with measurement errors, especially systematic errors, is to use multiple measures of the same construct. Especially if the different measures don't share the same systematic errors, you will be able to **triangulate** across the multiple measures in the data analysis and get a more accurate sense of what's happening.

triangulate Combining multiple independent measures to get at a more accurate estimate of a variable.

5.2b Theory of Reliability

Now that you have an understanding of true score theory and the two types of errors in measurement, we are ready to discuss reliability theory. What is reliability? We hear the term used a lot in research contexts, but what does it really mean? If you think about how we use the word reliable in colloquial language, you might get a hint. For instance, we often speak about a machine as reliable: "I have a reliable car." Or, media persons talk about a "reliable news source." In both cases, the word reliable usually means dependable or trustworthy. In research, the term reliable also means dependable in a general sense, but that's not a precise enough definition. The reason dependable is not a good enough description is that it can be confused too easily with the idea of a valid measure (see the section, "Construct Validity," later in this chapter). Certainly, when researchers speak of a dependable measure, we mean one that is both reliable and valid. So we have to be a little more precise when we try to define reliability.

In research, the term reliability means repeatability or consistency. A measure is considered reliable if it would give you the same observation over and over again (assuming that what you are measuring isn't changing).

Let's explore in more detail what it means to say that a measure is repeatable or consistent. We'll begin by defining a measure of a single construct that we'll arbitrarily label X. It might be a person's score on a math achievement test or a score on a severity of illness scale. It is the value (numerical or otherwise) that you observe in your study. Now, to see how repeatable or consistent an observation is, you can measure it twice. You use subscripts to indicate the first and second observation of the same measure, as shown in **Figure 5.9**. If you assume that what you're measuring doesn't change between the time of the first and second observation, you can begin to understand how you get at reliability.

If your observation is reliable, you should pretty much get the same result each time you measure it. Why would the results be essentially the same in these circumstances? If you look at Figure 5.9, you should see that the only thing that the two observations have in common is their true scores, T. How do you know that? Because the error scores (e_1 and e_2) have different subscripts indicating that they are different values. (You are likely to have different errors on different occasions.) However, the true score symbol (T) is the same for both observations. What does this mean? The two observed scores, X_1 and X_2, are related only to the degree that the observations share a true score. You should remember that the error score is assumed to be random (see the section "True Score Theory" earlier in this chapter). Sometimes errors will lead you to perform better on a test than your true ability (you had a good day guessing!) while other times they will lead you to score worse. The true score—your true ability on that construct—would be the same on both observations (assuming, of course, that your true ability didn't change between the two measurement occasions).

With this in mind, we can now define reliability more precisely. Reliability is a ratio or fraction. In layperson terms, you might define this ratio as shown in **Figure 5.10**.

You might think of reliability as the proportion of truth in your observation. Now, it makes no sense to speak of the reliability of a score for an individual; reliability is a characteristic of an observation that's taken across individuals. So, to get closer to a more formal definition, we'll restate the definition of reliability in terms of a set of observations. The easiest way to do this is to introduce the idea of the variance of the scores. The variance is a measure of the spread or distribution of a set of scores (see Chapter 11). So, we can now state the definition as shown in **Figure 5.11**.

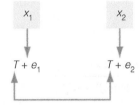

x_1　　　　x_2

$T + e_1$　　$T + e_2$

Figure 5.9 Reliability and true score theory.

True level on the measure
The entire observed measure (with error included)

Figure 5.10 Reliability can be expressed as a simple ratio.

The variance of the true score
The variance of the entire observed measure

Figure 5.11 The reliability ratio can be expressed in terms of variances.

$$\frac{var(T)}{var(X)}$$

Figure 5.12 The reliability ratio expressed in terms of variances in abbreviated form.

$$\frac{covariance(X_1, X_2)}{sd(X_1)^* \, sd(X_2)}$$

Figure 5.13 The formula for estimating reliability.

We might put this into slightly more technical terms by using the abbreviated name for the variance and our variable names (see **Figure 5.12**).

We're getting to the critical part now. If you look at the equation in Figure 5.12, you should recognize that you can easily determine or calculate the bottom part of the reliability ratio; it's just the variance of the set of observed scores. (You remember how to calculate the variance: It's the sum of the squared deviations of the scores from their mean, divided by the number of scores. If you're still not sure, see Chapter 11.) So how do you calculate the variance of the true scores? Remember, you can't see the true scores. (You only see X!) You never know the true score for a specific observation. Therefore, if you can't calculate the variance of the true scores, you can't compute the ratio, which means *you can't compute reliability!* Everybody got that? Here's the bottom line: *You can't compute reliability because you can't calculate the variance of the true scores!*

Great. So where does that leave you? If you can't compute reliability, perhaps the best you can do is to estimate it. Maybe you can get an estimate of the variability of the true scores. How do you do that? Remember your two observations, X_1 and X_2? You assume (using true score theory described earlier in this chapter) that these two observations would be related to each other to the degree that they share true scores. So, let's calculate the correlation between X_1 and X_2. **Figure 5.13** shows a simple formula for the correlation.

In Figure 5.13, the *sd* stands for the standard deviation which is the square root of the variance (see Chapter 11). If you look carefully at this equation, you can see that the covariance, which simply measures the shared variance between measures, must be an indicator of the variability of the true scores, because the true scores in X_1 and X_2 are the only things the two observations share! So, the top part is essentially an estimate of *var(T)* in this context. Additionally, since the bottom part of the equation multiplies the standard deviation of one observation with the standard deviation of the same observation at another time, you would expect that these two values would be the same (it is the same observation we're making) and that this is essentially the same thing as squaring the standard deviation for either observation. However, the square of the standard deviation is the same thing as the variance of the observation. So, the bottom part of the equation becomes the variance of the score, or *var(X)*. If you read this paragraph carefully, you should see that *the correlation between two observations of the same construct is an estimate of reliability*. Got that? We've just shown that a simple and straightforward way to estimate the reliability of an observation is to compute the correlation of the scores obtained from two administrations of the measure!

It's time to reach some conclusions. You know from this discussion that you cannot calculate reliability directly because you cannot measure the true score component of an observation. However, you can still estimate the true score component as the covariance between two observations of the same thing. With that in mind, you can estimate reliability as the correlation between two observations of the same thing. It turns out that there are several ways to estimate this reliability correlation. These are discussed in the section "Types of Reliability" later in this chapter.

There's only one other issue we want to address here. What is the range of a reliability estimate? What is the largest or smallest value a reliability estimate can be? To figure this out, let's go back to the equation given earlier (see Figure 5.12).

Remember, because X = T + e, you can substitute it in the bottom of the ratio as shown in **Figure 5.14**.

With this slight change, you can easily determine the range of a reliability estimate. If an observation is *perfectly* reliable, there is no error in measurement; everything you observe is true score. Therefore, for a perfectly reliable observation, *var(e)* is zero and the equation would reduce to the equation shown in **Figure 5.15**. Therefore, reliability = 1.

Now, if you have a perfectly unreliable measure, there is no true score; the measure is entirely error. In this case, the equation would reduce to the equation shown in **Figure 5.16**. Therefore, the reliability = 0. From this you know that, in principle, reliability will always range between 0 and 1, with higher values indicating higher levels of reliability. Though, theoretically, it doesn't make any sense; in practice, it is possible to have a negative estimate of reliability when the correlation between two observations of the same thing is negative. However, theoretically reliability should range from 0 to 1.

The value of a reliability estimate tells you the proportion of variability in the observation attributable to the true score. A reliability of .5 means that about half of the variance of the observed score is attributable to true score and half is attributable to error. A reliability of .8 means the variability is about 80 percent true ability and 20 percent error, and so on.

5.2c Types of Reliability

You learned in the section "Theory of Reliability," earlier in this chapter, that it's not possible to calculate reliability exactly. Instead, you have to estimate reliability, and by its very nature, this is always an imperfect endeavor. Here, we want to introduce the major reliability estimators and talk about their strengths and weaknesses.

There are four general classes of reliability estimators, each of which estimates reliability in a different way:

- **Inter-rater or inter-observer reliability** is used to assess the degree to which different raters/observers give consistent estimates of the same phenomenon.
- **Test-retest reliability** is used to assess the consistency of an observation from one time to another.
- **Parallel-forms reliability** is used to assess the consistency of the results of two tests constructed in the same way from the same content domain.
- **Internal consistency reliability** is used to assess the consistency of results across items within a test.

We'll discuss each of these in turn.

Inter-Rater or Inter-Observer Reliability

Whenever you use humans as a part of your measurement procedure, you have to worry about whether the results you get are reliable or consistent. People are notorious for their inconsistency. We are easily distracted. We get tired of doing repetitive tasks. We daydream. We misinterpret. So how do you determine whether two observers are being consistent in their observations? You calculate inter-rater reliability! Inter-rater or inter-observer reliability is the degree to which two or more raters or coders agree with each other when using the same instruments at the same time (**Figure 5.17**).

$$\frac{var(T)}{var(T) + var(e)}$$

Figure 5.14 The reliability ratio expressed in terms of variances with the variance of the observed score subdivided according to true score theory.

$$\frac{var(T)}{var(T)}$$

Figure 5.15 When there is no error in measurement you have perfect reliability and the reliability estimate is 1.0.

$$\frac{0}{var(e)}$$

Figure 5.16 When there is only error in measurement, you have no reliability and the reliability estimate is 0.

inter-rater or inter-observer reliability The degree of agreement or correlation between the ratings or codings of two independent raters or observers of the same phenomenon.

test-retest reliability The correlation between scores on the same test or measure at two successive time points.

parallel-forms reliability The correlation between two versions of the same test or measure that were constructed in the same way, usually by randomly selecting items from a common test question pool.

internal consistency reliability A correlation that assesses the degree to which items on the same multi-item instrument are interrelated. The most common forms of internal consistency reliability are the average inter-item correlation, the average item-total correlation, the split half correlation and Cronbach's Alpha.

Figure 5.17 Sometimes raters are unreliable or inconsistent!

Robert Michael/Corbis

You should establish this form of reliability outside of the context of the measurement in your study. After all, if you use data from your study to establish reliability, and you find that reliability is low, you're kind of stuck. Probably it's best to do this as a side study or pilot study. If your study continues for a long time, you may want to reestablish inter-rater reliability from time to time to ensure that your raters aren't changing.

Here are two of the many ways to actually estimate inter-rater reliability. First, if your measurement consists of categories—the raters are checking off which category each observation falls in—you can calculate the percentage of agreement between the raters. For instance, let's say you had 100 observations that were being rated by two raters. For each observation, the rater could check one of three categories. Imagine that on 86 of the 100 observations the raters checked the same category. In this case, the percent of agreement would be 86 percent. Okay, it's a crude measure, but it does give an idea of how much agreement exists, and it works no matter how many categories are used for each observation.

The main problem with percentage agreement is that the value obtained can be misleading because it fails to take into account the extent of agreement we would expect just by chance alone (when in actuality, the observations are completely unrelated). Using percentages can result in two raters appearing to be somewhat reliable and in agreement with each other, even if they assigned their scores completely randomly. For example, consider two researchers who are supposed to examine the behavior of the same child, but, by mistake, they start observing two different children. Even though the researchers are observing behaviors in two separate children, it is very likely that they record the same type of behavior. In this case, their observations will probably agree with each other to a large extent—however this agreement is simply due to sheer luck (Wood, 2007)! For this reason, percentage agreement has been widely criticized in recent years as an inferior measure.

Cohen's Kappa (Cohen, 1960) was introduced as a measure of agreement that avoids this problem by adjusting the agreement score to take account of the amount of agreement that would be expected by chance alone. While we might hope that this would be the end of the story of inter-rater agreement, there are

Cohen's Kappa A statistical estimate of inter-rater agreement or reliability that is more robust than percent agreement because it adjusts for the probability that some agreement is due to random chance.

other ways of estimating inter-rater agreement depending on the particular study (e.g., a weighted version of Cohen's Kappa is used when codes are ordered, and there are other statistics to use when you have multiple raters).

Another way to estimate inter-rater reliability is appropriate when the measure is a continuous one, rather than a categorical one. In such a case, all you need to do is calculate the correlation between the ratings of the two observers. For instance, they might be rating the overall level of activity in a classroom on a 1-to-7 scale. The correlation between the ratings of two observers would give you an estimate of the reliability or consistency between the raters (see **Figure 5.18**).

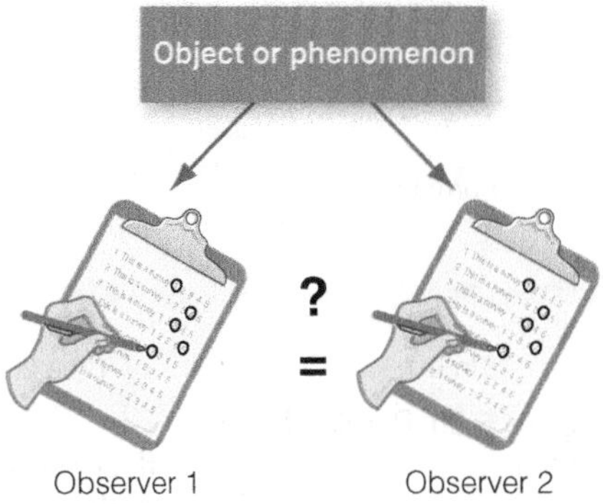

There are other things you could do to encourage reliability between observers, even if you don't estimate it. For instance, one of us used to work in a psychiatric unit where every morning a nurse had to do a ten-item rating of each patient on the unit. Of course, we couldn't count on the same nurse being present every day, so we had to find a way to ensure that all the nurses would give comparable (reliable) ratings. The way we did it was to hold weekly calibration meetings where we would have all of the nurses' ratings for several patients and discuss why they chose the specific values they did. If there were disagreements, the nurses would discuss them and attempt to come up with rules for deciding when they would give a 3 or a 4 for a rating on a specific item. Although this was not an estimate of reliability, more effective communication between raters (in this case, nurses) probably went a long way toward improving the reliability among them.

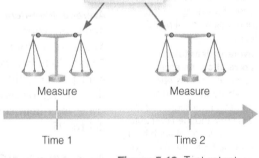

Figure 5.18 Inter-rater or inter-observer reliability.

Test-Retest Reliability

You estimate test-retest reliability when you administer the same test to the same (or a similar) sample on two different occasions (see **Figure 5.19**). This is a classic way to check on the stability of a measure. This approach assumes that there is no substantial change in the construct being measured between the two occasions. In estimating test-retest reliability, the focus is on analyzing the data collection instrument as a potential source of error. The amount of time allowed between measures is critical. You know that if you measure the same thing twice, the correlation between the two observations will depend in part on how much time elapses between the two measurement occasions. The shorter the time gap, the higher the correlation; the longer the time gap, the lower the correlation, because the two observations are related over time; the closer in time you get, the more similar the factors that contribute to error. Since this correlation is the test-retest estimate of reliability, you can obtain considerably different estimates depending on the time interval. Ideally, the interval between the two observations should be long enough so that values obtained the second time around are not affected by the previous measurement (for example, the respondent may just simply remember his or her response if the time interval is too short) but not so distant that knowledge of new things over time alters the way the study participants respond to the question.

Figure 5.19 Test-retest reliability.

Parallel-Forms Reliability

In **parallel-forms reliability**, you first have to create two parallel forms or different versions of an assessment tool. One way to accomplish this is to start with

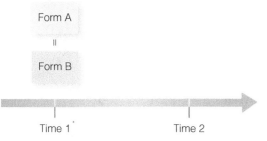

Figure 5.20 Parallel-forms reliability.

a relatively large set of questions that address the same construct and then randomly divide the questions into two sets. You administer both instruments to the same sample of individuals (see **Figure 5.20**). The correlation between the two parallel forms is the estimate of reliability. Sometimes, exam programs use parallel-forms reliability to check the consistency of student scores on multiple versions of an exam (like the ACT for example) that are similar in content and difficulty level. One major problem with this approach is that you have to be able to generate lots of items that reflect the same construct, which is often no easy feat. Furthermore, this approach makes the assumption that the alternate forms are parallel or equivalent. Even by chance, this will sometimes not be the case.

Internal Consistency Reliability

In internal consistency reliability estimation, the key idea is to check whether items on a single test or instrument appear to be measuring the same thing. For this, you use a single measurement instrument administered to a group of people on one occasion to estimate reliability. In effect, you judge the reliability of the instrument by estimating how well the items on the instrument that reflect the same construct yield similar results. You are looking at how consistent the results are for different items for the same construct within the measure. There are a wide variety of internal consistency measures you can use.

average inter-item correlation An estimate of internal consistency reliability that uses the average of the correlations of all pairs of items.

average item-total correlation An estimate of internal consistency reliability where you first create a total score across all items and then compute the correlation of each item with the total. The average inter-item correlation is the average of those individual item-total correlations.

Average Inter-item Correlation The **average inter-item correlation** uses all of the items on your instrument that are designed to measure the same construct. You first compute the correlation between each pair of items, as illustrated in **Figure 5.21**. For example, if you have six items, you will have fifteen different item pairings (fifteen correlations). The average inter-item correlation is simply the average or mean of all these correlations. In the example, you find an average inter-item correlation of .90 with the individual correlations ranging from .84 to .95.

Average Item-total Correlation The average item-total correlation involves computing a total score across the set of items on a measure and treating that total score as though it were another item, thereby obtaining all of the item to total score

Figure 5.21 The average inter-item correlation.

		I_1	I_2	I_3	I_4	I_5	I_6
Item 1							
Item 2	I_1	1.00					
Item 3	I_2	.89	1.00				
	I_3	.91	.92	1.00			
Item 4	I_4	.88	.93	.95	1.00		
Item 5	I_5	.84	.86	.92	.85	1.00	
Item 6	I_6	.88	.91	.95	.87	.85	1.00

Measure →

.89 + .91 + .88 + .84 + .88 + .92 + .93 + .86 + .91 + .92 + .95 + .85 + .87 + .85 = 13.41

13.41 / 15 = .90

Figure 5.22 Average item-total correlation.

	I_1	I_2	I_3	I_4	I_5	I_6
I_1	1.00					
I_2	.89	1.00				
I_3	.91	.92	1.00			
I_4	.88	.93	.95	1.00		
I_5	.84	.86	.92	.85	1.00	
I_6	.88	.91	.95	.87	.85	1.00
Total	.84	.88	.86	.87	.83	.82	1.00

Measure → Item 1, Item 2, Item 3, Item 4, Item 5, Item 6

.85

correlations. **Figure 5.22** shows the six item-to-total correlations at the bottom of the correlation matrix. They range from .82 to .88 in this sample analysis, with the average of these at .85.

Split-Half Reliability In **split-half reliability**, you randomly divide into two sets all items that measure the same construct. You administer the entire instrument to a sample and calculate the total score for each randomly divided half of the measure. The split-half reliability estimate, as shown **Figure 5.23**, is simply the correlation between these two total scores. In the example, it is .87. Note, the parallel-forms approach may seem similar to the split-half reliability, but the objectives of the two estimations are different. The major difference is that in parallel forms, two separate instruments are constructed so that equivalence of the two forms (that are to be used independently of each other) can be considered. For this you would actually create two versions of the same form rather than creating one long one and splitting it in half. With split-half reliability, the aim is to check for homogeneity within a single instrument.

Figure 5.23 Split-half reliability.

Cronbach's Alpha (α) You might see a problem in split-half reliability because you picked two halves at random. You might find yourself wondering, wouldn't it be better if you could take into account all possible split halves? That's the logic behind Cronbach's Alpha (α). Imagine that you compute one split-half reliability and then randomly divide the items into another set of split halves and recompute, and keep doing this until you have computed all possible split-half estimates of reliability. **Cronbach's Alpha** is mathematically equivalent to the average of all possible split-half estimates (although that's not how it's typically computed). Notice that when we say you compute all possible split-half estimates, we don't mean that for each time you measure a new sample! That would take forever. Instead, you calculate all split-half estimates from the same sample. Because you measured the entire sample on each of the six items, all you have to do is have the computer analysis do the random subsets of items and compute the resulting correlations. **Figure 5.24** shows several of the split-half estimates for our six-item example and lists them as SH with a subscript. Keep in mind that although Cronbach's Alpha is equivalent to the average of all possible split-half correlations, you would

split-half reliability An estimate of internal consistency reliability that uses the correlation between the total score of two randomly selected halves of the same multi-item test or measure.

Cronbach's Alpha One specific method of estimating the internal consistency reliability of a measure. Although not calculated in this manner, Cronbach's Alpha can be thought of as analogous to the average of all possible split-half correlations.

Figure 5.24 Cronbach's
Alpha estimate of reliability.

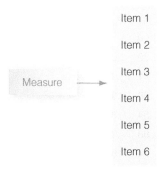

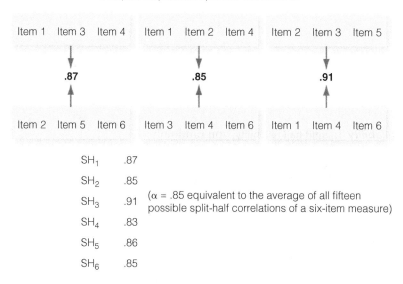

Compute all possible split-half correlations.

SH$_1$	.87
SH$_2$	.85
SH$_3$	.91
SH$_4$	.83
SH$_5$	.86
SH$_6$	.85

(α = .85 equivalent to the average of all fifteen possible split-half correlations of a six-item measure)

never actually calculate it that way. Some clever researcher (Cronbach, no doubt!) figured out a way to get the mathematical equivalent a lot more quickly.

Comparison of Reliability Estimators

Each of the reliability estimators has certain advantages and disadvantages. Inter-rater reliability is one of the best ways to estimate reliability when your measure is an observation done by a person. However, it is useful only if multiple raters or observers are present. As an alternative, you could look at the correlation of ratings of the same single observer repeated on two different occasions. For example, let's say you collected videotapes of child–mother interactions and had a rater code the videos for how often the mother smiled at the child. To establish inter-rater reliability, you could take a sample of videos and have two raters code them independently. You might use the inter-rater approach especially if you were interested in using a team of raters and you wanted to establish that they yielded consistent results. If you get a suitably high inter-rater reliability, you could then justify allowing them to work independently on coding different videos. (When you only have a single rater and cannot easily train others, you might alternatively use a test-retest approach by having this person rate the same sample of videos on two occasions and computing the correlation). On the other hand, in some studies it is reasonable to do both inter-rater and test-retest approaches, to help establish the reliability of the raters or observers. Inter-rater reliability can be strengthened by making the assessment criteria explicit among various raters. In fact, the more individual judgment is involved in an assessment, the more crucial it is that independent raters follow a set criterion to record their observations.

Use the parallel-forms estimator only in situations where you intend to use the two forms as alternate measures of the same thing. Both the parallel-forms and all of the internal consistency estimators have one major constraint: you should have lots of items designed to measure the same construct. This is relatively easy to achieve in certain contexts like achievement testing. (It's easy, for instance, to construct many similar addition problems for a math test.) However, for more complex or subjective constructs, this can be a real challenge. Cronbach's Alpha tends to be the most frequently used estimate of internal consistency.

The test-retest estimator is especially feasible in most experimental and quasi-experimental designs that use a no-treatment control group. In these designs, you typically have a control group that is measured on two occasions (pretest and posttest). The main problem with this approach is that you don't have any information about reliability until you collect the posttest and, if the reliability estimate is low, it's too late to fix the pretest. However, even in this case, you may be able to identify a subset of items that are reliable and use only those.

Each of the reliability estimators is likely to give a different value for reliability. In general, the test-retest and inter-rater reliability estimates will be lower in value than the parallel-forms and internal consistency estimates because the former involve measurement at different times or with different raters. Since reliability estimates are often used in statistical analyses of quasi-experimental designs (see the section "The Nonequivalent Groups Design" in Chapter 9, "Quasi-Experimental Design"), the fact that different estimates can differ considerably makes the analysis even more complex.

5.2d Validity

While reliability is necessary for ensuring the quality of any measurement, it alone is not sufficient. Validity is the other important dimension that is crucial for any kind of meaningful measurement, in particular, and research, in general. To give you an example, suppose you weigh yourself every day, but your scale is off by five pounds. This means that you will probably get the same reading every day (not considering the cheesecake you had last night!), thereby making the measurement reliable, but

"*How accurate is our misinformation?*"

Christopher Weyant/Conde Nast Collection

Figure 5.25 It is *always* important to think about validity.

it will not be an accurate measure. In other words, you are better off investing in a new weighing scale that provides you both a reliable as well as a valid measure of your weight—in fact, the latter is in some cases (for example, where weight issues are concerned!) more important than the former. In short, validity deals with accuracy or precision of measurement (**Figure 5.25**).

Construct Validity

In this text, as in most research methods texts, construct validity is presented in the section on measurement; however, it is typically presented as one of many different types of validity (for example, face validity, predictive validity, or concurrent validity) that are related to measurement. We don't see it that way at all. We see construct validity as the overarching category that contributes (with reliability) to the quality of measurement, with all of the other measurement validity labels falling beneath it. In fact, it is important to remember that construct validity is not limited only to measurement. It is as relevant to the independent variable—the program or treatment—as it is to the dependent variable. Before we are able to convince you of this, let us first begin by describing what construct validity means.

Construct validity is often defined as the extent to which your measure or instrument actually measures what it is theoretically supposed to measure. Formally speaking, it is defined as the degree to which inferences can legitimately be made from the operationalizations in your study to the theoretical constructs on which those operationalizations were based. What does this mean? In research, you "operationalize" a concept you want to measure when you describe exactly *how* you'll measure it. An operationalization is your translation of the idea into something real and concrete.

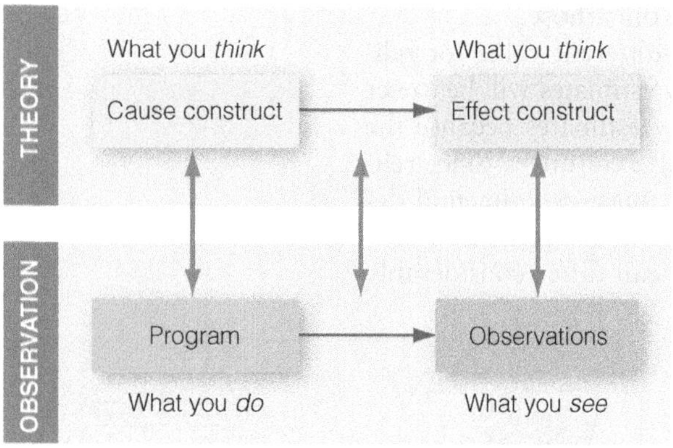

We find that it helps when thinking about construct validity to make a distinction between two broad territories that we call the *land of theory* and the *land of observation*, as illustrated in **Figure 5.26**. The land of theory is what goes on inside your mind, and your attempt to explain or articulate this to others. It is all of the ideas, theories, hunches, and abstract hypotheses you have about the world. In the land of theory, you think of the program or treatment as it should be. In the land of theory, you have the idea or construct of the outcomes or measures you are trying to affect. The land of observation, on the other hand, consists of what you actually see or measure happening in the world around you and the public manifestations of that world. In the land of observation, you find your actual program or treatment, and your actual measures or observational procedures. If you have construct validity, then you have constructed the land of observation based on your ideas in the land of theory. You developed the program to reflect the kind of program you had in mind. You created the measures to get at what you wanted to get at.

In other words, construct validity is an assessment of how well your actual programs or measures reflect your ideas or theories, how well the bottom of Figure 5.26 reflects the top. To give an example, imagine that you want to measure the construct of self-esteem. You have an idea of what self-esteem means—this is the land of theory. Perhaps you construct a ten-item paper-and-pencil instrument to measure that self-esteem concept. The instrument is the operationalization; it's the translation of the idea of self-esteem into something concrete. The construct validity question here would be how well the ten-item instrument (the operationalization) reflects the idea you had of self-esteem. Essentially, construct validity is all about representation, and it can be viewed as a *truth in labeling* issue—that is, when you measure what you call self-esteem, is that what you are really measuring?

Figure 5.26 The idea of construct validity.

Construct Validity of What?

As mentioned previously, when discussing construct validity, it's dumb to limit our scope only to measurement. The idea speaks to the validity of any operationalization. That is, any time you translate a concept or construct into a functioning and operating reality (*the operationalization*), you need to be concerned about how well the translation reflects what you intended. This issue is as relevant when talking about treatments or programs as it is when talking about measures. Come to think of it, you could also think of sampling in this way. The population of interest in your study is the construct, and the sample is your operationalization. If you think of it this way, you are essentially talking about the construct validity of sampling (see Chapter 4, "Sampling"). The construct validity question, "How well does my sample represent the idea of the population?" is directly related to the external validity question, "How well can I generalize from my sample to the population?" In fact, construct validity is intimately connected with the original concept of "generalizing," which is what the idea of external validity is about. We want to know how reasonable it is to "generalize" from our actual measure, program, or sample to the construct of each. Although we don't usually talk about sampling in terms of construct validity, it makes sense and helps to illustrate how important the idea is in research.

5.2e Construct Validity and Other Measurement Validity Labels

Now that we have a basic understanding of construct validity let's try to put it in the context of other measurement validity types. There's an awful lot of confusion in the methodological literature that stems from the wide variety of labels used to describe the validity of measures. Because construct validity refers to the general case of translating any construct into an operationalization, we view all of the other validity labels as subcategories to the overarching idea of construct validity. The validity labels, which we call "measurement validity types," reflect different ways you can demonstrate the various aspects of construct validity. Here's a list of the validity types that are typically mentioned in texts and research papers when talking about the quality of measurement and how we would categorize them:

Construct validity

- **Translation validity**
  - Face validity
  - Content validity
- **Criterion-related validity**
  - Predictive validity
  - Concurrent validity
  - Convergent validity
  - Discriminant validity

We have to warn you here that we made this list up. We've never heard of translation validity before, but we needed a good name to summarize what both face and content validity are getting at, and this title seemed sensible. (See how easy it is to be a methodologist?) All of the other labels are commonly known, but the way we've organized them is different from what we've seen elsewhere.

Let's see if we can make some sense out of this list. First, as mentioned previously, we would like to use the term construct validity to be the overarching category. Construct validity is the approximate truth of the conclusion or inference that your operationalization accurately reflects its construct. All of the other validity types essentially address some aspect of this general issue (which is why we've subsumed them under the general category of construct validity). Second, we make a distinction between two broad types of construct validity: translation validity and criterion-related validity. That's because these correspond to the two major ways you can assure/assess the validity of an operationalization.

In **translation validity,** you focus on whether the operationalization is a good translation of the construct. This approach is definitional in nature; it assumes you have a good, detailed definition of the construct and that you can check the operationalization against it. In **criterion-related validity,** you examine whether the operationalization or the implementation of the construct behaves the way it should, according to some theoretical criteria. This type of validity is a more relational approach to construct validity. It assumes that your operationalization should function in predictable ways in relation to other operationalizations, based upon your theory of the construct. (If all this seems a bit dense, hang in there until you've gone through the following discussion, and then come back and reread this paragraph.) Let's go through the specific validity types under each of the two headings.

Translation Validity

In essence, both of the translation validity types (face and content validity) attempt to assess the degree to which you accurately *translated* your construct into the operationalization, and hence the choice of name. Let's look at the two types of translation validity.

Face Validity

In **face validity,** you look at the operationalization and see whether *on its face* it seems like a good translation of the construct. In other words, does it *look like* your measure is measuring what you want it to? This is probably the weakest way to try to demonstrate construct validity. For instance, you might look at a measure of math ability, read through the questions, and decide that it seems like it is a good measure of math ability (that is, that the label *math ability* seems appropriate for this measure). Or, you might observe a teenage pregnancy-prevention program and conclude that it is indeed a teenage pregnancy-prevention program. However, this would clearly be weak evidence of validity because it is essentially a subjective judgment call. (Note that just because it is weak evidence doesn't mean that it is wrong. You need to rely on your subjective judgment throughout the research process. It's just that this form of judgment won't be especially convincing to others.) You can improve the quality of a face-validity assessment considerably by making it more systematic and checking that it "makes sense" as a measure of the construct to people who know the construct well.

Content Validity

In **content validity,** you essentially check the operationalization against the relevant content domain for the construct. The content domain is like a comprehensive checklist of the traits of your construct. This approach assumes that you have a good, detailed description of the content domain, something that's not always true. Let's look at an example where it is true. You might lay out all of the characteristics of a teenage pregnancy-prevention program. You would probably include in this domain specification the definition of the target group, a description of whether the program is preventive in nature (as opposed to treatment-oriented),

translation validity A type of construct validity related to how well you translated the idea of your measure into its operationalization.

criterion-related validity The validation of a measure based on its relationship to another independent measure as predicted by your theory of how the measures should behave.

face validity A validity that checks that "on its face" the operationalization seems like a good translation of the construct.

content validity A check of the operationalization against the relevant content domain for the construct.

and the content that should be included, such as basic information on pregnancy, the use of abstinence, birth control methods, and so on. Then, armed with these characteristics, you create a type of checklist to be used when examining your program. Only programs that have these characteristics can legitimately be defined as teenage pregnancy-prevention programs. This all sounds fairly straightforward, and for many operationalizations it may be. However, for other, more abstract constructs (such as self-esteem or intelligence), it may not be as easy to decide which characteristics constitute the content domain.

Criterion-Related Validity

In criterion-related validity, you check the performance of your operationalization against some criterion. An example of this idea would be to see how well graduate school admissions exams predict student performance in graduate school (the criterion). How is this different from translation validity? In translation validity, the question is, how well did you translate the idea of the construct into its manifestation? No other measure comes into play. In criterion-related validity, you usually make a prediction about how the operationalization will perform on some other measure, based on your theory of the construct. The differences among the criterion-related validity types are in the criteria they use as the standard for judgment.

For example, think again about measuring self-esteem. For content validity, you would try to describe all the things that self-esteem is in your mind and translate that into a measure. You might say that self-esteem involves how good you feel about yourself, that it includes things like your self-confidence and the degree to which you think positively about yourself. You could translate these notions into specific questions, a translation-validity approach. On the other hand, you might reasonably expect that people with high self-esteem, as you interpret it, would tend to act in certain ways. You might expect that you could distinguish them from people with low self-esteem. For instance, you might argue that high self-esteem people will volunteer for a task that requires self-confidence (such as speaking in public). Notice that in this case, you validate your self-esteem measure by demonstrating that it is correlated with some other independent indicator (raising hands to volunteer) that you theoretically expect high self-esteem people to exhibit. This is the essential idea of criterion-related validity: validating a measure based on its relationship to another independent measure. This is different from translational validity because it goes beyond the realm of definition.

Predictive Validity

In **predictive validity**, you assess the operationalization's ability to predict something it should theoretically be able to predict. For instance, you might theorize that a measure of math ability should be able to predict how well a person will do in an engineering-based profession. You could give your measure to experienced engineers and see whether there is a high correlation between scores on the measure and their salaries as engineers (which, hopefully, is related to how well they do as engineers). A high correlation would provide evidence for predictive validity; it would show that your measure can correctly predict something that you theoretically think it should be able to predict. However, as you may have already guessed, one challenge with predictive validity is that the two measures may not be available at the same time. For instance, you may have

predictive validity A type of construct validity based on the idea that your measure is able to predict what it theoretically should be able to predict.

scores of individuals on a measure of math ability now, but you will probably have to wait until they graduate and join the workforce to get their salaries. This problem goes away when assessing concurrent validity, as discussed below.

Concurrent Validity

In **concurrent validity**, you assess the operationalization's ability to distinguish between groups that it should theoretically be able to distinguish between. For example, if you come up with a way of assessing depression, your measure should be able to distinguish between people who are diagnosed as depressed and those diagnosed as paranoid schizophrenic. If you want to assess the concurrent validity of a new measure of empowerment, you might give the measure to both migrant farmworkers and to the farm owners, theorizing that your measure should show that the farm owners are higher in empowerment. As in any discriminating test, the results are more powerful if you are able to show that you can discriminate between two similar groups than if you can show that you can discriminate between two groups that are very different.

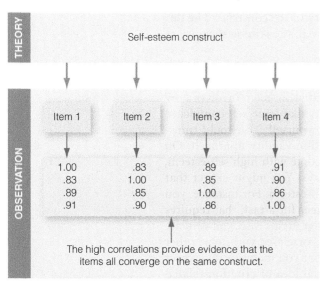

Figure 5.27 Convergent validity correlations.

concurrent validity An operationalization's ability to distinguish between groups that it should theoretically be able to distinguish between.

convergent validity The degree to which the operationalization is similar to (converges on) other operationalizations to which it should be theoretically similar.

Convergent and Discriminant Validity

Convergent and discriminant validity are both considered subcategories of construct validity. The important thing to recognize is that they work together; if you can demonstrate that you have evidence for both convergent and discriminant validity, you have by definition demonstrated that you have evidence for construct validity. However, neither one alone is sufficient for establishing construct validity. Because it is easiest to think about convergent and discriminant validity as two interlocking propositions, we will discuss them under one section.

Convergent Validity In **convergent validity**, you examine the degree to which the operationalization is similar to (converges on) other operationalizations to which it theoretically should be similar. That is, you have convergent validity if measures of constructs that theoretically *should* be related to each other are, in fact, observed to be related to each other. If you can demonstrate this, you have shown a correspondence between similar constructs. For instance, to show the convergent validity of a test of arithmetic skills, you might correlate the scores on your test with scores on other tests that purport to measure basic math ability, where high correlations would be evidence of convergent validity. Let's walk you through a step-by-step example of establishing convergent validity. In **Figure 5.27**, you see four scales that supposedly reflect the construct of self-esteem. For instance, Scale 1 might be a 1-to-5 scale measuring "how you feel about yourself," Scale 2 might be an observational scale of your self-esteem as rated by a parent, Scale 3 could be the same rating by a teacher, and Scale 4 might be a classroom observation rating scale of self-esteem-related behaviors. You theorize that all four scales reflect the idea of self-esteem (which is why we labeled the top part of the figure "Theory"). On the bottom part of the figure ("Observation"), you see the intercorrelations between the four scales. This might be based on measuring your scales for a sample of respondents. You should readily see that the item intercorrelations for all item pairings are extremely high. (Remember that correlations range from −1.00 to +1.00.) The correlations provide support for your theory that all four scales are related to the same construct.

Notice, however, that whereas the high intercorrelations demonstrate the four scales are probably related to the same construct, that doesn't automatically mean that the construct is self-esteem. Maybe there's some other construct to which all four items are related (more about this later). However, at least, you can assume from the pattern of correlations that the four scales are converging on the same thing, whatever you might label it.

Discriminant Validity In **discriminant validity**, you examine the degree to which the operationalization is not similar to (or diverges from) other operationalizations that it theoretically should not be similar to. That is, measures of constructs that theoretically should *not* be related to each other are, in fact, observed *not* to be related to each other (that is, you should be able to *discriminate* between dissimilar constructs). For instance, to show the discriminant validity of a test of arithmetic skills, you might correlate the scores on your test with scores on tests of verbal ability, where *low* correlations would be evidence of discriminant validity. To estimate the degree to which any two measures are related to each other, you would typically use the correlation coefficient discussed later in the book (see Chapter 11.) That is, you look at the patterns of intercorrelations among the measures. Correlations between theoretically similar measures should be "high"; whereas correlations between theoretically dissimilar measures should be "low." In **Figure 5.28**, you again see four scales or test scores. Here, however, two of the scales are thought to reflect the construct of self-esteem, whereas the other two are thought to reflect intelligence (IQ). The top part of the figure shows the theoretically expected relationships among the four scales. If you have discriminant validity, the relationship between measures from different constructs should be low. There are four correlations between measures that reflect different constructs, and these are shown on the bottom of the figure (Observation). You should see immediately that these four cross-construct correlations are low (nearer to zero) and certainly much lower than the convergent correlations in Figure 5.27.

As mentioned previously in the case of convergent validity, just because there is evidence that the two sets of two measures seem to be related to different constructs (because their intercorrelations are so low) doesn't mean that the constructs they're related to are self-esteem and intelligence. However the correlations do provide evidence that the two sets of measures are discriminated from each other.

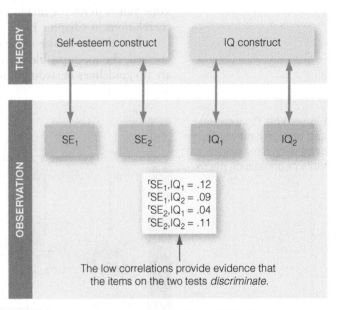

$r_{SE_1,IQ_1} = .12$
$r_{SE_1,IQ_2} = .09$
$r_{SE_2,IQ_1} = .04$
$r_{SE_2,IQ_2} = .11$

The low correlations provide evidence that the items on the two tests *discriminate*.

Figure 5.28 Discriminant validity correlations.

discriminant validity The degree to which the operationalization is not similar to (or diverges from) other operationalizations that it theoretically should not be similar to.

Putting It All Together

Okay, so where does this leave us? We've shown how to provide evidence for convergent and discriminant validity separately; but, as we said at the outset, to argue for construct validity, you really need to be able to show that both of these types of validity are supported. Given the previous discussions of convergent and discriminant validity, you should be able to see that you could put both principles together into a single analysis, to examine both at the same time. This is illustrated in **Figure 5.29**.

Figure 5.29 shows six scales or tests: three that are theoretically related to the construct of self-esteem and three that are thought to be related to intelligence. The top part of the figure shows this theoretical arrangement. The bottom of the figure shows what a correlation matrix based on a pilot sample might show. To understand this table, first you need to be able to identify the convergent correlations and the discriminant ones. The two sets or blocks of convergent coefficients appear in regular type: one 3×3 block for the self-esteem intercorrelations in the upper left of the table, and one 3×3 block for the intelligence correlations in the lower right. Additionally, two 3×3 blocks of discriminant coefficients appear in blue and italics, although if you're really sharp you'll recognize that they are the same values in mirror image. (Do you know why? You might want to read up on correlations in Chapter 11.)

How do you make sense of the correlations' patterns? As you might have guessed, the main problem with this convergent-discriminant idea is that there are no guidelines regarding how high correlations need to be to provide evidence

Figure 5.29 Convergent and discriminant validity correlations in a single table or correlation matrix.

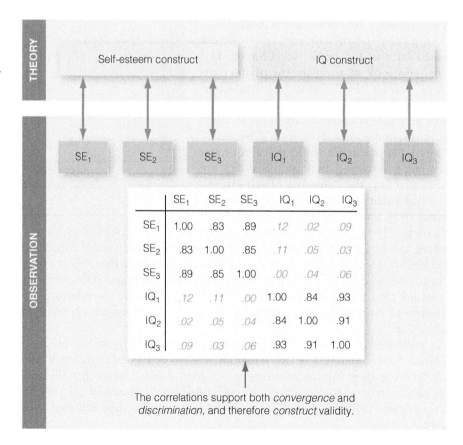

	SE₁	SE₂	SE₃	IQ₁	IQ₂	IQ₃
SE₁	1.00	.83	.89	*.12*	*.02*	*.09*
SE₂	.83	1.00	.85	*.11*	*.05*	*.03*
SE₃	.89	.85	1.00	*.00*	*.04*	*.06*
IQ₁	*.12*	*.11*	*.00*	1.00	.84	.93
IQ₂	*.02*	*.05*	*.04*	.84	1.00	.91
IQ₃	*.09*	*.03*	*.06*	.93	.91	1.00

The correlations support both *convergence* and *discrimination*, and therefore *construct* validity.

for convergence and how low they need to be to provide evidence for discrimination. In general, convergent correlations should be as high as possible and discriminant ones should be as low as possible, but there is no hard-and-fast rule. Well, let's not let *that* stop us. One thing you can assume to be true is that the convergent correlations should always be higher than the discriminant ones. At least that helps a bit. Take a good look at the table and you will see that in this example the convergent correlations are always higher than the discriminant ones. We would conclude from this that the correlation matrix provides evidence for both convergent and discriminant validity, all in one table!

It's true the pattern supports discriminant and convergent validity, but does it show that the three self-esteem measures actually measure self-esteem or that the three intelligence measures actually measure intelligence? Of course not. That would be much too easy.

So, what good is this analysis? It does show that, as you predicted, the three self-esteem measures seem to reflect the same construct (whatever that might be). The three intelligence measures also seem to reflect the same construct (again, whatever that is), and the two sets of measures seem to reflect two different constructs (whatever they are). That's not bad for one simple analysis.

Okay, so how do you get to the really interesting question? How do you show that your measures are actually measuring self-esteem or intelligence? We hate to disappoint you, but there is no simple answer to that. (We bet you knew that was coming!) You can do several things to address this question. First, you can use other ways to address construct validity to help provide further evidence that you're measuring what you say you're measuring. For instance, you might use a face-validity or content-validity approach to demonstrate that the measures reflect the constructs you say they are.

One of the most powerful approaches is to include even more constructs and measures. Your test of construct validity becomes even stronger when you make sure that the variables you use for discriminant validity are closer in meaning to the construct you are interested in. For instance, if your theory is that self-esteem is something different from self-worth, showing that measures of each of these have lower correlations than within-construct correlations would be an even more powerful demonstration of discriminant validity. Put in other terms, it is easier to show discrimination with constructs that are less similar to the target construct than to ones that are more similar, but the evidence for construct validity would be much stronger with more similar constructs. In general, the more complex your theoretical model (if you find confirmation of the correct pattern in the correlations), the more evidence you are providing that you know what you're talking about (theoretically speaking). Of course, it's also harder to get all the correlations to give you the exact right pattern as you add more measures. In many studies, you simply don't have the luxury of adding more and more measures because it's too costly or demanding. Despite the impracticality, if you can afford to do it, adding more constructs and measures enhances your ability to assess construct validity.

5.2f Threats to Construct Validity

Before launching into a discussion of the most common threats to construct validity, take a moment to recall what a threat to validity is. In a research study, you are likely to reach a conclusion that your program was a good operationalization of what you wanted and that your measures reflected what you wanted them to reflect. Would you be correct? How will you be criticized if you make these types of claims? How might you strengthen your claims? The kinds of questions and issues your critics will raise are what we mean by **threats to construct validity**.

We base the list of threats from the discussion in Cook and Campbell's text (Cook & Campbell, 1979). Although we love their discussion, we do find some of their terminology a bit cumbersome; much of what we'll do here is try to translate their terms into words that are more understandable to us normal human beings. Here, then, are the major threats to construct validity (drum roll, please):

Inadequate Preoperational Explication of Constructs

Huh? Okay, this is a fairly daunting phrase, but don't panic. Breathe in. Breathe out. Let's decipher this phrase one part at a time. It isn't nearly as complicated as it sounds. You know what inadequate means, don't you? (If you don't, we'd say you're pretty inadequate!) Here, *preoperational* is what you were thinking about before you developed your measures or treatments. *Explication* is just a fancy word for explanation. Put it all together, what this phrase means is that you didn't do a good enough job defining what you meant by the construct before you tried to translate it into a measure or program. In other words, you weren't thinking carefully. How is this a threat? Imagine that your program consisted of a new type of approach to rehabilitation. A critic comes along and claims that, in fact, your program is neither *new* nor a true *rehabilitation* program. You are being accused of doing a poor job of thinking through your constructs. Here are some possible solutions:

- **Think through your concepts better.**
- **Use structured methods (for example, concept mapping) to articulate your concepts.**
- **Get experts to critique your operationalizations.**

Mono-Operation Bias

Mono-operation bias occurs when you use only one version of the treatment or program in your study. Note that it is only relevant to the independent variable, the cause, program, or treatment in your study. It does not pertain to measures or outcomes (see mono-method bias in the following section). If you only use a single version of a program in a single place at a single point in time, you may not be capturing the full breadth of the concept of the program. For example, suppose a nonprofit agency plans a program that intends to provide training and equipment to doctors and nurses in developing countries—but in its implementation, it ends up providing training only to doctors, leaving out other allied health professionals. In this case, the operationalization suffers from mono-operation bias. Every operationalization is a flawed, imperfect reflection of the construct on which it is based. If you conclude that your program reflects the construct of the program, your critics are likely to argue that the results of your study reflect only the peculiar version of the program that you implemented, and not the full breadth of the construct you had in mind. Solution: try to implement multiple versions of your program.

Mono-Method Bias

Mono-method bias occurs when you only use one measure of a construct. Note that it is only relevant to your measures or observations, not to your programs or causes. Otherwise, it's essentially the same issue as mono-operation bias. In the

threats to construct validity Any factor that causes you to make an incorrect conclusion about whether your operationalized variables (e.g., your program or outcome) reflect well the constructs they are intended to represent.

mono-operation bias A threat to construct validity that occurs when you rely on only a single implementation of your independent variable, cause, program, or treatment in your study.

mono-method bias A threat to construct validity that occurs because you use only a single method of measurement.

nonprofit example above, say you want to measure the change in knowledge and practices of medical professionals after you have implemented your program. You might consider interviewing a sample of doctors about their overall training experience at the end of program implementation. However, your critics will suggest that you aren't measuring the actual change in knowledge and practices, because the doctors' views are likely to be subjective and bound by need for future investment from the nonprofit at their institution. Another way to measure change in knowledge and practices would be to administer a pre- and post-questionnaire on the subject matter covered by the training. Even with this, your critics may still argue that you're only measuring part of the construct—in this case, theoretical knowledge—and not practices. So, you might consider sending an independent evaluator to check on their surgical skills before and after the program. With only a single measure, you can't provide much evidence that you're really measuring change in knowledge and practices. Solution: try to implement multiple measures of key constructs and try to demonstrate (perhaps through a pilot or side study) that the measures you use behave as you theoretically expect them to behave.

Interaction of Different Treatments

Suppose you design a new program that encourages high-risk teenage girls to go to school and not become pregnant. The results of your program show that the girls in your treatment group have higher school attendance and lower pregnancy rates. You're feeling pretty good about your program until your critics point out that the targeted at-risk treatment group in your study is also likely to be involved simultaneously in several other programs designed to have similar effects. Can you really claim that the program effect is a consequence of your program? The real program that the girls received may actually be the *combination* of the separate programs in which they participated. Though this seems more like a threat to internal validity (did your program actually cause the outcome that you observe?), it is more properly considered a construct validity threat because it deals with a labeling issue— you think what was actually implemented was the construct you had in mind, but that was really not the case. What can you do about this threat? One approach is to try to isolate the effects of your program from the effects of any other treatments. You could do this by creating a research design that uses a control group. (This is discussed in detail in Chapter 8, "Introduction to Research Design.") In this case, you could randomly assign some high-risk girls to receive your program and some to a no-program control group. Even if girls in both groups receive some other treatment or program, the only systematic difference between the groups would be your program. If you observe differences between them on outcome measures, the differences must be due to the program. By using a control group that makes your program the only thing that differentiates the two groups, you control for the potential confusion or "confounding" of multiple treatments.

Interaction of Testing and Treatment

Does testing or measurement itself make the groups more sensitive or receptive to the treatment? If it does, the testing is essentially a part of the treatment; it's inseparable from the effect of the treatment. This is also a labeling issue (and, hence, a concern of construct validity), because you want to use the label *program* to refer to the program alone, but in fact it also includes the testing. As in the previous threat, one way to control for this is through research design. If you are worried that a pretest makes your program participants more sensitive or receptive to the treatment, randomly assign your program participants into two groups, one of which gets the pretest while the other does not. If there are differences on outcomes between these groups, you have evidence that there is an effect of the testing. If not, the testing doesn't matter.

Restricted Generalizability Across Constructs

This is what we like to refer to as the *unintended consequences* threat to construct validity. You do a study and conclude that Treatment X is effective. In fact, Treatment X is effective, but only on the outcome you measured. What you failed to anticipate is that the treatment may have drastic negative consequences or side effects on other outcomes. For example, a nonprofit that trains medical professionals in developing countries does so with the end goal of developing individual capacity (or at least that's what they have in mind in the land of theory). But, it is possible that because of the training you provided, medical professionals decide to leave their jobs for more lucrative offers elsewhere. This actually ends up hurting the overall institutional capacity of the organization where you implemented your program. When you say that Treatment X is effective, you have defined *effective* in regard to only the outcomes you measured. But, in fact, significant unintended consequences might affect constructs you did not measure and cannot generalize to. This threat should remind you that you have to be careful about whether your observed effects (Treatment X is effective) would generalize to other potential outcomes. How can you deal with this threat? The critical issue here is to try to anticipate the unintended and measure a broad range of potential relevant outcomes.

Confounding Constructs and Levels of Constructs

Imagine a study to test the effect of a new drug treatment for cancer. A fixed dose of drug X is given to a randomly assigned treatment group and a placebo to the other group. No treatment effects are detected. But perhaps the observed result is only true for a certain dosage level. Slight increases or decreases of the dosage of drug X may radically change the results. In this context, it is not fair for you to label the treatment as "drug X" because you only looked at a narrow range of dosage of the drug. Like the other construct validity threats, this threat is essentially a labeling issue; your label is not a good description for what you implemented. What can you do about it? If you find a treatment effect at a specific dosage, be sure to conduct subsequent studies that explore the range of effective doses. Note that, although we use the term "dose" here, you shouldn't limit the idea to medical studies. If you find an educational program effective at a particular dose—say one hour of tutoring a week—conduct subsequent studies to see if dose responses change as you increase or decrease from there. Similarly, if you don't find an effect with an initial dose, don't automatically give up. It may be that at a higher dose the desired outcome will occur.

5.2g The Social Threats to Construct Validity

The remaining major threats to construct validity can be distinguished from the ones discussed previously, because they are all related to the social and human nature of research.

Hypothesis Guessing

Most people don't just participate passively in a research project. They engage in **hypothesis guessing** and try to determine the real purpose of the study. Therefore, they are likely to base their behavior on what they guess, not just on your treatment. In an educational study conducted in a classroom, students might guess that the key dependent variable has to do with class participation levels. If they increase their participation not because of your program but because they think that's what you're studying, you cannot label the outcome as an effect of the program. It is this labeling issue that makes this a construct validity threat. This is a difficult threat to eliminate. In some studies, researchers try to hide the real purpose of the study, but this may be unethical, depending on the circumstances. In some instances, they eliminate the need for participants to guess by telling them the real purpose (although who's to say that participants will believe them). If this is a potentially serious threat, you may think about trying to control for it explicitly through your research design. For instance, you might have multiple program groups and give each one slightly different explanations about the nature of the study, even though they all get exactly the same treatment or program. If they perform differently, it may be evidence that they were guessing differently and that this was influencing the results.

Evaluation Apprehension

Many people are anxious about being evaluated. Some are even phobic about testing and measurement situations. If their apprehension makes them perform poorly (and not your program conditions), you certainly can't label that as a treatment effect. Another form of evaluation apprehension concerns the human tendency to want to look good, or look smart, and so on (often referred to as Social Desirability). If, in their desire to look good, participants perform better (and not as a result of your program), you would be wrong to label this as a treatment effect. In both cases, the apprehension becomes confounded with the treatment itself, and you have to be careful about how you label the outcomes. Researchers take a variety of steps to reduce apprehension. In any testing or measurement situation, it is probably a good idea to give participants some time to get comfortable and adjusted to their surroundings. You might ask a few warm-up questions, (knowing that you are not going to use the answers), trying to encourage the participant to get comfortable responding. In many research projects, people misunderstand what you are measuring. If it is appropriate, you may want to tell them that there are no right or wrong answers and that they aren't being personally judged or evaluated based on what they say or do.

Researcher Expectancies

The researcher can bias the results of a study in countless ways, both conscious and unconscious. Sometimes the researcher can communicate what the desired outcome for a study might be (and the participants' desire to look good leads them to react that way). For instance, the researcher might look pleased when participants give a desired answer. If researcher feedback causes the response, it would be wrong to label the response a treatment effect. As in many of the previous threats, probably the most effective way to address this threat is to control for it through your research design. For instance, if resources allow, you can have multiple researchers who differ in their characteristics. Or, you can address the threat through measurement; you can measure expectations prior to the study and use the information in that analysis to attempt to adjust for expectations.

hypothesis guessing A threat to construct validity and a source of bias in which participants in a study guess the purpose of the study and adjust their responses based on that.

5.3 Integrating Reliability and Validity

Now that we have covered both aspects of measurement quality in detail, let's discuss how they are linked to each other. We often think of reliability and validity as separate ideas but, in fact, they're intimately interconnected. One of our favorite metaphors for the relationship between reliability and validity is that of a target. Think of the center of the target as the concept or construct you are trying to measure. Imagine that for each person you are measuring, you are taking a shot at the target. If you measure the concept perfectly for a person, you are hitting the center of the target. If you don't, you are missing the center. The more "off" you are for that person, the further you are from the center (see **Figure 5.30**).

Figure 5.30 shows four possible situations. In the first one, you are hitting the target consistently, but you are missing the center of the target. That is, you are consistently and systematically measuring the wrong value for all respondents. This measure is reliable, but not valid. (It's consistent but wrong.) The second shows hits that are randomly spread across the target. You seldom hit the center of the target, but, on average, you are getting the right answer for the group (but not very well for individuals). In this case, you get a valid group average, but you are inconsistent. Here, you can clearly see that reliability is directly related to the variability of your measure. The third scenario shows a case where your hits are spread across the target and you are consistently missing the center. Your measure in this case is neither reliable nor valid. Finally, the figure shows the Robin Hood or William Tell scenario; you consistently hit the center of the target. Your measure is both reliable and valid. (We bet you never thought of Robin Hood or William Tell in those terms before.)

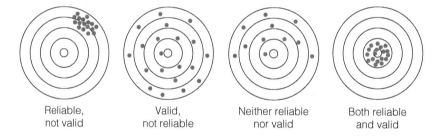

Figure 5.30 The target metaphor for reliability and validity of measurement.

| Reliable, not valid | Valid, not reliable | Neither reliable nor valid | Both reliable and valid |

SUMMARY

This chapter laid the foundation for the idea of measurement. Three broad topics were considered. First, the level of measurement describes the relationship implicit among that measure's values and determines the type of statistical manipulations that are sensible. Second, reliability refers to the consistency or dependability of your measurement. Reliability is based upon true score theory, which holds that any observation can be divided into two values—a true score and an error component. Reliability is defined as the ratio of the true score variance to the observed variance in a measure. There are a variety of methods for estimating reliability. Reliability is a necessary but not sufficient condition for validity. And finally, construct validity refers to the degree to which you are measuring what you intended to measure. Construct validity is divided into translation validity (the degree to which you translated the construct well) and criterion-related validity (the degree to which your measure relates to or predicts other criteria as theoretically predicted). With these three ideas—level of measurement, reliability, and construct validity—as a foundation, you can now move on to some of the more practical and useful aspects of measurement in the next few chapters.

Key Terms

average inter-item correlation
   p. 124
average item-total correlation
   p. 124
Cohen's Kappa p. 122
concurrent validity p. 132
content validity p. 130
convergent validity p. 132
criterion-related validity p. 130
Cronbach's Alpha p. 125
discriminant validity p. 133
face validity p. 130
hypothesis guessing p. 139

inter-rater or inter-observer
   reliability p. 121
internal consistency reliability p. 121
interval level of measurement
   p. 114
level of measurement p. 113
mono-method bias p. 136
mono-operation bias p. 136
nominal level of measurement
   p. 114
ordinal level of measurement
   p. 114
parallel-forms reliability p. 121

predictive validity p. 131
random error p. 117
ratio level of measurement
   p. 115
split-half reliability p. 125
systematic error p. 117
test-retest reliability p. 121
threats to construct validity
   p. 136
translation validity p. 130
triangulate p. 118
true score theory p. 116

Suggested Websites

The U.S. Government Accountability Office

http://www.gao.gov/search?q=reliability

The U.S. GAO is often called the "watchdog" of government because it is dedicated to independent objective assessment of how the government spends money. You can get a sense of just how integral the concepts of reliability and validity are in daily life by browsing this website. For example, go to the website and do keyword searches on "reliability" and "validity" to see the vast range and centrality of measurement in objective analyses conducted by the GAO.

The National Council on Measurement in Education (NCME)

http://ncme.org/

The NCME is a national organization dedicated to all aspects of measurement in education, including technical, but also practical and ethical aspects of use of measures in education. Their website includes a library as well as links to the journals they publish and other measurement resources.

Review Questions

1. What level of measurement is a person's political party affiliation?

a. nominal
b. ordinal
c. interval
d. ratio
(Reference: 5.1a)

2. Cronbach's coefficient alpha is an internal consistency estimate of scale reliability that is mathematically equivalent to

a. the average item-total correlation.
b. all possible parallel-forms reliability estimates.
c. all possible split-half estimates of reliability.
d. none of the above.
(Reference: 5.2)

3. Which of the following is the broadest category of validity?

a. content validity
b. construct validity
c. criterion-related validity
d. translation validity
(Reference: 5.2d)

4. Claiming that a general psychology chapter test was unfair, a student argued that the test was heavily weighted with material related to four key concepts, rather than the possible twelve presented by the textbook author. This student was making an argument based on his awareness of what kind of validity?

a. content validity
b. construct validity
c. criterion-related validity
d. translation validity
(Reference: 5.2e)

5. What type of validity is assessed if an "integrity scale" is given to a group of prisoners and contrasted with the performance of a group of Rotary Club members?

a. content validity
b. face validity

c. criterion-related validity
d. translation validity
(Reference: 5.2e)

6. A newly created graduate school performance prediction exam is given to a set of students. All are admitted to graduate school, and their performance is tracked. After five years, scores of the group who successfully completed the program are compared with scores of students who failed to graduate. This would be an example of what type of validity?

a. face validity
b. predictive validity
c. convergent validity
d. translation validity
(Reference: 5.2e)

7. In the best of all worlds, we want convergent correlation coefficients to be as ___ as possible and discriminant correlation coefficients to be as ___ as possible.

a. high, low
b. high, high
c. low, high
d. low, low
(Reference: 5.2e)

8. What type of threat to construct validity exists if a single set of measures is used to assess a program?

a. mono-construct bias
b. mono-operation bias
c. mono-method bias
d. monolithic bias
(Reference: 5.2f)

9. What type of threat to construct validity exists when a researcher consciously or unconsciously communicates the desired response in her or his approach?

a. hypothesis confirmation
b. evaluation apprehension
c. experimenter expectancy
d. social desirability
(Reference: 5.2g3)

10. Like _____ validity, construct validity is related to generalizing.

a. external
b. discriminant
c. convergent
d. content

(Reference: 5.2d)

11. Reliance on tried-and-true methods like self-report poses no particular issues in terms of threats to validity.

a. True
b. False

(Reference: 5.2f)

12. Inadequate preoperational explication of constructs is most likely to occur when a researcher charges hastily into a program of measurement of a new construct without critically thinking about the construct.

a. True
b. False

(Reference: 5.2f)

13. The overarching category of measurement validity (the one that supersedes and connects the others) is construct validity.

a. True
b. False

(Reference: 5.2e)

14. The strongest way to demonstrate *construct* validity is through *face* validity.

a. True
b. False

(Reference: 5.2e)

15. Examining whether test performance is correlated with job performance in a particular field is a form of predictive validity.

a. True
b. False

(Reference: 5.2e)

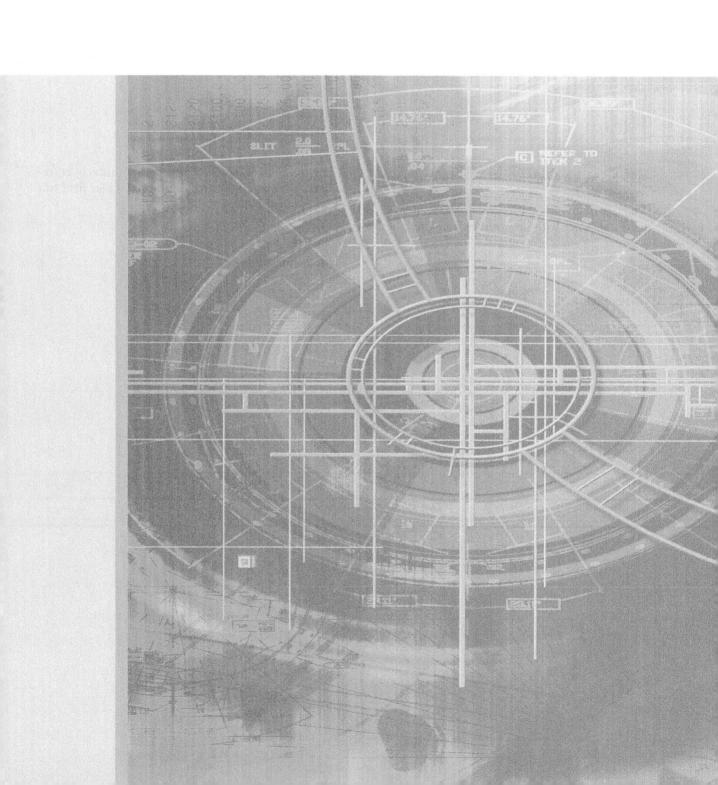

6

Scales, Tests, and Indexes

6.1 Foundations of Scales, Tests, and Indexes

In this chapter, we discuss the three most common approaches for quantitatively measuring a construct: scales, tests, and indexes. These three measures are sometimes difficult to distinguish from each other, and there are conflicting views about how they should be defined. A scale is usually developed to measure an abstract concept like an attitude—say, satisfaction—that might be important in a research study or an opinion poll. A test is typically developed to assess knowledge or skill. Finally, indexes are typically used to assess behaviors of people, organizations, or processes, and they often combine different quantitative variables into a single score. In their simplest manifestation, indexes are basically descriptive metrics. For instance, a count of the number of people who visit an emergency room each day can be considered an index. In more complex cases, the variables pooled together in an index often include different types of constructs, each of which might be measured in a different way.

We begin with a discussion of general issues in scaling, including the distinction between a scale and a response format. We also explain the difference between unidimensional and multidimensional scaling, concluding with an in-depth look at three types of unidimensional scales: Thurstone, Likert, and Guttman. Then, we move on to something you're very familiar with as a student—tests. We will show how tests themselves are evaluated and will briefly consider some of the ethical issues involved in testing. In the last part of this chapter, we'll discuss indexes. Specifically, we'll focus on what indexes are used for and how they are constructed. We'll also talk about some common indexes that you may be familiar with already.

From the discussions presented in this chapter, we hope that you will gain a comprehensive understanding of scales, tests, and indexes, including when each type of measure is most appropriate to use.

6.2 Scales and Scaling

Scaling involves the construction of a measure by combining qualitative judgments about a construct with quantitative metric units. A scale is typically designed to yield a single numerical score that represents the construct of interest. In many ways, scaling remains one of the most mysterious and misunderstood aspects of social research measurement. It attempts to do one of the most difficult of research tasks—measure abstract concepts.

The basic idea of scaling is described in the following section, "General Issues in Scaling." The discussion includes the important distinction between a scale and a response format. Scales are generally divided into two broad categories: unidimensional and multidimensional. The unidimensional scaling methods were developed in the first half of the twentieth century and are generally named after their inventors. We'll look at three types of unidimensional scaling methods here:

- **Thurstone or Equal-Appearing Interval Scaling**
- **Likert or Summative Scaling**
- **Guttman or Cumulative Scaling**

In the late 1950s and early 1960s, measurement theorists developed advanced techniques for creating multidimensional scales. Although these techniques are

scaling The branch of measurement that involves the construction of an instrument that associates qualitative constructs (i.e., objects) with quantitative metric units.

outside the scope of this text, an understanding of the most common unidimensional scaling methods will provide a good foundation for more complex variations.

6.2a General Issues in Scaling

In our opinion, S. S. Stevens (1946) came up with the simplest and most straightforward definition of scaling (**Figure 6.1**). He said, "Scaling is the assignment of objects to numbers according to a rule (1959, p. 25)." What does that mean? In most scaling, the objects are text statements, usually statements regarding a particular attitude or belief. In **Figure 6.2**, three statements describe attitudes toward immigration. To scale these statements, you have to assign numbers to them. Usually, you would like the result to be on at least an interval scale (see "Levels of Measurement" in Chapter 5, "Introduction to Measurement"), as indicated by the ruler in the figure. What does "according to a rule" mean? If you look at the statements, you can see that as you read down, the attitude towards immigration becomes more restrictive; if people agree with any one statement on the list, it's likely that they will also agree with all other statements on the list that are higher than that particular statement. In this case, the rule is a cumulative one.

So what does it all come down to? Scaling basically allows you to meaningfully assign numbers to objects. The following sections introduce several approaches to scaling.

First, we have to clear up one of our pet peeves. People often confuse the idea of a scale and a response scale. A **response scale** is the way you collect responses from people on an instrument. You might use a **dichotomous response** scale like Agree/Disagree, True/False, or Yes/No; or, you might use an **interval response scale** like a 1-to-5 or 1-to-7 rating. However, if all you are doing is attaching a response scale to an object or statement, you can't call that scaling. As you will see, scaling involves procedures that you perform independently of the respondent so that you can come up with an overall numerical value for the object. In true scaling research,

Figure 6.1 Harvard psychologist S. S. Stevens, who contributed much to experimental design and measurement.

response scale A sequential-numerical response format, such as a 1-to-5 rating format.

dichotomous response A measurement response that has two possible options (e.g., true/false or yes/no).

interval response scale A measurement response format with multiple response options that are set at equal intervals (e.g., 1-to-5 or 1-to-7).

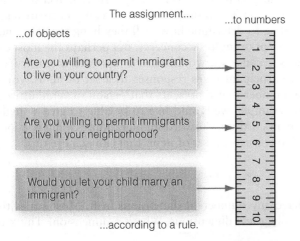

The assignment...

...of objects

...to numbers

Are you willing to permit immigrants to live in your country?

Are you willing to permit immigrants to live in your neighborhood?

Would you let your child marry an immigrant?

...according to a rule.

Figure 6.2 Scaling as the assignment of numbers according to a rule.

Table 6.1 Differences between scaling and response scales.

Scaling	Response Scale
Results from a process of assigning numbers to objects	Used to collect the response for an item
Each item on a scale has a scale value	Item not associated with a scale value
Refers to a set of items	Used for a single item

you use a scaling procedure to develop your instrument (scale) and a response scale to collect the responses from participants. Simply assigning a 1-to-5 response scale for an item is *not* scaling! The differences are illustrated in Table 6.1.

Also, it's important to realize that while a scale is an instrument that can be used alone, it is often integrated into a larger and more complex instrument such as a survey. Many surveys are designed to assess multiple topics of interest and to collect data that enable us to study the interrelationships among these topics. In many surveys, we often embed one or more scales as separate sections of the survey. When the data are collected, the analyst will compute the various scale scores by combining the responses to items from each scale according to the rules for that scale. However, just because a survey asks a set of questions on a single topic and asks you to respond on a similar response scale (such as a 1-to-5 disagree-agree response scale), you cannot automatically conclude that the set of questions constitutes a scale. Got that? Please reread that sentence until you're sure you get the distinction! The set of questions on a survey cannot be considered a scale unless a scaling process was followed to identify the questions and determine how the responses would be combined. So, just because a set of questions on a survey looks like a scale, it collects data using the same response scale, and it is even analyzed like a scale, it isn't a real scale unless some type of scaling process was used to create it. We'll present several of the most famous scaling processes later in this chapter.

6.2b Purposes of Scaling

Why do scaling? Why not just create text statements or questions and use response formats to collect the answers? First, sometimes you do scaling to test a hypothesis. You might want to know whether the construct or concept is unidimensional or multidimensional (more about dimensionality later). Sometimes, you do scaling as part of exploratory research. You want to know which dimensions underlie a set of ratings. For instance, if you create a set of questions, you can use scaling to determine how well they hang together and whether they measure one concept or multiple concepts. But perhaps the most common reason we do scaling is because we would like to represent a construct using a single score. When a participant gives responses to a set of items, you often want to assign a single number that represents that person's overall attitude or belief. In Figure 6.2, for example, we would like to be able to give a single number that describes a person's attitudes toward immigration. Scaling is a formal procedure that helps you to construct a set of items that can achieve this.

6.2c Dimensionality

A scale can have any number of dimensions in it. Most scales that researchers develop have only a few dimensions. What's a dimension? Think of a dimension as a number line, as illustrated in **Figure 6.3**. If you want to measure a construct, you have to decide whether the construct can be measured well with one number line or whether it may need more. For instance, height is a concept that is unidimensional, or one-dimensional. You can measure the concept of height well with only a single number line (a ruler). Weight is also unidimensional; you can measure it with a single scale. Thirst might also be considered a unidimensional concept; you are either more or less thirsty at any given time. It's easy to see that height and weight are unidimensional; but what about a concept like self-esteem? If you think you can measure a person's self-esteem well with a single ruler that goes from low to high, you probably have a unidimensional construct.

What would a two-dimensional concept be? Many models of intelligence or achievement postulate two major dimensions: mathematical and verbal ability. In this type of two-dimensional model, a person can be said to possess two types of achievement, as illustrated in **Figure 6.4**. Some people will have high verbal skills but lower math ones. For others, it will be the reverse. If a concept is truly two-dimensional, it is not possible to depict a person's level on it using only a single number line. In other words, to describe achievement you would need to locate a person as a point in two-dimensional (x, y) space as shown in Figure 6.4.

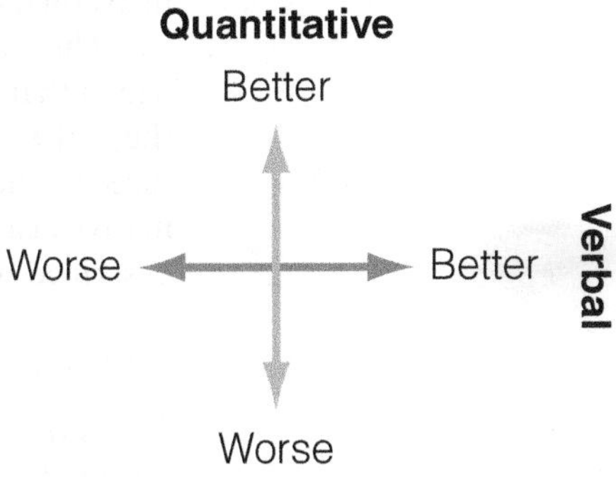

Okay, let's push this one step further: how about a three-dimensional concept? Psychologists who study the idea of meaning theorized that the meaning of a term could be well described in three dimensions. Put in other terms, any object can be distinguished or differentiated from another along three dimensions. They labeled these three dimensions activity, evaluation, and potency. They called this general theory of meaning the **semantic differential**. Their theory essentially states that you can rate any object along those three dimensions. For instance, think of the idea of ballet. If you like the ballet, you would probably rate it high on activity, favorable on evaluation, and powerful on potency. On the other hand, think about the concept of a book like a novel. You might rate it low on activity (it's passive), favorable on evaluation (assuming you like it), and about average on potency. Now, think of the idea of going to the dentist. Most people would rate it low on activity (it's a passive activity), unfavorable on evaluation, and powerless on potency. (Few routine activities make you feel as powerless!) The theorists who came up with the idea of the semantic differential thought that the meaning of any concept could be described well by rating the concept on these three dimensions. In other words, to describe the meaning of an object you have to locate it as a dot somewhere within the cube (three-dimensional space), as shown in **Figure 6.5**.

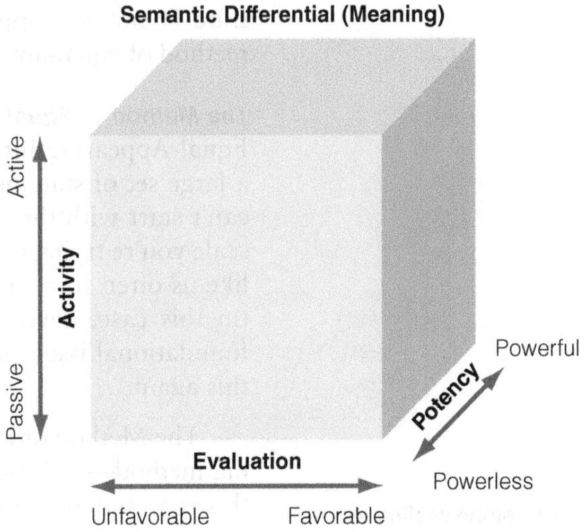

6.2d Unidimensional or Multidimensional?

What are the advantages of using a unidimensional model? Unidimensional concepts are generally easier to understand. You have either more or less of it, and

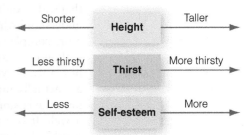

Figure 6.3 Unidimensional scales.

Quantitative

Better

Worse ← → Better

Verbal

Worse

Figure 6.4 A two-dimensional scale.

Semantic Differential (Meaning)

Active / Passive — **Activity**

Evaluation

Unfavorable / Favorable

Potency

Powerful

Powerless

Figure 6.5 A three-dimensional scale.

semantic differential A scaling method in which an object is assessed by the respondent on a set of bipolar adjective pairs.

that's all. You're either taller or shorter, heavier or lighter. It's also important to understand what a unidimensional scale is as a foundation for comprehending the more complex multidimensional scales. But the best reason to use unidimensional scaling is because you believe the concept you are measuring is unidimensional in reality. As you've seen, many familiar concepts (height, weight, temperature) are actually unidimensional. However, if the concept you are studying is in fact multidimensional in nature, a unidimensional scale or number line won't describe it well. If you try to measure academic achievement on a single dimension, you would place every person on a single line ranging from low to high achievers. How would you score someone who is a high math achiever and terrible verbally, or vice versa? A unidimensional scale can't capture that more general type of achievement. You would need at least two unidimensional scales or one multidimensional one.

There are three major types of unidimensional scaling methods. They are similar in that they each measure the concept of interest on a number line. However, they differ considerably in how they arrive at scale values for different items. The three methods are Thurstone or Equal-Appearing Interval Scaling, Likert or Summative Scaling, and Guttman or Cumulative Scaling. Each of these approaches is described in the following sections.

Thurstone Scaling

Thurstone was one of the first and most productive scaling theorists. He actually invented three different methods for developing a unidimensional scale that may all be referred to as **Thurstone scaling**: the *method of equal-appearing intervals,* the *method of successive intervals,* and the *method of paired comparisons.* The three methods differed in how the scale values for the items were constructed, but in all three cases, the resulting scale was rated the same way by respondents. To illustrate Thurstone's approach, we'll show you the easiest method of the three—the method of equal-appearing intervals.

The Method of Equal-Appearing Intervals Developing the Focus: The Method of Equal-Appearing Intervals starts like almost every other scaling method—with a large set of statements to which people respond. Oops! We did it again! You can't start with the set of statements; you have to first define the focus for the scale you're trying to develop. Let this be a warning to all of you: methodologists like us often start our descriptions with the first objective, methodological step (in this case, developing a set of statements) and forget to mention critical foundational issues like the development of the focus for a project. So, let's try this again....

The Method of Equal-Appearing Intervals starts like almost every other scaling method—with the development of the focus for the scaling project. Because this is a unidimensional scaling method, you have to be able to assume that the concept you are trying to scale is reasonably thought of as one-dimensional. The description of this concept should be as clear as possible so that the person creating the statements has a clear idea of what you are trying to measure. We like to state the focus for a scaling project in the form of an open-ended statement to give to the people who will create the candidate statements. You want to be sure that everyone who is generating statements has some idea of what you are after in this focus command. You especially want to be sure that technical language and acronyms are spelled out and understood.

Thurstone scaling A class of scaling methods (the method of equal-appearing intervals, the method of successive intervals, and the method of paired comparisons) that were designed to yield unidimensional, interval-level, multi-item scales.

Generating Potential Scale Items: In this phase, you're ready to create statements. Who should create the statements for a scale? That depends. You might have experts who know something about the phenomenon you are studying. Since the people affected are likely to be experts about what they're experiencing, you might sample them to generate statements. For instance, if you are trying to create a scale for quality of life for people who have a certain type of health condition, you might want to ask them to create potential items. Finally, you can make up the items. Obviously, each of these approaches has advantages and disadvantages, so in many situations you may want to use some or all of them.

You want a large set of candidate statements, usually as many as 80–100, because you are going to select your final scale items from this pool. You also want to be sure that all of the statements are worded similarly—that they don't differ in grammar or structure. For instance, you might want them each to be worded as a statement with which respondents agree or disagree. You don't want some of them to be statements while others are questions.

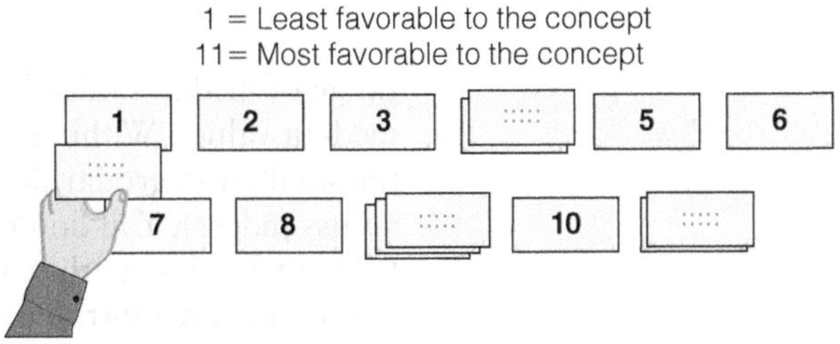

Rating the Scale Items: So now you have a set of items or statements. The next step is to have a group of people called judges rate each statement on a 1-to-11 scale in terms of how much each statement indicates a *favorable* attitude toward the construct of interest. Pay close attention here! You *don't* want the judges to tell you what their attitudes on the statements are, or whether they would agree with the statements. You want them to rate the favorableness of each statement in terms of the construct you are trying to measure, where 1 = extremely unfavorable attitude toward the construct and 11 = extremely favorable attitude toward the construct. One easy way to actually accomplish this is to type each statement on a separate index card and have each judge rate them by sorting them into eleven piles, as shown in **Figure 6.6**. Who should the judges be? As with generating the items, there is no simple answer. Generally you want to have people who are "experts" on the construct of interest do this. But there are many kinds of expertise, ranging from academically trained and credentialed experts to the people who are most directly experienced with the phenomenon.

1 = Least favorable to the concept
11 = Most favorable to the concept

Figure 6.6 Rating the candidate statements on a 1-to-11 scale by sorting them manually.

Computing Scale Score Values for Each Item: The next step is to analyze the rating data. For each item or statement, you need to compute the median and the interquartile range. The median is the value above and below which 50 percent of the ratings fall. The first quartile (Q1) is the value below which 25 percent of the cases fall and above which 75 percent of the cases fall—in other words, the 25th percentile. The median is the 50th percentile. The third quartile, Q3, is the 75th percentile. The **interquartile range** is the difference between the third and first quartile, or Q3–Q1. **Figure 6.7** shows a histogram for a single item and indicates the median and interquartile range.

You can compute these values easily with any introductory statistics program or with most spreadsheet programs. To facilitate the final selection of items for your scale, you might want to sort the table of medians and interquartile ranges in ascending order by median, and, within that, in descending order by interquartile range.

Selecting the Final Scale Items: Now you have to select the final statements for your scale. You should select statements that are at equal intervals across the

interquartile range The difference between the 75th (upper quartile) and 25th (lower quartile) percentile scores on the distribution of a variable. The interquartile range is an estimate of the spread or variability of the measure.

Figure 6.7 Histogram displaying the median and interquartile ranges of the attitude data.

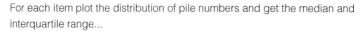

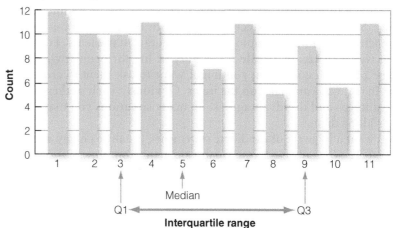

range of medians. Ideally, one statement would be selected for each of the eleven median values. Within each value, you should try to select the statement that has the smallest interquartile range (the statement with the least amount of variability across judges). You don't want the statistical analysis to be the only deciding factor here. Look over the candidate statements at each level and select the statement that makes the most sense. If you find that the best statistical choice is a confusing statement, select the next-best choice.

Administering the Scale: You now have a scale—a yardstick you can use for measuring the construct of interest. Each of your final scale items has a scale score—the median value. And the item scores should range across the spectrum of potential attitudes or beliefs on this construct (because you selected items from across the median range). You can now give the final set of items to respondents and ask them to agree or disagree with each statement. To get an individual's final scale score, average only the scale scores of all the items that person agreed with. When you average the scale items for the statements with which the respondent agreed, you get an average score that has to range between 1 and 11. If they agreed with scale items that were low in favorableness to the construct, then the average of the items they agreed to should be low. If they agreed with items that your judges had said were highly favorable to the construct, then their final score will be on the higher end of the scale.

You should see a couple of things from this discussion. First, you use the judges to create your scale. Think of the scale as a ruler that ranges from 1 to 11, with one scale item or statement at each of the 11 points on the ruler. Second, when you give the set of scale items to a respondent and ask them to tell you which ones they agree with, you are essentially trying to measure them with that ruler. Their scale score—where you would mark the individual on your 11-point ruler—is the average value for the items the respondent agreed with.

The Other Thurstone Methods The other Thurstone scaling methods are similar to the Method of Equal-Appearing Intervals. All of them begin by focusing on a concept that is assumed to be unidimensional and involve generating a large set of potential scale items. All of them result in a scale consisting of relatively few items that the respondent rates on an Agree/Disagree basis. The major differences are in how the data from the judges are collected. For instance, the method of paired comparisons requires each judge to make a judgment about each pair of statements. With lots of statements, this can become time consuming.

Likert Scaling

Another unidimensional scaling method is the very widely used method known as **Likert scaling**. If you found the Thurstone methods complex, then you have a friend in Rensis Likert, who developed his method because he thought there may be an easier and more efficient way to develop a scale. Perhaps because his method was so successful and popular, it has become unfortunately common to hear people refer to all kinds of scales as "Likert Scales." Remember that a true Likert scale can only be developed by the method described here and is always a measure of agreement with a statement that reflects a particular attitude or idea. By the way, you might impress your friends (your future grad school friends anyway) with correct pronunciation of Professor Likert's name: It is pronounced "*Lick*-ert", not "*Like*-ert," and definitely not "Like-*it*" or "Lik*er*"!

Here, we'll explain the basic steps in developing a Likert or a **summative scale**. You will learn about the term Likert response scale in Chapter 7. A Likert scale is a type of response scale and is different from Likert Scaling (see the discussion in Table 6.1 for the differences between response scales and scaling).

Defining the Focus: As in all scaling methods, the first step is to define what it is you are trying to measure. Because this is a unidimensional scaling method, it is assumed that the concept you want to measure is one-dimensional in nature. You might operationalize the definition as an instruction to the people who are going to create or generate the initial set of candidate items for your scale.

Generating the Items: Next, you have to create the set of potential scale items. These should be items that can be rated on a 1-to-5 or 1-to-7 Disagree-Agree response scale. Sometimes you can create the items by yourself based on your intimate understanding of the subject matter. More often than not, though, it's helpful to engage a number of people in the item creation step. For instance, you might use some form of brainstorming to create the items. It's desirable to have as large a set of potential items as possible at this stage; about 80–100 would be best.

Rating the Items: The next step is to have a group of judges rate the items. Usually you would use a 1-to-5 rating scale where:

1 = Strongly unfavorable to the concept
2 = Somewhat unfavorable to the concept
3 = Undecided
4 = Somewhat favorable to the concept
5 = Strongly favorable to the concept

Notice that, as in other scaling methods, the judges are not telling you what they believe; they are judging how favorable each item is with respect to the construct of interest.

Likert or summative scaling A method of scaling in which the items are assigned interval-level scale values and the responses are gathered using an interval-level response format.

Who should the judges be? As in any scaling method, that's not an easy question to answer. Some argue that experts familiar with the construct should be used. Others suggest that you should use a random sample of the same types of people who are ultimately your respondents of interest for the scale. There are advantages and disadvantages to each.

Selecting the Items: The next step is to compute the intercorrelations between all pairs of items, based on the ratings provided by the judges. In making judgments about which items to retain for the final scale, any items that have a low correlation with the total (summed) score across all items. In most statistics packages, it is relatively easy to compute this type of Item-Total correlation. First, you create a new variable that is the sum of all of the individual items for each respondent. Then, you include this variable in the correlation-matrix computation. (If you include it as the last variable in the list, the resulting Item-Total correlations will all be the last line of the correlation matrix and will be easy to spot.) How low should the correlation be for you to throw out the item? There is no fixed rule here; you might eliminate all items that have a correlation with the total score less than .6, for example. (Interpreting a correlation is covered in the discussion on correlation in Chapter 11, "Introduction to Analysis.")

Note that if an item is a reversal item (described below) it should have a high negative correlation. For reversal items the rule would be to discard items if the correlation with the total score is not lower than −0.6.

Administering the Scale: You're now ready to use your Likert scale. Each respondent is asked to rate each item on some response scale. For instance, respondents could rate each item on a 1-to-5 response scale where:

1 = Strongly disagree
2 = Disagree
3 = Undecided
4 = Agree
5 = Strongly agree

There are a variety of possible response scales (1-to-7, 1-to-9, 0-to-4). All of these odd-numbered scales have a middle value, which is often labeled Neutral or Undecided. It is also possible to use a response scale with an even number of responses and no middle neutral or undecided choice. This is called a **forced-choice response scale**. In this situation, respondents are forced to decide whether they lean more toward the agree- or disagree-end of the scale for each item.

The final score for the respondent on the scale is the sum of his or her ratings for all of the items. (This is why this is sometimes called a summative scale.) On some scales, you will have items that are reversed in meaning from the overall direction of the scale. These are called **reversal items**. You will need to reverse the response value for each of these items before summing for the total. That is, if the respondent gave a 1, you make it a 5; if a respondent gave a 2, you make it a 4; 3 = 3; 4 = 2; and, 5 = 1. Researchers disagree about whether you should have a "neutral" or "undecided" point on the scale (an odd number of responses) or whether the response scale should be a "forced choice" one with no neutral point and even number of responses (as in a 1-to-4 scale).

Example: The Rosenberg Self-Esteem Scale Table 6.2 shows an example of a ten-item Likert scale that attempts to estimate the level of self-esteem a person has

forced-choice response scale A response scale that does not allow for a neutral or undecided value. By definition, a forced-choice response scale has an even number of response options. For example, the following would be a forced-choice response scale: 1 = strongly disagree; 2 = disagree; 3 = agree; 4 = strongly agree.

reversal items Items on a multi-item scale whose wording is in the opposite direction of the construct of interest. Reversal items must have their scores reversed prior to computing total scale scores. For example, if you had a multi-item self-esteem scale, an item such as "I generally feel bad about myself as a person" would be a reversal item.

Table 6.2 The Rosenberg scale.

Please record the appropriate answer for each item, depending on whether you Strongly agree, agree, disagree, or strongly disagree with it.
1 = Strongly agree
2 = Agree
3 = Disagree
4 = Strongly disagree
_____ 1. On the whole, I am satisfied with myself.
_____ 2. At times I think I am no good at all.
_____ 3. I feel that I have a number of good qualities.
_____ 4. I am able to do things as well as most other people.
_____ 5. I feel I do not have much to be proud of.
_____ 6. I certainly feel useless at times.
_____ 7. I feel that I'm a person of worth.
_____ 8. I wish I could have more respect for myself.
_____ 9. All in all, I am inclined to think that I am a failure.
_____ 10. I take a positive attitude toward myself.

(Rosenberg, 1979). Notice that this instrument has no center or neutral point in the response scale; the respondent has to declare whether he/she is in agreement or disagreement with the item. Notice also that items 2, 5, 6, 8, and 9 are reversal items. The total self esteem score is obtained by summing the individual item scores after reversing the values for the negative items (i.e., making 1 = 4, 2 = 3, 3 = 2, and 4 = 1).

Guttman Scaling

Guttman scaling is also sometimes known as **cumulative scaling**. In Chapter 5, we introduced the term Guttman scale in section 5-1a. Again, a Guttman scale is a type of response scale and is different from Guttman Scaling (see Table 6.1 for the differences between response scales and scaling). The purpose of Guttman scaling is to establish a one-dimensional continuum for a concept you want to measure. What does that mean? Essentially, you would like a set of items or statements so that a respondent who agrees with any specific item in the list will also tend to agree with all previous questions. Put more formally, you would like to be able to predict item responses perfectly, knowing only the total score for the respondent. For example, imagine a ten-item cumulative scale. If the respondent scores a four, it should mean that he/she agreed with the first four statements. If the respondent scores an eight, it should mean he/she agreed with the first eight. The object is to find a set of items that perfectly matches this pattern. In practice, you would seldom expect to find this cumulative pattern perfectly. So, you use scalogram analysis to examine how closely a set of items corresponds with this idea of cumulativeness. Here, we'll explain how you develop a Guttman scale.

Guttman or cumulative scaling A method of scaling in which the items are assigned scale values that allow them to be placed in a cumulative ordering with respect to the construct being scaled.

Define the Focus: As in all of the scaling methods, you begin by defining the focus for your scale. Let's imagine that you want to develop a cumulative scale that measures U.S. citizen attitudes toward immigration. You would want to be sure to specify in your definition whether you are talking about any type of immigration (legal and illegal) from anywhere (Europe, Asia, Latin and South America, Africa).

Develop the Items: Next, as in all scaling methods, you would develop a large set of items that reflects the concept. You might do this yourself or you might engage a knowledgeable group to help. Of course, you would want to come up with many more statements (about 80–100 is desirable) than you will ultimately need.

Rate the Items: Next, you would want to have a group of judges rate the statements or items in terms of how favorable they are to the concept of interest. They would give a Yes if the item is favorable toward the construct and a No if it is not. Notice that you are not asking the judges whether they personally agree with the statement. Instead, you're asking them to make a judgment about how the statement is related to the construct of interest.

Develop the Cumulative Scale: The key to Guttman scaling is in the analysis. You construct a matrix or table that shows the responses of all the respondents on all of the items. You then sort this matrix so that respondents who agree with more statements are listed at the top and those who agree with fewer are at the bottom. For respondents with the same number of agreements, sort the statements from left to right, from those that most agreed to, to those that fewest agreed to. You might get a table something like the one in **Figure 6.8.** Notice that the scale is nearly cumulative when you read from left to right across the columns (items). Specifically, a person who agreed with Item 7 always agreed with Item 2. Someone who agreed with Item 5 always agreed with Items 7 and 2. The matrix shows that the cumulativeness of the scale is not perfect, however. While, in general, a person agreeing with Item 3 tended to also agree with 5, 7, and 2, there are several exceptions to that rule.

Although you can examine the matrix if there are only a few items in it, if there are many items, you need to use a data analysis procedure called **scalogram analysis**

Figure 6.8 Developing a cumulative scale with Guttman scaling.

scalogram analysis A method of analysis of a set of scale items used when constructing a Guttman or cumulative scale. In scalogram analysis, one attempts to determine the degree to which responses to the set of items allows the items to be ordered cumulatively in one dimension with respect to the construct of interest.

When sorted by row and column, it will show whether there is a cumulative scale.

Respondent	Item 2	Item 7	Item 5	Item 3	Item 8	Item ...
7	Y	Y	Y	Y	Y	Y
15	Y	Y	Y	–	Y	–
3	Y	Y	Y	Y	–	–
29	Y	Y	Y	Y	–	–
19	Y	Y	Y	–	–	–
32	Y	Y	–	Y	–	–
41	Y	Y	–	–	–	–
6	Y	Y	–	–	–	–
14	Y	–	–	Y	–	–
33	–	–	–	–	–	–

Exceptions

to determine the subsets of items from the pool that best approximate the cumulative property. Then, you review these items and select your final scale elements. There are several statistical techniques for examining the table to find a cumulative scale. Because there is seldom a perfectly cumulative scale, you usually have to test how good it is. These statistics also estimate a scale score value for each item. This scale score is used in the final calculation of a respondent's score.

Administering the Scale: After you've selected the final scale items, it's relatively simple to administer the scale. You simply present the items and ask respondents to check items with which they agree.

Each scale item has a scale value associated with it (obtained from the scalogram analysis). To compute a respondent's scale score you simply sum the scale values of every item the respondent agrees with. In this example, the final value should be an indication of the respondent's view on the construct of interest.

6.3 Tests

For better or worse, we are all familiar with tests. In this section, we'll describe the major uses and some of the measurement operations behind **tests**. You wouldn't believe how many different kinds of people are involved in the world of testing, nor what a big business it has become (well, you might believe that last part). Besides the test taker, there are test makers, test administrators, test scorers, test publishers, test marketers, test reviewers, and, of course, test researchers. Many of these people utilize tests for making decisions about things like who is the best candidate for a job, what kind of illness a person might have, how much students have learned in a course, and other decisions related to evaluating performance or classifying individuals. Others use tests as measures in a research study. As research methodologists, our main concern is the latter kind of test use, but we will touch on other applications as well as some of the social and ethical issues related to testing.

The history of testing actually goes back to 2200 BCE, when Chinese government officials were selected based on their performance on a common examination (Provenzo, 2008). This had such major social implications, it seems, that it opened the door to employment opportunities across classes and helped to mitigate circumstances of wealth and power. This goal of objective assessment, to facilitate accurate and fair matching of people to positions, remains with us today. But the social consequences of testing programs continue to include opportunity as well as controversy in modern society. You may not immediately associate the words "justice" and "test," and you may have good reasons from your own experience, but the fact remains that decision making based on reliable and valid measurement (i.e., through tests) is likely to be fairer than the alternatives. Did you notice that important assumption in the last sentence— "*reliable and valid measurements*"? This is the core issue in both the science and ethics of test use.

Like the scales we've discussed previously, test data have become central to research and evaluation in modern society. In the United States, this happened as a result of the field of psychology moving from philosophical theorizing and laboratory experimenting in the late 1800s to the practical problems of the workplace, classroom, and battlefield. As society grappled with these new challenges,

tests Measurement instruments designed to assess a respondent's knowledge, skill, or performance.

testing became a centerpiece of psychological methods. The emergence of statistical methods, communication and data analysis technologies, a societal focus on child development, and competitive forces within and between countries, all provided fuel for the testing culture that we see today.

In modern usage, the term "test" is most accurately used to describe the assessment of an aptitude or ability, or some form of learning or achievement (APA, 1999). However, in everyday use, the term is routinely applied to all kinds of measures, from medical diagnostic tests to personality inventories to the performance of a program (e.g., a literacy improvement program). How are tests different from scales? While scales are constructed to measure subjective opinions or attitudes in a given population, tests strive to be objective measures of knowledge or ability. Tests are very relevant to our study of research methods because they are based on measurement methodology and are often used in scholarly inquiry and evaluation.

Here is an example of a study that used tests to measure the driving abilities of teenagers with attention deficit problems, a group that is worse than teen drivers in general in terms of accidents, injuries, costs, arrests, and other problems. Dr. Greg Fabiano and his colleagues at the University at Buffalo developed a comprehensive driver-training program for adolescents with ADHD (Fabiano et al., 2011). The program included both a driving simulator (Figure 6.9) and measurement devices installed in the teen's car that would record such things as abrupt and extreme braking, amount of time driving over 70 miles an hour, and other objective indicators of driving. These tests provided very specific estimates of the level of ability of participants prior to a study, and enabled the researchers to see what difference the program made from the participant's baseline. Objective tests (whether they are paper and pencil, or behavioral, like the driving measurement systems) can also be used to make sure that groups of participants are equivalent prior to the study, and to statistically adjust for pre-program differences when the data is analyzed.

Figure 6.9 This driving simulator was used as part of a driver-training study for adolescents with attention deficit problems (Fabiano et al., 2011).

Kevin F. Hulme

6.3a Validity, Reliability, and Test Construction

The process of test construction begins with the familiar idea of defining the construct of interest—with construct validity as our primary concern. Similar to the scale development processes you've read about, this means that we define the domain as a construct (e.g., critical thinking or driving ability), and then devise items that should theoretically reflect variability in levels of the construct. As with other kinds of measurements, we should be able to demonstrate that the test (usually the total score) is a strong indicator of the construct. The process of validating a test means that several specific kinds of evidence are gathered, including evidence that the content of the test represents the domain (content validity), and that the test produces scores that are useful in prediction (predictive or criterion-related validity). Content validity of a test is very often studied by having experts review the items in a test to be sure they cover the domain comprehensively. As you've already learned, construct validity can be demonstrated by studying the pattern of correlations of a construct with others that are thought to be similar and different. One kind of construct validity, predictive validity, can be evaluated by seeing how well the tests identify performance on a future activity such as graduate school or a job. For example, football fans know that every year there is a testing program for college football players called "The Combine" in which an enormous number of measurements are obtained to try to predict professional playing ability.

Validity can also be studied using advanced statistical procedures like **factor analysis** that help us understand the structure of a construct (like vocational interests). For instance, let's say researchers think theoretically that vocational interests seem to be composed of three distinct subconstructs (i.e., a three-factor structure) that they call "people, data, and things." Factor analysis uses the intercorrelations among test items to explore whether three factors are present. In this way, we can use actual data, instead of just our theoretical judgment, to see whether a large number of potential test items can reasonably be synthesized and summarized in terms of a fewer number of factors or constructs (e.g., a vocational interest test obtains data on preferences for hundreds of different activities that are all related to the main factors like "people, data, and things").

Another kind of validity was proposed specifically because of concern about the social impact of tests. Samuel Messick (1996, 1998), a famous psychometrician, called this "**consequential validity**" to bring attention to the fact that there are personal and social consequences of test scores. He was particularly concerned about the potential for unfortunate consequences when a test systematically produces biased decisions and social injustices. From a measurement perspective, the source of the trouble is that a test may include irrelevant variance in scores (e.g., socioeconomic status that leads to variation on an intelligence test) or underrepresents the construct (as in **Figure 6.10**).

As we have discussed in previous chapters, reliability refers to the consistency of a measure. When a measure includes a significant degree of error, then measures are inconsistent and therefore unreliable. When you think of how your own test performance can be affected by factors like effort, fatigue, the time of the semester, or your health, you are identifying factors that impact your score by reducing the reliability of a test from occasion to occasion. This is typically referred to as test-retest reliability, represented by the correlation between one occasion and another on the same test (or a very similar or "parallel" one). Test reliability is also assessed by estimating internal consistency, as with any other kind of measure, and once again Cronbach's coefficient alpha provides a summary of internal consistency. At the individual item level, we want to look at

factor analysis A multivariate statistical analysis that uses observed correlations (variability) as input and identifies a fewer number of unobserved variables, known as factors, that describe the original data more efficiently.

consequential validity The approximate truth or falsity of assertions regarding the intended or unintended consequences of test interpretation and use.

Figure 6.10 What validity issue do you see here?

the correlation of a particular item with the total subtest (e.g., the verbal versus quantitative scores in the SAT) or test score. This process is known as **item analysis** and can be conducted at any level, from individual classes like the ones you are in, to nationally representative samples for large-scale testing programs like the SAT, ACT, or GRE.

6.3b Standardized Tests

At present, standardized testing is a growth industry, with many millions of dollars invested annually. Oversight of testing standards is primarily conducted by professional organizations (such as the American Psychological Association, American Educational Research Association, American Evaluation Association, and National Council on Measurement in Education) who publish standards and guidelines, as well as the academic departments in universities that train researchers and practitioners in proper test evaluation and use.

In order to meet the goal of strong evidence of reliability and validity, standardization is used to attempt to control factors that are irrelevant to the construct being measured, and to gather evidence of validity, so that accurate interpretation, evaluation, and prediction can be accomplished. Just as the laboratory scientist attempts to control environmental and other variables that may affect study results, the test developer must attend to any factor that could have an impact on scores that is not relevant to the thing being measured (e.g., disabilities not related to the test, but related to ability to complete the test as intended). This is an example of what we think of as enhancing the signal of the variable we want to measure, relative to the noise of irrelevant variance that can be captured in our measures.

If we asked you to use your new knowledge about reliability and validity of measurement to design an assessment of research methods learning that could be used in your class, you might start by thinking about things like making sure that the content was relevant to what had been studied, controlling sources of error, and assuring that any scores resulting from your assessment would really be related to what your classmates had learned, not to something else. Then if we asked you if

item analysis The systematic statistical analysis of items on a scale or test undertaken to determine the properties of the items, especially for purposes of selecting final items or deciding which items to combine. An item-to-total-score correlation is an example of an item analysis.

your assessment could be used in other classes, you might want to write some guidelines for how to administer, score, and interpret the results of the assessment. In essence, you'd be doing (of course, in much briefer form!) what the makers of the SAT, ACT, GRE and other standardized tests do. Your guidelines might provide some descriptive statistics showing what the mean, median, and standard deviation of your assessment were, and perhaps correlations with other known measures of research methods knowledge. Eventually, with enough data, you might have norms that could facilitate interpretation of new scores by allowing comparison with percentiles in the larger database. All of these activities would be the kind that test developers would engage in to develop reliable and valid measures. In this light, test development might sound interesting and actually kind of pro-social. As you know, the translation to real-world practice can complicate matters, as it has since test scores were first used in decision making. One of the most hotly debates issues is fairness, discussed next.

6.3c Test Fairness

Test fairness means that everyone who takes the test should have the same chance of demonstrating his or her ability, knowledge, or aptitude. Yet critics of standardized tests have pointed out correlations of test scores with variables that are not related to the construct, but instead are related to opportunities in life, many associated with socioeconomic circumstances. This might be a good example of what Messick described in discussing consequential validity. These concerns have become so significant that some colleges and universities are eliminating the requirement of standardized test scores for admission. Many others have reconfigured their decision-making processes to make sure that decisions are made on the basis of multiple indicators, as recommended by most researchers and professional organizations concerned with these issues.

In actual practice, one of the less desirable outcomes of testing programs happens when the definition of the construct becomes a function of the test, instead of the other way around. For example, in the world of intelligence assessment, there is a saying that "intelligence is what the test measures." This statement reflects the fact that once a test is established and scores become widely used, the practical definition of the construct becomes a test score (e.g., intelligence = SAT score). This not only compromises construct validity, it also creates fairness issues, because opportunities to learn the needed material are not equal. Since no test is perfect, we must be very careful about this kind of pattern. In addition to continual refinement of tests, we should use multiple indicators, especially when stakes are high, as in school and employment decision making.

6.3d How to Find a Good Test

Similar to rating scales, sometimes we find that a good test has already been developed. So, before you go trying to develop one on your own, look around to see what's already out there that you could use or adapt. Researchers can turn to test publishers, the primary research literature, and to test reviews for help in determining if a good test already exists. Test reviews are published in academic journals as well as by the Buros Center for Testing at the University of Iowa. You can follow the link at the end of the chapter to explore the center's website. It is very likely that your library will have both printed and online versions of the Buros reviews. Every test selected for review is examined by two experts who independently evaluate the test on the kind of criteria discussed here, as well as the technical aspects described in the Standards for Educational and Psychological

Testing (APA, 1999). When a suitable test does not exist, or when the evidence for a particular population or use has not been established, then you might decide to construct or adapt a test. The test construction or adaptation process is similar to scale development, with the additional considerations related to reliability, validity, and score interpretation outlined above. Note that we've only scratched the surface of the field of psychometrics, which includes many clinical and other applied issues beyond our scope.

In summary, tests are a particular kind of measurement most often used in the evaluation of performance for the purpose of decision making. Tests are also useful in many kinds of studies in which a statistical relationship or intervention is to be evaluated as a study outcome. In any application of tests, the quality of the test depends on the same reliability and validity ideas as other kinds of measures. In practice, the use of standardized tests is intended to "level the playing field," but has created much controversy over the potential for harm.

6.4 Indexes

An **index** is a quantitative score that is typically repeatedly measured over time, and is often constructed by applying a set of rules to combine two or more variables in order to reflect a more general construct. So, what does this mean? First of all, an index is a score, a numerical value that purportedly measures something. Second, an index is usually measured repeatedly so we can track the construct we are interested in. Sometimes the index values are adjusted relative to some expected or baseline value that we use as a comparison. In this sense, we "index" the construct we are interested in by seeing how the measure performs over time compared to some standard or expected value. Third, an index is often (but not always) a composite. It puts different variables together. Sometimes these variables are very different kinds of things and may even be measured in different ways and on different scales. Fourth, indexes that are combinations of multiple measures put them together using a rule or set of rules. Sometimes the rule is as simple as just adding up or averaging the scores of each variable to get a total index score. Sometimes the rule is actually a formula or set of procedures that describe how the variables are combined. Finally, we usually construct an index because we want to measure something that none of the individual components alone does as good a job of measuring. An index score is typically trying to get at something that cuts across the variables that are combined, something that is more general than its separate parts.

6.4a Some Common Indexes

You are probably already familiar with several famous indexes. One of the best known is the Consumer Price Index or CPI that is collected every month by the Bureau of Labor Statistics of the U.S. Department of Labor (U.S. Department of Labor, 2004). Each month the CPI index is reported and is considered to be a reflection of generally how much consumers have to pay for things. To construct this single score each month, the government identified eight major categories of spending for the typical consumer: food and beverages; housing; apparel; transportation; medical care; recreation; education and communication; and other goods and services. Then, they (the government staff) break down these eight areas into over 200 specific categories. For each of these, they sample from the many

index A quantitative score that measures a construct of interest by applying a formula or a set of rules that combines relevant data.

items that reflect each category. For example, to represent the "apple" category that is in the "food and beverages" area, they might sample a "particular plastic bag of golden delicious apples, U.S. extra fancy grade, weighing 4.4 pounds" (U.S. Department of Labor, 2004). Each month, people call all over the country to get the current price for over 80,000 items. Through a rather complicated weighting scheme that takes into account things like the location and the probability that the item will be purchased, these prices are combined. That is, there is a series of formulas and rules that are used each month to combine the prices into an index score. Actually, they compute thousands of different Consumer Price Index scores each month, to reflect different groups of consumers and different locations, although one of these is typically reported in the news as *the* CPI. The CPI is considered an index of consumer costs, and therefore is a general economic indicator. It illustrates one of the most important reasons for creating an index—to track a phenomenon and its ups and downs over time (**Figure 6.11**).

Figure 6.11 Maybe we should construct an index of indexes that would summarize life on planet Earth.

"Which should I be worrying about? The wholesale price index, the consumer price index, or the industrial price index?"

A second well-known type of index is the Socioeconomic Status (SES) index. Unlike the CPI, the SES almost always involves the combination of several very different types of variables. Traditionally the SES is a combination of three constructs: income, education, and occupation. Income would typically be measured in dollars. Education might be measured in years or degree achieved. And occupation typically would be classified into categories or levels by status. Then, these very different elements would need to be combined to get the SES score. In one of the early classic studies in this area (Duncan, 1981), the researchers used the degree to which education and income predicted occupation as the basis for constructing the index score. This SES measure is now typically referred to as the Duncan Socioeconomic Index (SEI). For this index, an SEI score has been created for each of hundreds of occupations. The score is a weighted combination of "occupational education" (the percent of people in that occupation who had one or more years of college education) and "occupational income" (the percentage of people in the occupation who earned more than a specific annual income). With the SEI, all you need to know is the occupation of a person, and you can look up the SEI score that presumably reflects the status of the occupation as related to both education and income. Almost from its inception, the measurement of socioeconomic status has been controversial, and different researchers attempt to accomplish it in a great variety of different ways (Hauser & Warren, 1996; Stevens & Cho, 1985).

6.4b Constructing an Index

There are several steps that are typically followed in constructing an index. We'll go over them briefly here, but you should know that in practice each one of these steps is considerably more complex than we're able to convey in this brief description. Each step can involve sophisticated methods and considerable effort, when accomplished well. Here are the basic steps:

Conceptualize the index. It probably won't surprise you that the first thing you need to decide is what you would like the index to measure. This may seem like a simple issue at first. However, for almost anything you would like to measure with an index, different people might reasonably disagree about what it means. What is socioeconomic status? Does it include income, education, and occupation? If you measure education and occupation, won't that be highly related to income? If so, do you need a separate component that reflects income, or will just the two components be sufficient? If you were trying to measure a construct like "quality of life," what components would you need to include in order to capture the construct? Even with a well-established measure like the CPI, researchers worry about defining basic terms like "consumption by whom?" and "prices of what?". To begin composing an index you first need to identify the construct you are trying to reflect in the index, and describe the variables that are components of the construct. There are a wide variety of ways to accomplish this step. You can make it up using your own hunches and intuitions (a surprisingly large amount of social research happens this way). You can review the literature and use current theory as a guide. You can engage experts or key stakeholders in formal processes for conceptualizing, using approaches like brainstorming, concept mapping, or interviewing, to identify what the key concept you are trying to measure means to different people. Think about several conceptual issues at this stage. What is the purpose of the index? How will it be used and by whom? Is this a one-time or short-term measure or one that you would like to use over a long period of time?

Operationalize and measure the components. It is one thing to say that you would like to measure education and occupation as major components of socio-economic status. But it is quite another to figure out how to measure each one. If you are trying to measure education as it relates to status, do you simply count the number of years in school? If two people spend the same number of years in college majoring in very different subjects, should they get the same numerical value on educational status? Or should we give more "points" for someone majoring in one field than another? Should all bachelor's degrees be counted the same? If you are trying to look at occupation as it relates to status, how do you even classify occupations? How do you decide the numerical value for each occupation as it relates to status? Over time, do occupations change in status? Do new occupations get created (there weren't any web programmers before the Internet!). If so, how do you accommodate this in an index that tries to measure changes in status over years or decades? What is the unit you are measuring? Are you measuring educational levels of individuals? Or are you looking at some other unit like the community or an organization? For example, if you want to measure the educational level of a community, and you know the number of years a representative sample of community members went to school, it may be reasonable simply to average the number of years for the community estimate. But, if you only have a coding by level of education (e.g., 1 = grade school, 2 = some high school, 3 = some college, 4 = associate's degree, 5 = bachelor's degree, etc.) you cannot average these values. In this case you may need to calculate the proportion of the community that achieved a particular level (e.g., the proportion of high school graduates) as an estimate of the community educational level. In any event, you need to figure out how you will measure each component of an index before you can move on to calculating the composite index score.

Develop the rules for calculating the index score. Once you have the components that you think make up the construct of interest, you need to figure out how to combine these component scores to create a single index score. There are many complications here. In the simplest case, you might be able to combine the component scores just by adding or averaging them. In essence, this is what the CPI does. This can be done for the CPI because each consumer item is measured in the same way—its price. But what if each component is measured in entirely different ways? In SES measures, you can't measure income the same way you measure education. So, you're not likely to be able to add or average the scores for income and education in any straightforward way. Even if the components are measured in a similar manner, what if you think different components should be given different emphasis in measuring the construct? For example, what if you believe that income should be considered more important than education when trying to measure socioeconomic status? How much more important? There are several things you can do when combining the components of an index. It helps if you can develop a model of the index that shows the index score, each of the components, and how you think theoretically these are related. You then need to develop precise rules for how to combine the components in the model. Sometimes these rules can be stated as a set of procedures that you follow to compute the index, almost like a recipe. In other cases, the rules are essentially a formula or set of formulas (a simple average of several components is essentially a formula).

One of the most important questions you have to address when constructing an index is whether you are giving each component equal weight or you are constructing a weighted index score. A **weighted index** is one where you combine different components of the index in different amounts or with different emphasis. You're almost certainly familiar with a weighted index, because most of you

weighted index A quantitative score that measures a construct of interest by applying a formula or a set of rules that combines relevant data where the data components are weighted differently.

have probably had a teacher that at one time or another used a weighting scheme to come up with your grade for a class. For example, imagine that your teacher measures you on three characteristics: test scores, class participation, and a class project. For the sake of argument (and this is no simple matter in itself), let's assume that you get scored on each of those on a 0-to-100 scale. If you score perfectly on all your tests you get a 100 on the test score, if your project is perfect, you get 100 on the project score. One way to get a total index score for your course performance would be to average these three components. But what if you believe (or, more to the point, your professor believes) that these components should not receive equal weight? For instance, maybe participation should only be weighted half as much as the test or project component. You might reason that it doesn't matter how much you participate as long as you can do well on the tests and project. To construct this index score, you would need a formula that weights the test and project components twice as high as the participation. Here's one:

$$\text{Performance} = [(2 \times \text{Test}) + (2 \times \text{Project}) + (1 \times \text{Participation})]/5$$

Why divide by 5? We want the final index score to be on a scale of 1 to 100. Notice that the idea of weighting in index construction can get rather confusing (so what's new?). For example, your professor could have measured both your test and project performance on a 1-to-40 scale (where the best performance gets a 40) and your attendance on a 1-to-20 scale. Then, to construct your index score you might simply add the three component scores! It looks like this is not a weighted index, and technically it isn't because you're simply adding the scores. But the truth is that you built the weighting into the measurement of each component.

Validate the index score. Once you have constructed the index you will need to validate it. This is essentially accomplished in the same way any measure is validated (see Construct Validity in Chapter 5). If the index score is going to be used over time, it is especially important to do periodic validation studies because it's quite possible that the components and/or how they relate to the index score have changed in important ways over time. For instance, the classification of occupations today differs in important ways from classifications used in 1950. And the selection of consumer goods continually changes over time. Consequently, indexes like the CPI and SES have to be recalibrated or adjusted periodically if they are to be valid reflections of the construct of interest.

Indexes are essential in social research. They range from formal, complex, sophisticated national indexes that track phenomena over years or decades, to simple measures developed for use in a single study (or to compute your grade for a course!).

SUMMARY

A lot of territory was covered in this chapter. We began with a discussion of what a scale is and we described the basic univariate scale types: Thurstone, Likert, and Guttman. You saw that scales can be used as stand-alone instruments, but they can also be integrated into a larger survey. Then we looked at tests as a specific kind of measure, once again with reliability and validity as central issues, and we also touched on some of the social and cultural aspects of testing in society. We also learned about indexes. We looked briefly at two of the most prominent indexes, the Consumer Price Index (CPI) and Socioeconomic Status (SES) index. We then went through the basic steps for constructing an index score.

At this point you should have a much clearer sense of how scales, tests, and indexes are similar to and different from each other. One clear commonality is that a scale, a test, and an index all yield a single numerical score or value that is designed to reflect the construct of interest. But there are lots of ways in which they are different. Scales typically involve rating a set of similar items on the same response scale, as in the 1-to-5 Likert response format. Tests can provide important data about the knowledge or skills of study participants and can function as an outcome or dependent variable. Indexes often combine numerical values that are counts or that are more objectively observable (like prices). Scales very often are constructed to get at more subjective and judgmental constructs like attitudes or beliefs. Tests are most often built to examine an ability, aptitude, or some form of learning or achievement. Indexes very often are used to combine component scores that are very different from each other and are measured in different ways, like income, occupation, and education in the SES.

There is considerable disagreement among researchers about whether and how scales, tests, and indexes can be defined and distinguished. Some researchers argue that scales and tests comprise a particular type or subset of an index. Others argue that they are very different things altogether. Some maintain that a unique feature of scaling is the sophistication of the methodology used to select the items. Others contend that good index development can get as sophisticated and advanced as any scaling procedure. And so it goes. However we define them, it should be clear to you that scales, tests, and indexes are essential tools in social research.

Based on this chapter, you should have a feel for what would be involved in creating and using a scale, test, or index. The next chapter introduces you to the general topic of survey research, which incorporates both qualitative and quantitative measurement. You'll see, for example, how scales, as discussed in this chapter, can be embedded in a larger questionnaire or interview.

Key Terms

consequential validity p. 159	index p. 162	reversal items p. 154
dichotomous response p. 147	interquartile range p. 151	scaling p. 146
factor analysis p. 159	interval response scale p. 147	scalogram analysis p. 156
forced-choice response scale p. 154	item analysis p. 160	semantic differential p. 149
Guttman or cumulative scaling p. 155	Likert or summative scaling p. 153	tests p. 157
	response scale p. 147	Thurstone scaling p. 150
		weighted index p. 165

Suggested Websites

Educational Testing Service

http://www.ets.org/?WT.ac=grehome
_etshome_121017

GRE fairness and validity

http://www.ets.org/gre/revised_general/about
/fairness/

Buros Center for Testing

http://buros.org/

U.S. Department of Labor Bureau of Labor Statistics: Consumer Price Index

http://www.bls.gov/cpi/

Review Questions

1. The Thurstone, Likert, and Guttman scales are all

a. unidimensional scales.
b. multidimensional scales.
c. based on the exact same procedures.
d. qualitative and unobtrusive forms of measurement.
(Reference: 6.2d)

2. A/An _____ is a quantitative score constructed by applying a set of rules to combine two or more variables in order to reflect a more general construct.

a. variable
b. scale
c. index
d. attitude
(Reference: 6.4)

3. The four steps in developing a new index, in order, are

a. conceptualization, validation, calculation, operationalization.
b. operationalization, conceptualization, calculation, validation.
c. calculation, validation, operationalization, conceptualization.
d. conceptualization, operationalization, calculation, validation.
(Reference: 6.4b)

4. A teacher who uses a formula like: total score = .3 (quiz scores) + .4 (term paper) + .1 (class participation) + .2 (class presentation) is using which kind of index?

a. Likert scale
b. weighted index
c. scalogram index
d. Guttman or cumulative scaling
(Reference: 6.4b)

5. According to S. S. Stevens, "_____is the assignment of objects to numbers according to a rule."

a. Science
b. Empiricism
c. Scaling
d. Quantitative analysis
(Reference: 6.2a)

6. Messick's concept of the "consequential validity" of tests has to do with:

a. The social impact of tests
b. How students learn best
c. Predictive power of a test for job performance
d. The impact of varying test length on reliability of the test
(Reference: 6.3a)

7. Thurstone's method of equal-appearing intervals allows the researcher to

a. identify whether a construct is unidimensional or multidimensional.
b. determine the attitudes of students toward fraternities and sororities.
c. follow a well-defined set of steps to construct a unidimensional scale.
d. determine the internal consistency of a set of scale items.
(Reference: 6.2d)

8. Which method of scale construction includes studying the item-total correlations in order to make decisions about which items best reflect the construct?

a. Thurstone's method of equal-appearing intervals
b. Likert scales
c. Guttman's scalogram analysis
d. semantic differential
(Reference: 6.2d)

9. Which scaling method is based on the idea that attitudes can be measured as cumulative?

a. Thurstone's method of equal-appearing intervals
b. Likert scales
c. Guttman's scalogram analysis
d. semantic differential
(Reference: 6.2d)

10. All scaling methods require first and foremost that you

a. understand the concept of correlation.
b. are able to calculate internal reliability.
c. generate a large pool of scale items.
d. define the focus.
(Reference: 6.2d)

11. The terms *scale* and *index* are synonymous.

a. True
b. False
(Reference: 6.1)

12. The primary difference between tests and scales is that tests are generally intended to objectively measure constructs like achievement while scales are generally used to measure subjective constructs like attitudes.

a. True
b. False
(Reference: 6.1)

13. The operationalization step in index construction basically involves figuring out how to translate the construct into specific observables.

a. True
b. False
(Reference: 6.4)

14. If a survey asks a set of questions on a single topic and asks you to respond on a similar response scale (such as a 1-to-5 disagree-agree response scale), you can automatically conclude that the set of questions constitutes a scale.

a. True
b. False
(Reference: 6.2a)

15. A Likert scale is any scale that has the respondent rate things from 1 to 7.

a. True
b. False
(Reference: 6.2)

7
Survey Research

7.1 Foundations of Survey Research

Ultimately, much of applied social science research concerns itself with understanding and measuring people's knowledge, opinions, perceptions, and behaviors. So how do we collect this type of data? A simplistic—and perhaps a little glib—response would be: We ask them! That's the point of survey research, which is the focus of this chapter.

A **survey** is a systematic way of asking people to volunteer information regarding their opinions and behaviors (**Figure 7.1**). For this reason, survey research is one of the most important areas of measurement in applied social research. In this chapter, we'll begin broadly by defining surveys, the different ways in which surveys are administered (questionnaires and interviews), and what factors to consider when selecting a particular survey method. Then, we will get more specific and discuss how to design a survey instrument. We'll present a number of issues, including the different types of questions, decisions about question content and wording, thinking about response format, and question placement and sequencing in your instrument. Finally, we will turn to some of the special issues involved when there is an interviewer collecting the responses.

© iStockphoto.com/wdstock

Figure 7.1 A survey is a systematic way of asking people to volunteer information.

7.2 Types of Survey Research

A survey is a data collection tool that is used to gather information from a population of interest. The information is collected from a sample of respondents, usually in a way that makes it possible to generalize results to the population within a certain degree of error.

A survey can be administered in two key ways: the questionnaire and the interview. Questionnaires are typically instruments that the respondent completes—in other words, questionnaires are self-administered. In the interview method, a researcher asks the participants questions and then goes on to complete the instrument based on their answers. An interview instrument looks very much like a questionnaire except that it usually also includes instructions for the interviewer and space for the interviewer to record observations about how the interview progressed. A common misconception about surveys is that questionnaires always ask short, closed-ended questions while interviews always ask broad open-ended ones. This is not always the case; it is quite common to see questionnaires with open-ended questions (although they do tend to be shorter than interview questions), and there will often be a series of closed-ended questions asked in an interview.

survey A measurement tool used to gather information from people by asking questions about one or more topics.

Survey research has changed dramatically in the last ten years. Automated telephone surveys use random dialing methods. Web-based surveys enable people to respond anytime and from anywhere in the world. A new variation of the group interview has evolved as focus group methodology. With the increasing use of mobile technology, many researchers are also using smartphones and tablets to collect and store survey data.

Below we will describe the major types of questionnaires and interviews. We will also discuss the relative advantages and disadvantages of these different survey types.

7.2a Questionnaires

When most people think of questionnaires, they think of the **mail survey**. All of us have, at some time, probably received a questionnaire in the mail. Mail surveys have many advantages. They are relatively inexpensive to administer—the cost mainly includes the cost of postage and the printed questionnaire. You can send the exact same instrument to a wide number of people, so there is wide geographical coverage. The respondent can fill it out at their own convenience. However, there are some disadvantages as well. Response rates from mail surveys are often low (Hager, Wilson, Thomas & Rooney (2003)) and untimely, and, mail questionnaires are not the best vehicles for receiving detailed written responses.

A second type of survey is the **group-administered questionnaire**. A sample of respondents is brought together and asked to respond to a structured sequence of questions. Traditionally, questionnaires have been administered in group settings for convenience. The researcher can give the questionnaire to those who are present and be fairly sure that there will be a high response rate. If the respondents don't understand the meaning of a question, they can ask for clarification. Additionally, in many organizational settings, it is relatively easy to assemble the group (in a company or business, for instance).

A less familiar type of questionnaire is the **household drop-off survey**. In this approach, a researcher goes to the respondent's home or business and hands the respondent the instrument. In some cases, the respondent is asked to mail it back; in others, the interviewer returns to pick it up. This approach attempts to blend the advantages of the mail survey and the group-administered questionnaire. Like the mail survey, the respondent can work on the instrument in private, when it's convenient. Like the group-administered questionnaire, the interviewer makes personal contact with the respondent; they don't just send an impersonal survey instrument. Additionally, respondents can ask questions about the study and get clarification on what they are being asked to do. Generally, this increases the percentage of people willing to respond. Finally, as compared to the mail-back option, if the researcher returns to pick up a completed survey, there is likely to be a higher response rate (Allred & Ross-Davis, 2010). However, drop-off surveys are relatively less economical, since the cost of administering a survey includes travel and wait time.

Another type of survey that is becoming more and more common in recent times is the **point-of-experience survey**. This method is especially useful in engaging people at the "point-of-experience"—such as right after a training event or outside a polling booth—to ensure timely and accurate data collection. Many kinds of customer-satisfaction surveys can be included under this category. Increasingly, your hotel room has a survey on the desk or your waiter presents a short survey with your check. You might also be asked to complete a survey as you come to the end

mail survey A paper-and-pencil survey that is sent to respondents through the mail.

group-administered questionnaire A survey that is administered to respondents in a group setting. For instance, if a survey is administered to all students in a classroom, we would describe that as a group-administered questionnaire.

household drop-off survey A paper-and-pencil survey that is administered by dropping it off at the respondent's household and, either picking it up at a later time, or having the respondent return it directly. The household drop-off method assures a direct personal contact with the respondent while also allowing the respondent the time and privacy to respond to the survey on their own.

point-of-experience survey A survey that is delivered at or immediately after the experience that the respondent is being asked about. A customer-satisfaction survey is often a point-of-experience survey.

of a hospital stay. Even though these point-of-experience surveys differ from each other in many aspects, they share the common need to achieve quick responses. A rapid response is needed in many situations, especially because as time passes, respondents are likely to forget details about the experience they are being asked about. It is important to note that all customer-satisfaction surveys are not point-of-experience surveys. Often, when you check out at a retail store, you are reminded to go online to fill out a short survey rating your shopping experience. In other cases, you might receive a questionnaire in the mail asking you to provide information on the quality of a particular service delivery. Customer-satisfaction surveys utilize online and mail methods in order to decrease social desirability bias. For example, if we were to interview patients on the quality of their hospital stay, we might justifiably worry that we are being given answers that are socially desirable (telling us what we want to hear), rather than their true feelings or beliefs, because the respondents are completing the survey in the presence of the very people who treated them.

Perhaps the most significant change in survey research in recent years has been the rise of the **electronic survey** or e-surveys. The main advantage of this method is the complete elimination of paper, postage, mail-out, and data entry costs. It also holds the potential for encouraging timely responses (for instance, while with a mail survey, the turnaround time is likely to be at least a few days, with an electronic survey it could be as small as a matter of a few hours). Additionally, with an electronic survey, a researcher can also reach out to many international respondents—which would be extremely expensive with a mail survey.

There are a wide and increasing variety of technologies for electronic surveys, but most of them fall into one of two categories: **email surveys** or **web surveys**. Although these two are similar in the technology used for delivery of the survey (email versus web, of course) perhaps the more important difference is that email surveys are "pushed" directly to the respondent's computer, whereas with web surveys you have to "pull" the respondent to a website. This distinction has important implications for how the respondent perceives the survey and for response rates. At the same time, email surveys are relatively limited in their interaction capabilities and graphics as compared to web surveys. However, that being said, email surveys are much easier to create as compared to their web-based cousins.

Electronic surveys also raise important questions about who can be reached. Although computer access is becoming more ubiquitous, there are still many people who have limited access or none at all. Where this is the case, you may need to conduct **dual-media surveys,** where you make the survey available through multiple channels (e.g., e-survey and mail survey) and allow respondents to select their preferred method of response. Another potential problem with electronic surveys is related to sampling. Because one individual can have several different email addresses, or each member of the household can have his/her own email address, it is often difficult to access a random sample of individuals or households, respectively, when it comes to using electronic survey methods. In addition, some people share email addresses so it can be difficult to know who has actually seen or replied to the survey.

7.2b Interviews

Interviews are a far more personal form of research than questionnaires. In the **personal interview,** the interviewer works directly with the respondent. In contrast to mail surveys, the interviewer has the opportunity to probe or ask follow-up questions,

---

electronic survey A survey that is administered via a computer program, typically distributed via email and/or a web site.

email survey Any survey that is distributed to respondents via email. Generally, the survey is either embedded in the email message and the respondent can reply to complete it, is transmitted as an email attachment that the respondent can complete and return via email, or is reached by providing a link in the email that directs the respondent to a website survey.

web survey A survey that is administered over a website (either Intranet or Internet). Respondents use their web browser to reach the website and complete the survey.

dual-media surveys A survey that is distributed simultaneously in two ways. For instance, if you distribute a survey to participants as an attachment they can print, complete, and fax back, or they can complete directly on the web as a web form, you can describe this as a dual-media survey.

personal interview A one-on-one interview between an interviewer and respondent. The interviewer typically uses an interview guide that provides a script for asking questions and follow-up prompts.

Figure 7.2 Interviews are personal.

and interviews are generally easier for the respondent, especially if you are seeking opinions or impressions (**Figure 7.2**). However, interviews can be time consuming and resource intensive. The interviewer is considered a part of the measurement instrument and interviewers have to be well trained to respond to any possible situation.

An increasingly important type of interview is the **group interview** or **focus group**. In a focus group, the interviewer is essentially a facilitator of the group discussion. Small groups of five to ten people are asked to discuss one or more focus questions. For example, focus groups are often used in marketing research—mainly to get feedback regarding new products that are about to be introduced. In social science research, a researcher might assemble a focus group to get people's opinions and views regarding a particular social issue like gun control. The facilitator strives to assure that each person has an opportunity to give their opinion. Focus groups enable deeper consideration of complex issues than many other survey methods. When people hear the points others make, they often will trigger ideas or responses they wouldn't have thought of by themselves (much like in brainstorming). But you always have to be concerned about how respondents in a group might be constrained from saying what they believe because others are present (social desirability bias). You also need to have ground rules regarding confidentiality. Even with ground rules in place, you have to be concerned about the impact of discussion of sensitive issues when participants have ongoing relationships outside the group (e.g., a worker and supervisor).

What's the difference between a group-administered questionnaire (discussed in the previous section) and a group interview or a focus group? In the group-administered questionnaire, each respondent is handed an instrument and asked to complete it while in the room. In a focus group, the interviewer facilitates the session. People work as a group, listening to each other's comments and answering the questions. Someone takes notes for the entire group; people don't complete the interview individually.

Almost everyone is familiar with the **telephone interview**. Telephone interviews enable a researcher to gather information rapidly. These interviews are usually conducted using a landline or a cell phone. Most of the major public opinion polls that are reported are based on telephone interviews. Like personal interviews, they allow for some personal contact between the inter-

group interview An interview that is administered to respondents in a group setting. A focus group is a structured form of group interview.

focus group A qualitative measurement method where input on one or more focus topics is collected from participants in a small-group setting where the discussion is structured and guided by a facilitator.

telephone interview A personal interview that is conducted over the telephone.

viewer and the respondent. They also allow the interviewer to ask follow-up questions; but they have some major disadvantages. Many people don't have publicly listed telephone numbers; some don't have telephones; people often don't like the intrusion of a call to their homes; and telephone interviews have to be relatively short or people will feel imposed upon. Additionally, cell phone interviews are relatively more expensive than a traditional landline survey (Pew Research Center, 2013). Another complication with cell phone surveys is that unlike a landline, cell phones are used virtually anywhere. For example, there might be legal and safety implications if an interviewee is reached while driving. Finally, if cell phones are more likely to be used in public places where respondents don't have enough privacy, there is a concern about the veracity of responses received through such a survey. However, because more and more people are moving exclusively to cell phones these days, it is common for researchers to include them in their sample along with landline phone surveys.

7.3 Selecting the Survey Method

Selecting the type of survey you are going to use is one of the most critical decisions in many social research contexts. Decisions regarding what type of survey method to use usually depend on the target population, the kind of information that is being sought, and the availability of resources—including budget and time. It's hard to compare the advantages and disadvantages of the major different survey types. Even though each type has some general advantages and disadvantages, there are exceptions to almost every rule. Table 7.1 shows our general assessment. A few simple rules will help you make the decision; you have to use your judgment to balance the strengths and weaknesses of different survey types. Here, we want to give you a number of questions you might ask to guide your decision.

7.3a Population Issues

The first set of considerations has to do with the population and its accessibility.

- *Can the population units be identified?* For some populations, you have a complete listing of the units to be sampled. For others, such a list is difficult or impossible to compile. For instance, there are complete listings of registered voters or persons with active drivers' licenses; but no one keeps a complete list of homeless people. If you are doing a study that requires input from homeless persons, it's likely that you'll need to go and find the respondents personally. In such contexts, you can pretty much rule out the idea of mail surveys or telephone interviews.
- *Is the population literate?* Questionnaires require that your respondents read. While initially this might seem like a reasonable assumption for most adult populations, recent research suggests that the instance of adult illiteracy is alarmingly high (National Assessment of Adult Literacy, 2014, http://nces. ed.gov/naal/index.asp). Even if your respondents can read to a certain degree, your questionnaire might contain difficult or technical vocabulary. Clearly, you would expect some populations to be illiterate. Young children would not be good targets for questionnaires.

Table 7.1 Advantages and disadvantages of different survey methods.

Issue	Questionnaire				Interview		
	Group	Mail	Electronic	Drop-Off	Personal	Phone	Focus Group
Are visual presentations possible?	Yes	Yes	Yes	Yes	Yes	No	Yes
Are long response categories possible?	Yes	Yes	???	Yes	???	No	???
Is privacy a feature?	No	Yes	Yes	No	Yes	???	No
Is the method adaptable on the spot?	No	No	No	No	Yes	Yes	Yes
Are longer, open-ended questions feasible?	No	No	No	No	Yes	Yes	Yes
Are reading and writing needed?	???	Yes	Yes	Yes	No	No	No
Can you judge quality of response?	Yes	No	No	???	Yes	???	Yes
Are high response rates likely?	Yes	No	No	Yes	Yes	No	Yes
Can you explain the study in person?	Yes	No	No	Yes	Yes	???	Yes
Is it low cost?	Yes	Yes	Yes	No	No	No	No
Are staff and facilities needs low?	Yes	Yes	Yes	No	No	No	No
Does it give access to dispersed samples?	No	Yes	Yes	No	No	No	No
Does respondent have time to formulate answers?	No	Yes	Yes	Yes	No	No	No
Is there personal contact?	Yes	No	No	Yes	Yes	No	Yes
Is a long survey feasible?	No	No	No	No	Yes	No	No
Is there quick turnaround?	No	Yes	Yes	No	No	Yes	???

- *Are there language issues?* We live in a multilingual world. Virtually every society has members who speak a language other than the predominant language. Some countries (like Canada) are officially multilingual, and our increasingly global economy requires us to do research that spans countries and language groups. Can you produce multiple versions of your questionnaire? For mail instruments, can you know in advance which language your respondent speaks, or do you need to send multiple translations of your instrument? Can you be confident that important connotations in your instrument are not culturally specific? Could some of the important nuances get lost in the process of translating your questions? Do you need to include pictures to facilitate an easier interpretation of your questions? If so, a web-based interview may be a better option than an email survey.

- *Will the population cooperate?* People who do research on illegal immigration have a difficult methodological problem. They often need to speak with illegal immigrants or people who may be able to identify others who are. Why would those respondents cooperate? Although the researcher may mean no harm, the respondents are at considerable risk legally if information they divulge should get into the hands of the authorities. The same can be said for any target group that is engaging in illegal or unpopular activities.
- *What are the geographic restrictions?* Is your population of interest dispersed over too broad a geographic range for you to study feasibly with a personal interview? It may be possible for you to send a mail or an electronic instrument to a nationwide sample. You may be able to conduct phone interviews with them; but it will almost certainly be less feasible to do research that requires interviewers to visit directly with respondents if they are widely dispersed.

7.3b Sampling Issues

The sample is the actual group you will have to contact in some way. When doing survey research you need to consider several important sampling issues.

- *What data is available?* What information do you have about your sample? Do you have current addresses? Current phone numbers? Are your contact lists, such as email addresses, up to date?
- *Can respondents be found?* Can your respondents be located? Some people are very busy. Some travel a lot. Some work the night shift. Even if you have an accurate phone, address, or email, you may not be able to locate or make contact with your sample.
- *Who is the respondent?* Who is the respondent in your study? Let's say you draw a sample of households in a small city. A household is not a respondent. Do you want to interview a specific individual? Do you want to talk only to the head of the household (how is that person defined)? Are you willing to talk to any member of the household? Do you decide that you will speak to the first adult member of the household who opens the door? What if that person is unwilling to be interviewed but someone else in the house is willing? How do you deal with multifamily households? Similar problems arise when you sample groups, agencies, or companies. Can you survey any member of the organization? Or do you want to speak only to the Director of Human Resources? What if the person you would like to interview is unwilling or unable to participate? Do you use another member of the organization? If so, how do you decide who to survey?
- *Can all members of the population be sampled?* If you have an incomplete list of the population (sampling frame) you may not be able to sample every member of the population. Lists of various groups are extremely hard to keep up to date. People move or change their names. Even though they are on your sampling frame listing, you may not be able to get to them. It's also possible they are not even on the list.
- *Are response rates likely to be a problem?* Even if you are able to solve all of the other population and sampling problems, you still have to deal with the issue of response rates. Some members of your sample will simply refuse to respond. Others have the best of intentions, but can't seem to find the time to send in your questionnaire by the due date. Still others misplace the instrument or forget about the appointment for an interview. Low response rates are among the most difficult of problems in survey research. They can ruin

an otherwise well-designed survey effort. Often, electronic surveys are used to improve response rates.

- *Will incentives for participation help?* In addition to following the guidelines for effective survey construction, you may be able to increase your response rate by using carefully chosen incentives. An incentive is a way of increasing motivation of your respondents to complete the survey. Incentives may be very concrete rewards such as money, gift cards, or lottery chances. Other kinds of incentives might be apparent if you carefully consider your population and what might be most meaningful to them. For example, if you are conducting research on pets, you might be able to offer to make a donation to the SPCA in an amount that corresponds to the survey completion rate.

7.3c Question Issues

Sometimes the nature of what you want to ask respondents determines the type of survey you select.

- *What types of questions can you ask?* Are you going to be asking personal questions? If so, group interviews are probably not going to be appropriate. Are you going to need to get lots of detail in the responses? Are the questions on the survey complex and technical? If so, it may be better to conduct an interview so that you can provide explanations if needed, rather than send out a questionnaire and risk receiving answers based on improperly understood questions. Can you anticipate the most frequent or important types of responses and develop reasonable closed-ended questions?
- *Will filter questions be needed?* A filter question may be needed to determine whether the respondent is qualified to answer your question(s) of interest. For instance, you wouldn't want to ask for respondents' opinions about a specific computer program without first screening to find out whether they have any experience with the program. Sometimes you have to filter on several variables (for example, age, gender, and experience). The more complicated the filtering, the less likely it is that you can rely on paper-and-pencil instruments without confusing the respondent.
- *Can question sequence be controlled?* Is your survey one in which you can construct a reasonable sequence of questions in advance? Or, are you doing an initial exploratory study in which you may need to ask follow-up questions that you can't easily anticipate? If you need to ask follow-up questions, then interviews are a more appropriate survey method.
- *Will lengthy questions be asked?* If your subject matter is complicated, you may need to give the respondent some detailed background for a question. Can you reasonably expect your respondent to sit still long enough in a phone interview to listen to your question?
- **Will long response scales be used?** If you are asking people about the different computing equipment they use, you may have to have a lengthy response list (tablets, smartphones, laptops, netbooks, desktop computers, and so on). Clearly, it may be difficult to ask about each of these in a short phone interview.

7.3d Content Issues

The content of your study can also pose challenges for the different survey types you might use.

- *Can the respondents be expected to know about the issue?* If respondents do not keep up with the news (for example, by reading the newspaper, watching television news, or talking with others), they may not even know of the news issue you want to ask them about. Or, if you want to do a study of family finances and you are talking to the spouse who doesn't pay the bills on a regular basis, he or she may not have the information to answer your questions. In such cases, a mail survey seems like an inappropriate choice. A personal interview might work the best here. You could also try a phone interview where you first ascertain whether you have the right kind of respondent before beginning the survey.

- *Will the respondent need to consult records?* Even if the respondents understand what you're asking about, you may need to allow them to consult their records to get an accurate answer. For instance, if you ask them how much money they spent on food in the past month, they may need to look up their personal check and credit card records. In this case, you don't want to be involved in an interview where they would have to go look things up while they keep you waiting (and they wouldn't be comfortable with that either). So, a mail or an electronic questionnaire seems like a better option.

7.3e Bias Issues

People come to the research endeavor with their own sets of biases and prejudices. Sometimes, these biases will be less of a problem with certain types of survey approaches.

- *Can social desirability be avoided?* Respondents generally want to look good in the eyes of others. None of us likes to look like we don't know an answer. We don't want to say anything that would be embarrassing. If you ask people about information that may put them in this kind of position, they may not tell you the truth, or they may spin the response so that it makes them look better. This may be more of a problem in a face-to face interview situation or a phone interview.

- *Can interviewer distortion and subversion be controlled?* Interviewers may distort an interview as well. They may not ask difficult questions or ones that make them uncomfortable. They may ask the question in a way that distorts the meaning of the question. They may not listen carefully to respondents on topics for which they have strong opinions. They may make the judgment that they already know what the respondent would say to a question based on their prior responses, even though that may not be true. If you are using multiple interviewers, you might also face inter-rater reliability bias (see topic on reliability in Chapter 5, "Introduction to Measurement").

- *Can false respondents be avoided?* With mail surveys, it may be difficult to know who actually responded. Did the head of household complete the survey or someone else? Did the CEO actually give the responses or instead pass the task off to a subordinate? Are the people you're speaking with on the phone actually who they say they are? At least with personal interviews, you have a reasonable chance of knowing to whom you are speaking. In mail surveys or phone interviews, this may not be the case. Similarly, while emails are usually personal, respondents may forward any electronic surveys they receive to others to complete on their behalf.

7.3f Administrative Issues

Last, but certainly not least, you have to consider the feasibility of the survey method for your study.

- *Costs.* Cost is often the major determining factor in selecting survey type. You might prefer to do personal interviews, but can't justify the high cost of training and paying for the interviewers. You may prefer to send out an extensive mailing but can't afford the postage to do so. So, you might consider sending out an electronic survey or doing a group survey or interview to cut down on the costs.
- *Facilities.* Do you have the facilities (or access to them) to process and manage your study? In phone interviews, do you have well-equipped phone surveying facilities? For focus groups, do you have a comfortable and accessible room to host the group? Do you have the equipment needed to record and transcribe responses?
- *Time.* Some types of surveys take longer than others. Do you need responses immediately (as in an overnight public opinion poll)? Have you budgeted enough time for your study to send out mail surveys and follow-up reminders, and to get the responses back by mail? Have you allowed for enough time to get enough personal interviews to justify that approach?
- *Personnel.* Different types of surveys make different demands of personnel. Interviews require well-trained and motivated interviewers. Group-administered surveys require people who are trained in group facilitation. Some studies may be in a technical area that requires some degree of expertise in the interviewer.

Clearly, there are lots of issues to consider when you are selecting which type of survey to use in your study, and there is no easy way to make this decision in many contexts because it might be that no single approach is the best. You may have to make trade-offs and weigh advantages and disadvantages of each approach. There is judgment involved. Two expert researchers might, for the same problem or issue, select entirely different survey methods; but, if you select a method that isn't appropriate or doesn't fit the context, you can doom a study before you even begin designing the instruments or questions themselves.

7.4 Survey Design

Before you begin constructing your survey, it is useful to reflect on the full list of issues that need to be explored through this instrument. Be very specific and clear regarding the purpose of the survey. Remember, it is important to focus on what you "need to know," as opposed to what is "nice to know." It is also important to have some idea about the kind of analysis you will perform on the survey responses—this will indicate, up-front, the level of detail that is required from the survey instrument. Finally, you should also research prior publications to see if there is an existing questionnaire already available that can be used to serve your purpose, or at least one that can be partly used as the basis for your survey.

Constructing a survey instrument is an art in itself. You must make numerous small decisions—about content, wording, format, and placement—that can have important consequences for your entire study. Although there's no one perfect way to accomplish this job, it is possible to increase your chances of developing a better final product by following the guidelines detailed below. There are three primary issues involved in writing a question:

- **Determining the question content, scope, and purpose**
- **Choosing the response format that you use for collecting information from the respondent**
- **Figuring out how to word the question to get at the issue of interest**

After you have your questions written, there is also the issue of how best to place them in your survey. You'll see that although many aspects of survey construction are just common sense, if you are not careful, you can make critical errors that have significant effects on your results.

7.4a Types of Questions

Survey questions can be divided into two broad types: *structured* and *unstructured*. While unstructured questions are open-ended, where the respondents need to create their own answer, structured questions are laid out with various response options, and the respondent simply has to choose or circle an option. For instance, if you were to simply ask, "What type of new meal options would you like to see in the future?" you would be asking an unstructured question. On the other hand, if you were to ask the same question and provide the respondent with a number of response options, like "Vegetarian, Asian, Mexican, Indian," or if you were to only ask, "Would you like to see any new meal options in the future?" with "yes" or "no" as response options, you would be asking a structured question.

From an instrument design point of view, structured questions pose the greater difficulties (see the section "Response Format" later in this chapter). From a content perspective, it may actually be more difficult to write good unstructured questions. Here, we'll discuss the variety of structured questions you can consider for your survey. (We discuss unstructured questioning more in the section "Interviews" later in this chapter.)

Dichotomous Response Formats

dichotomous response format A question response format that allows the respondent to choose between only two possible responses.

When a question has two possible responses, it has a **dichotomous response format**. Surveys often use dichotomous questions that ask for a Yes/No, True/False, or Agree/Disagree response (see **Figure 7.3**). There are a variety of ways to lay these questions out on a questionnaire:

Figure 7.3 Dichotomous response formats for a survey question.

> Do you believe that the death penalty is ever justified?
>
> ____ Yes
>
> ____ No
>
> Please enter your gender:
>
> ☐ Male ☐ Female

Questions Based on Level of Measurement

We can also classify questions in terms of the level of measurement used in the question's response format. (The idea of level of measurement is covered in

Chapter 5, "Introduction to Measurement.") For instance, you might measure occupation using a nominal response format, as in **Figure 7.4**. In a **nominal response format,** the number next to each response has no meaning except as a placeholder for that response; the choices in the example are a 2 for a lawyer and a 1 for a truck driver. From the numbering system used you can't infer that a lawyer is twice something that a truck driver is. The primary reason you might number responses in this manner is to make data entry more efficient. The person entering the data from this survey would only need to enter a single number rather than a longer category name like "truck driver."

When you ask respondents to rank-order their preferences, you are using an **ordinal response format.** For example, in **Figure 7.5** the respondent is asked to rate their dessert preferences by ranking them.

In this example, you want the respondent to put a 1, 2, 3, or 4 next to the dessert, where 1 is the respondent's first choice. Note that this could get confusing. The respondents might check their favorite dessert instead of entering a number, or assign higher numbers to desserts they prefer more instead of understanding that you want rank ordering where a higher number means a lower rank. Notice, in the example, we stated the prompt (question) explicitly so the respondent knows that we want a number from 1 to 4 and we indicated that a 1 meant the dessert was more favored than a 4.

You can also construct survey questions using an **interval level response format.** One of the most common of these types is the traditional 1-to-5 rating (or 1-to-7, or 1-to-9, and so on). This is sometimes referred to as a **Likert-type response scale.** In **Figure 7.6,** you see how you might ask an opinion question using a 1-to-5 bipolar scale. (It's called bipolar because there is a neutral point, and the two ends of the scale are at opposite positions of the opinion.) You should note that just because you use equal-spaced intervals doesn't automatically mean that you are measuring on an interval scale. For instance, in a traditional Likert response scale (1 = Strongly Disagree, 2 − Disagree, 3 = Neutral, 4 − Agree, 5 = Strongly Agree) we cannot assume that the distance between "Strongly Agree" and "Agree" is equal to the distance between "Agree" and "Neutral" even though the intervals between there respective numbers are equal.

Occupational class

1 = Truck driver

2 = Lawyer

3 = etc.

Figure 7.4 A nominal-level response format for a survey question.

Figure 7.5 An ordinal-level response format for a survey question.

nominal response format A response format that has a number beside each choice where the number has no meaning except as a placeholder for that response.

ordinal response format A response format in which respondents are asked to rank the possible answers in order of preference.

interval level response format A response measured using numbers spaced at equal intervals where the size of the interval between potential response values is meaningful. An example would be a 1-to-5 response scale.

Likert-type response scale A response format where responses are gathered using numbers spaced at equal intervals.

Rank the desserts in order of personal preference, from your most favorite (1) ...to your least favorite... (4).

____ Ice cream

____ Cookies

____ Cake

____ Candy

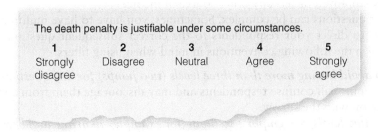

The death penalty is justifiable under some circumstances.

1	2	3	4	5
Strongly disagree	Disagree	Neutral	Agree	Strongly agree

Figure 7.6 A Likert-type scale is an interval level scale.

Figure 7.7 A semantic differential response format for a survey question.

Please state your opinions on national health insurance on the scale below.

	Very much	Somewhat	Neither	Somewhat	Very much	
Interesting	☐	☐	☐	☐	☐	Boring
Simple	☐	☐	☐	☐	☐	Complex
Uncaring	☐	☐	☐	☐	☐	Caring
Useful	☐	☐	☐	☐	☐	Useless

Please check each item with which you agree.

_____ Are you willing to permit immigrants to live in your country?

_____ Are you willing to permit immigrants to live in your community?

_____ Are you willing to permit immigrants to live in your neighborhood?

_____ Would you be willing to have an immigrant live next door to you?

_____ Would you let your child marry an immigrant?

Figure 7.8 A cumulative response format for a survey question.

Another example of how an interval response format is used is with a measurement approach called the semantic differential, as shown in **Figure 7.7**. Here, an item is assessed by the respondent on a set of bipolar (two-option) adjective pairs (in this example a 5-point rating interval-level response format is used).

Finally, another type of interval response format used to collect responses in the cumulative or Guttman scale. Here, the respondents check each item with which they agree. The items themselves are constructed so that they are cumulative; if you agree with one item, you probably agree with all of the ones above it in the list (see **Figure 7.8**), because that's the way such scales are constructed. Each item also has a scale score that is not shown with the item. A respondent's score is the highest scale score of an item with which they agreed.

Filter or Contingency Questions

Sometimes you have to ask respondents a preceding question to determine whether they are qualified or experienced enough to answer a subsequent one. This is called using a **filter or contingency question**. For instance, you may want to ask one question if the respondent has ever smoked marijuana and a different question if he or she reports they have not. In this case, you would have to construct a filter question to determine whether the respondent has ever smoked marijuana (see **Figure 7.9**). In another example, suppose you want to know how much the respondent earns, then you need to first ask whether the respondent is currently employed or has any other source of income.

Filter questions can be complex. Sometimes, you have to have multiple filter questions to direct your respondents to the correct subsequent questions. You should keep the following conventions in mind when using filters:

filter or contingency question A question you ask the respondents to determine whether they are qualified or experienced enough to answer a subsequent one.

- *Try to avoid having more than three levels (two jumps) for any question.* Too many jumps will confuse respondents and may discourage them from continuing on with the survey.
- *If only two levels, use graphics to jump (for example, an arrow and box).* The example in Figure 7.9 shows how you can make effective use of an arrow

and box to help direct the respondent to the correct subsequent question.

- **If possible, jump to a new page.** If you can't fit the response to a filter on a single page, it's probably best to be able to say something like, *If YES, please turn to page 4*, rather than *If YES, please go to Question 38*, because the respondent will generally have an easier time finding a page than a specific question.

Most web survey systems have built-in capabilities for allowing filtering. The advantage of using filtering in web contexts is that, unlike in paper instruments, the software can only show questions that are appropriate based on responses to one or more filters. In a paper instrument you have to include all responses and help the respondent navigate through the form.

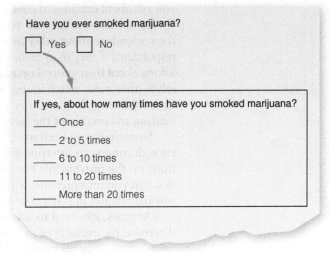

Figure 7.9 A filter or contingency question.

7.4b Question Content

For each question in your survey, you should ask yourself how well it addresses the content you are trying to get at. The following sections cover key content-related issues that you need to consider when writing your questions.

Is the Question Necessary and Useful?

Examine each question to determine whether you need to ask it at all, and whether you need to ask it at the level of detail you currently have, as in the following examples:

- **Do you need the age of each child or just the number of children under 16?**
- **Do you need to ask income or can you estimate?**

Are Several Questions Needed?

There might be occasions where additional questions are required. Sometimes we develop a question where we try to ask about too many things at once. For example, consider the following examples:

- **What are your feelings towards African Americans and Hispanic Americans?**
- **What do you think of proposed changes in benefits and hours?**

It's hard, often impossible, for a respondent to answer such questions by providing a single response, because it is likely that some have conflicting opinions about different parts of the question. They may feel very differently about African Americans as compared to Hispanic Americans, or about changes in benefits versus changes in hours. You can often spot these kinds of problem questions by looking for the conjunction *and* in your question. We refer to this classic question-writing problem as the **double-barreled question**. If you realize that your question has multiple subparts, you should split the question into separate ones.

You might also want to use more than one question to cover a single content area when the one question you ask does not cover all possibilities. For instance, if

double-barreled question A question in a survey that asks about two issues but only allows the respondent a single answer. For instance, the question "What do you think of proposed changes in benefits and hours in your workplace?" asks simultaneously about two issues but treats it as though they are one.

you ask about earnings, the respondent might end up mentioning only wage income. To get an accurate answer, you might need to include a follow-up probe asking the respondent about other income such as that from dividends or gifts. If you ask respondents if they're in favor of public TV, they might not understand that you're asking about their general opinion. They may not be in favor of public TV for themselves (they never watch it), but might favor it for their children (who watch *Sesame Street* regularly). You might be better off asking two questions: one about their own viewing and one about the viewing habits of other members of their households.

Sometimes you need to ask additional questions in order to provide you with enough context to interpret the respondents' answers on some of the other questions on the instrument. For instance, if you ask about attitudes toward Catholics, can you interpret this without finding out about your respondents' attitudes toward religion in general or toward other religious groups?

At times, you need to ask additional questions because your question does not determine the intensity of the respondent's attitude or belief. For example, if respondents say they support public TV, you probably should also ask whether they ever watch it or if they would be willing to have their tax dollars spent on it. It's one thing for respondents to tell you they support something; but the intensity of that response is greater if they are willing to back their sentiment of support with their behavior.

Do Respondents Have the Needed Information?

Look at each question in your survey to see whether the respondent is likely to have the necessary information to be able to answer the question. For example, let's say you want to ask the following question:

Do you think Dean Rusk acted correctly in the Bay of Pigs crisis?

The respondents won't be able to answer this question if they have no idea who Dean Rusk was or what the Bay of Pigs crisis was (most of you reading this weren't even born when Rusk was alive or the Bay of Pigs happened. Thankfully, your generation has Google!). In surveys of television viewing, you cannot expect the respondent to answer questions about shows they have never watched. You should ask a filter question first (such as, "Have you ever watched the show *Modern Family*?") before asking for opinions about it.

Does the Question Need to Be More Specific?

Sometimes researchers ask their questions too generally and the information they obtain is difficult to interpret. For example, let's say you want to find out a respondent's opinions about a specific book. You could ask the following question:

How well did you like the book?

and offer some scale ranging from Not At All to Extremely Well; but what would the response mean? What does it mean to say you liked a book extremely well? Instead, you might ask questions designed to be more specific:

Did you recommend the book to others?

or

Did you look for other books by that author?

Is the Question Sufficiently General?

You can err in the other direction as well by being too specific. For instance, if you ask people to list the television programs they liked best in the past week, you could get a different answer than if you asked them which show they've enjoyed most over the past year. Perhaps a show they don't usually like had a great episode in the past week, or their favorite show was preempted by another program, but if asked over the course of a year rather than just the past week, you would be able to get at their real preferences.

Is the Question Biased or Loaded?

One danger in question writing is that your own biases and blind spots may affect the wording (see the section "Question Wording" in this chapter). For instance, you might generally be in favor of tax cuts. If you ask the following question:

> **What do you see as the benefits of a tax cut?**

you're only asking about one side of the issue. You might get a different picture of the respondents' positions if you also asked about the disadvantages of tax cuts. The same thing could occur if you are in favor of public welfare and you ask

> **What do you see as the disadvantages of eliminating welfare?**

without also asking about the potential benefits.

Will the Respondent Answer Truthfully?

For each question on your survey, ask yourself whether respondents will have any difficulty answering the question truthfully. If there is some reason why they may not, consider rewording the question. For instance, some people are sensitive about answering questions about their exact age or income. In this case, you might give them **response brackets** to choose from (such as between 30 and 40 years old, or between $50,000 and $100,000 annual income). Sometimes even bracketed responses won't be enough. Some people do not like to share how much money they give to charitable causes. (They may be afraid of opening themselves up to even more solicitations.) No matter how you word the question, they would be unlikely to tell you their contribution rate. Sometimes you can work around such problems by posing the question in hypothetical terms. For example, they might respond if you ask how much money "people you know" typically give in a year to charitable causes, and that may be a reasonable proxy for how much they give (although they could be hanging out with people who are richer or poorer than they are). Finally, you can sometimes dispense with asking a question at all if you can obtain the answer unobtrusively. (This is covered in Chapter 3, "Qualitative Approaches to Research.") If you are interested in finding out which magazines the respondents read, you might instead tell them you are collecting magazines for a recycling drive and ask if they have any old ones to donate. Of course, you have to consider the ethical implications of such deception!

7.4c Response Format

The **response format** is how you collect the answer from the respondent. Let's start with a simple distinction between what we call *structured response formats* and *unstructured response formats*.

Structured Response Formats

Structured response formats help the respondent to respond more easily and help the researcher accumulate and summarize responses more efficiently; but they can also constrain the respondent and limit the researcher's ability to understand what the

response brackets A question response format that includes groups of answers, such as between 30 and 40 years old, or between $50,000 and $100,000 annual income.

response format The format you use to collect the answer from the respondent.

structured response format Provides a specific format for the respondent to choose their answer. For example, a checkbox question lists all of the possible responses.

Figure 7.10 A common fill-in-the-blank item.

Please enter your gender:

____ Male

____ Female

Figure 7.11 A checklist like this is another common structured item.

One of President Lincoln's most famous speeches, the _____ Address, only lasted a few minutes when delivered.

Figure 7.12 The fill-in-the-blank test item.

Please check the items that you have on the computer you use most:

☐ external hard drive

☐ USB

☐ DVD drive

☐ monitor

☐ printer

Figure 7.13 The check-box format is useful when you want respondents to select more than one item.

multi-option or multiple-response variable A question format in which the respondent can pick multiple variables from a list.

respondent really means. There are many different structured response formats, each with their own strengths and weaknesses. We'll review the major ones here.

- **Fill-in-the-Blank**. One of the simplest response formats is a blank line which can be used to collect data for a number of different response types. For instance, asking your name, as shown in **Figure 7.10**, is one of the simplest fill-in-the-blank formats.

Blanks are also used for checking responses in dichotomous questions, as illustrated in **Figure 7.11**.

Here, the respondent would probably put a check mark or an X next to the response. As described earlier in this chapter, this is also an example of a dichotomous response because it only has two possible values. Other common dichotomous responses are True/False and Yes/No. Another common use of a fill-in-the-blank response format is in preference ranking, as described earlier in Figure 7.5, where respondents entered their rank preferences for four desserts into the blank line in front of each category. Notice that in this case you expect the respondent to place a number on every blank, whereas in the previous example, you expect the respondent to choose only one. And there's always the classic fill-in-the-blank test item (see **Figure 7.12**).

- **Check the Answer.** The respondent places a check next to the response(s). The simplest form would be the example given previously that asks the respondents to indicate their gender. Sometimes, you supply a box that the person can fill in with an X, which is sort of a variation on the check mark. **Figure 7.13** shows a check-box format.

Notice that in this example, it is possible to check more than one response. By convention, you usually use the check-mark format when you want to allow the respondent to select multiple items.

This type of question is sometimes referred to as a **multi-option or multiple-response variable**. You have to be careful when you analyze data from a multioption variable. Because the respondent can select any of the options, you have to treat this type of variable in your analysis as though each option were a separate variable. For instance, for each option you would normally enter into a computer either a 0 if the respondent did not check it or a 1 if the respondent did check it. For the previous example, if the respondent had only a smartphone and tablet, you would enter the sequence 1, 1, 0, 0, 0 in five separate variables. There is an important reason why you should code this variable as either 0 or 1 when you enter the data. If you do, and you want to determine what percent of your sample has a printer, all you have to do is compute the average of the 0's and 1's for the printer variable. For instance, if you have ten respondents and only four have a printer, the average would be 4/10 = .40 or 40 percent, which is the percent who checked that item.

The previous example is also a good example of a checklist item. Whenever you use a checklist, you want to be sure that you ask the following questions:

- **Are all of the alternatives covered?**
- **Is the list of reasonable length (not too long)?**
- **Is the wording impartial?**
- **Is the form of the response easy, uniform?**

Sometimes you may not be sure that you have covered all of the possible responses in a checklist. If that is the case, you should probably allow the respondent to write in any other options that apply.

- **Circle the Answer.** Sometimes respondents are asked to circle an item to indicate their response. Usually, you are asking them to circle a number. For instance, you might have the example shown in **Figure 7.14**.

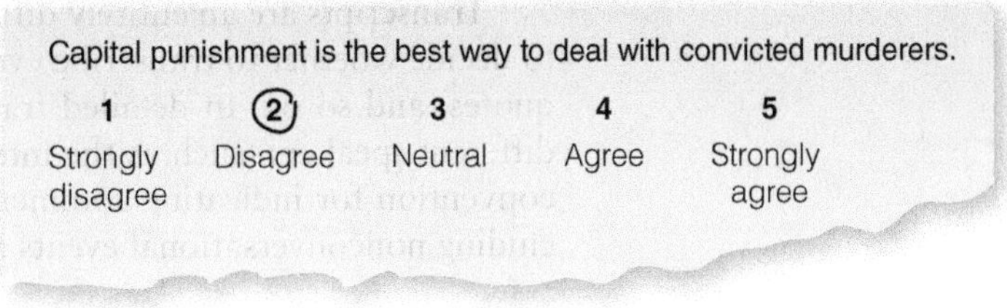

If respondents are answering questions on a computer, it's not feasible to have them circle a response. In this case, you would most likely use an option button as shown in **Figure 7.15**.

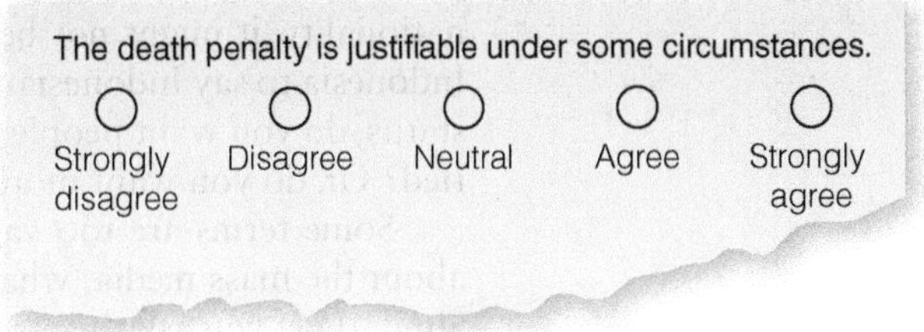

Only one option at a time can be checked. The rule of thumb is that you ask people to circle an item or click a button when you only want them to be able to select one of the options. In contrast to the multioption variable described previously, this type of item is referred to as a **single-option variable**; even though the respondents have multiple choices, they can only select one of them. You would analyze this as a single variable that can take the integer values from 1 to 5.

Unstructured Response Formats

A wide variety of structured response formats exist; however, there are relatively few unstructured ones. What is an **unstructured response format**? Generally, it is written text. If the respondent (or interviewer) writes down text as the response, you have an unstructured response format. These can vary from short comment boxes to the transcript of an interview.

In almost every questionnaire, there are usually one or more short text field questions. One of the most frequent is shown in **Figure 7.16**.

Actually, there's really not much more to text-based response formats of this type than writing the prompt and allowing enough space for a reasonable response. Such questions allow the respondent to answer in detail.

Figure 7.14 A circle-the-answer response format.

Figure 7.15 An option button response format on the web.

single-option variable A question response list from which the respondent can check only one response.

unstructured response format A response format that is not predetermined and that allows the respondent or interviewer to determine how to respond. An open-ended question is a type of unstructured response format.

Figure 7.16 The most common unstructured response format allows the respondent to add comments.

Transcripts are an entirely different matter. In those cases, the transcriber has to decide whether to transcribe every word or only record major ideas, thoughts, quotes, and so on. In detailed transcriptions, you may also need to distinguish different speakers (such as the interviewer and respondent) and have a standard convention for indicating comments about what's going on in the interview, including nonconversational events that take place and thoughts of the interviewer.

7.4d Question Wording

One of the major difficulties in writing good survey questions is getting the wording right. Even slight wording differences can confuse the respondent or lead to incorrect interpretations of the question. Here, we outline some key issues regarding the wording of survey questions:

Can the Question Be Misunderstood?

The survey author always has to be on the lookout for questions that could be misunderstood or confusing. For instance, if you ask a person for his or her nationality, it might not be clear what you want. (Do you want someone from Indonesia to say Indonesian, Asian, or Pacific Islander?) Or, if you ask for marital status, do you want people to say simply that they are either married or not married? Or, do you want more detail (like divorced, widow/widower, and so on)?

Some terms are too vague to be useful. For instance, if you ask a question about the mass media, what do you mean? The newspapers? Radio? Television? Similarly, if you ask the respondent whether they go for *regular* preventive health checkups, it is likely that you will get an answer, but it may not always be useful. An enormous amount of variation might exist in what the respondents mean by their answers because *regular* might mean different things to different people. Here's another one of our favorite questions that is often prone to misunderstandings. Let's say you want to know the following:

What kind of headache remedy do you use?

Do you want to know what brand-name medicine respondents take? Do you want to know about home remedies? Are you asking whether they prefer a pill, capsule, or caplet?

What Assumptions Does the Question Make?

Sometimes you don't stop to consider how a question will appear from the respondent's point of view. You don't think about the assumptions behind the questions. For instance, if you ask what social class someone's in, you assume that they know what social class is and that they think of themselves as being in one. In this case, you may need to use a filter question first to determine whether either of these assumptions is true.

Is the Time Frame Specified?

Whenever you use the words *will*, *could*, *might*, or *may* in a question, you might suspect that the question asks a time-related question. Be sure that, if it does, you have specified the time frame precisely. For instance, you might ask:

Do you think Congress will cut taxes?

or something like

Do you think Congress could successfully resist tax cuts?

Neither of these questions specifies a time frame. This causes the question to be vague. You can specify a time frame for the first question, such as:

Do you think Congress will cut taxes during its current session?

How Personal Is the Wording?

By changing just a few words, a question can go from being relatively impersonal to a probing, private one. Consider the following three questions, each of which asks about the respondent's satisfaction with working conditions:

- **Are working conditions satisfactory or not satisfactory in the plant where you work?**
- **Do you feel that working conditions are satisfactory or not satisfactory in the plant where you work?**
- **Are you personally satisfied with working conditions in the plant where you work?**

The first question is stated from a fairly detached, objective viewpoint. The second asks how you feel. The last asks whether you are personally satisfied. Be sure the questions in your survey are at an appropriate level for your context, and be sure that level is consistent across questions in your survey.

Is the Wording too Direct?

At times, asking a question too directly may be threatening or disturbing for respondents. This might encourage people to skip this question altogether. For instance, consider a study in which you want to discuss battlefield experiences with former soldiers who experienced trauma. Examine the following three question options:

- **How did you feel about being in the war?**
- **How well did the equipment hold up in the field?**
- **How well were new recruits trained?**

The first question may be too direct. For this population, it may elicit powerful, negative emotions based on individual recollections. The second question is a less direct one. It asks about equipment in the field; but, for this population, it may also lead the discussion toward more difficult issues to discuss directly. The last question is probably the least direct and least threatening. The question is likely to get the respondent talking and recounting anecdotes, without eliciting as much stress. Of course, all of this may simply be begging the question. If you are doing a study where the respondents might experience high levels of stress because of the questions you ask, you should reconsider the ethics of doing the study.

Other Wording Issues

The nuances of language guarantee that the task of the question writer is endlessly complex. Without trying to generate an exhaustive list, here are a few other guidelines to keep in mind:

- **Questions should not contain difficult or unclear terminology.**
- **Questions should make each alternative clear.**
- **Question wording should not be objectionable.**
- **Question wording should not be loaded or slanted. That is, they should not steer the respondent to a particular response (Otherwise, why ask them the question?).**

7.4e Question Placement

One of the most difficult tasks facing the survey designer involves the ordering of questions. Which topics should be introduced early in the survey and which later? If you leave your most important questions until the end, you may find that your respondents are too tired to give them the kind of attention you would like. If you introduce them too early, they may not yet be ready to address the topic, especially if it is a difficult or disturbing one. At the same time, you should ensure that the questions asked at the beginning of the questionnaire pique the respondent's interest and motivate them to continue on with the questionnaire.

There are no easy answers to the problems of question placement; you have to use your judgment. Whenever you think about question placement, consider the following potential issues:

- **The answer may be influenced by prior questions.**
- **The question may come too early or too late to arouse interest.**
- **The question may not receive sufficient attention because of the questions around it.**

The Opening Questions

Just as in other aspects of life, first impressions are important in survey work. The first few questions you ask will determine the tone for the survey and can help put your respondent at ease. With that in mind, the opening few questions should, in general, be easy to answer. You might start with some simple descriptive questions that will get the respondent rolling. You should never begin your survey with sensitive or threatening questions.

Sensitive Questions

In much of your social research, you will have to ask respondents about difficult or uncomfortable subjects. Before asking such questions, you should attempt to develop some trust or rapport with the respondent. Often, preceding sensitive questions with some easier warm-up ones will help; but, you have to make sure that the sensitive material does not come up abruptly or appear unconnected to the rest of the survey. The respondents will also know that the question is coming if you mention the purpose of the questionnaire clearly in the cover letter or in the introduction, in case you are conducting an interview. It is often helpful to have a transition sentence between sections of your instrument to give the respondent some idea of the kinds of questions that are coming. For instance, you might lead into a section on personal material with the following transition: "In this next section of the survey, we'd like to ask you about your personal relationships. Remember, we do not want you to answer any questions if you are uncomfortable doing so."

Guidelines for Question Sequencing

The survey-design business has lots of conventions or rules of thumb. You can use the following suggestions when reviewing your instrument:

- **Start with easy, nonthreatening questions.**
- **Put more difficult, threatening questions near the end.**
- **Never start a mail survey with an open-ended question.**

- For historical demographics, follow chronological order.
- Ask about one topic at a time.
- When switching topics, use a transition.
- Reduce response set (the tendency of respondent to just keep checking the same response).
- For filter or contingency questions, make a flowchart.

7.4f The Golden Rule

You are imposing in the life of your respondents. You are asking for their time, their attention, their trust, and often, for their personal information. Therefore, you should always keep in mind the golden rule of survey research (and, we hope, for the rest of your life as well!): Do unto your respondents as you would have them do unto you! To put this in more practical terms, you should keep the following in mind:

- Thank the respondent at the beginning for allowing you to conduct your study.
- Keep your survey as short as possible—only include what is absolutely necessary.
- Be sensitive to the needs of the respondent.
- Be alert for any sign that the respondent is uncomfortable.
- Thank the respondent at the end for participating.
- Assure the respondent that you will send a copy of the final results—and make sure you do.

7.5 Interviews

Interviews (McCracken, 1988) are among the most challenging and rewarding forms of measurement. They require a personal sensitivity and adaptability as well as the ability to stay within the bounds of the designed protocol. Here, we describe the preparation you typically need to do for an interview study and the process of conducting the interview itself. Keep in mind that the distinction between an interview and a questionnaire is not always clear-cut. Interviewers typically use a type of questionnaire instrument as the script for conducting the interview. It often has both structured and unstructured questions on it. This type of interview questionnaire would also have instructions for the interviewer that are not seen by the respondent and may include space for the interviewer to record any observations about the progress and process of the interview. These features would not be present in a mailed questionnaire.

7.5a The Role of the Interviewer

The interviewer is really the jack-of-all-trades in survey research. The interviewer's role is complex and multifaceted. It includes the following tasks:

- *Locate and enlist cooperation of respondents.* The interviewer has to find the respondent. In door-to-door surveys, this means being able to locate specific addresses. Often, the interviewer has to work at the least desirable times (like immediately after dinner or on weekends) because that's when respondents are most readily available.

- *Motivate respondents to do a good job.* If the interviewer does not take the work seriously, why would the respondent? The interviewer has to be motivated and has to be able to communicate that motivation to the respondent. Often, this means that the interviewer has to be convinced of the importance of the research.
- *Clarify any confusion/concerns.* Interviewers have to be able to think on their feet. Respondents may raise objections or concerns that were not anticipated. The interviewer has to be able to respond candidly and informatively.
- *Observe quality of responses.* Whether the interview is personal or over the phone, the interviewer is in the best position to judge the quality of the information that is being recorded. Even a verbatim transcript will not adequately convey how seriously the respondent took the task, or reveal any gestures or body language that were observed.
- *Conduct a good interview.* Last, and certainly not least, the interviewer has to conduct a good interview! Every interview has a life of its own. Some respondents are motivated and attentive; others are distracted or disinterested. The interviewer also has good or bad days. Assuring a consistently high-quality interview is a challenge that requires constant effort.

7.5b Training the Interviewers

One of the most important aspects of any interview study is the training of the interviewers themselves. In many ways, the interviewers are your measures, and the quality of the results is totally in their hands. Even in small studies involving only a single researcher-interviewer, it is important to organize in detail and rehearse the interviewing process before beginning the formal study.

Here are some of the major topics that you should consider during interviewer training:

- *Describe the entire study.* Interviewers need to know more than simply how to conduct the interview itself. They should learn about the background for the study, previous work that has been done, and why the study is important.
- *State who is the sponsor of the research.* Interviewers need to know who they are working for. They—and their respondents—have a right to know not only what agency or company is conducting the research, but also who is paying for the research.
- *Teach enough about survey research.* While you seldom have the time to teach a full course on survey-research methods, the interviewers need to know enough that they respect the survey method and are motivated. Sometimes it may not be apparent why a question or set of questions was asked in a particular way. The interviewers will need to understand the rationale behind the way you constructed the instrument.
- *Explain the sampling logic and process.* Naive interviewers may not understand why sampling is so important. They may wonder why you go through the difficulty of selecting the sample so carefully. You will have to explain that sampling is the basis for the conclusions that will be reached and for the degree to which your study will be useful.
- *Explain interviewer bias.* Interviewers need to know the many ways they can inadvertently bias the results. They also need to understand why it is important that they not bias the study. This is especially a problem when you are investigating political or moral issues on which people have strongly held convictions. While the interviewers may think they are doing

good for society by slanting results in favor of what they believe, they need to recognize that doing so could jeopardize the entire study in the eyes of others.

- *Walk through the interview.* When you first introduce the interview, it's a good idea to walk through the entire protocol so that the interviewers can get an idea of the various parts or phases and how they interrelate.
- *Explain respondent selection procedures, including the following:*

  - *Reading maps.* It's astonishing how many adults don't know how to follow directions on a map. In personal interviews, interviewers may need to locate respondents spread over a wide geographic area. They often have to navigate by night (respondents tend to be most available in evening hours) in neighborhoods they're not familiar with. Teaching basic map-reading skills and confirming that the interviewers can follow maps is essential (or, in these days, providing interviewers with GPS devices and making sure they know how to use them).
  - *Identifying households.* In many studies, it is impossible in advance to say whether every sample household meets the sampling requirements for the study. In your study, you may want to interview only people who live in single-family homes. It may be impossible to distinguish townhouses and apartment buildings in your sampling frame. The interviewer must know how to identify the appropriate target household.
  - *Identify respondents.* Just as with households, many studies require respondents who meet specific criteria. For instance, your study may require that you speak with a male head of household between the ages of 30 and 40 who has children under 18 living in the same household. It may be impossible to obtain statistics in advance to target such respondents. The interviewer may have to ask a series of filtering questions before determining whether the respondent meets the sampling needs.
  - *Rehearse the interview.* You should probably have several rehearsal sessions with the interview team. You should especially go over filter/contingency questions so the interviewer understands the basic flow of the interview. You might even videotape rehearsal interviews to discuss how the trainees responded in difficult situations. The interviewers should be familiar with the entire interview before ever facing a respondent.
  - *Explain supervision.* In most interview studies, the interviewers will work under the direction of a supervisor. In some contexts, such as university research, the supervisors may be faculty advisors; in others, such as a business context, they may be the boss. To assure the quality of the responses, the supervisor may have to observe a subsample of interviews, listen in on phone interviews, or conduct follow-up assessments of interviews with the respondents. This practice can be threatening to the interviewers. You need to develop an atmosphere in which everyone on the research team— interviewers and supervisors—feel like they're working together toward a common end.
  - *Explain scheduling.* The interviewers have to understand the demands being made on their schedules and why these are important to the study. In some studies, it will be imperative to conduct the entire set of interviews within a certain time period. In most studies, it's important to have the interviewers available when it's convenient for the respondents, not necessarily the interviewer.

7.5c The Interviewer's Kit

It's important that interviewers have all of the materials they need to do a professional job. Usually, you will want to assemble an interviewer kit that can be easily carried and that includes all of the important materials such as following:

- **A professional-looking 3-ring notebook (this might even have the logo of the company or organization conducting the interviews)**
- **Maps**
- **Sufficient copies of the survey instrument**
- **Official identification (preferably a picture ID)**
- **A cover letter from the principal investigator or sponsor**
- **A phone number the respondent can call to verify the interviewer's authenticity**

7.5d Conducting the Interview

Once all the preparation is complete, the interviewers, with their kits in hand, are ready to proceed. It's finally time to do an actual interview. Each interview is unique, like a small work of art (and sometimes the art may not be very good). Each interview has its own ebb and flow—its own pace. To the outsider, an interview looks like a fairly standard, simple, ordinary event; but to the interviewer, it can be filled with special nuances and interpretations that aren't often immediately apparent. Every interview includes some common components. There's the opening, where the interviewer gains entry and establishes the rapport and tone for what follows. There's the middle game, the heart of the process, which consists of the protocol of questions and the improvisations of the probe. Finally, there's the endgame, the wrap-up, during which the interviewer and respondent establish a sense of closure. Whether it's a two-minute phone interview or a personal interview that spans hours, the interview is a bit of theater, a mini-drama that involves real lives in real time.

- *Opening remarks.* In many ways, the interviewer has the same initial problem that a salesperson has. The interviewers have to get the respondents' attention initially for a long enough period that they can sell them on the idea of participating in the study. Many of the remarks here assume an interview that is being conducted at a respondent's residence; but the similarities to other interview contexts should be straightforward.
- *Gaining entry.* The first thing the interviewer must do is gain entry. Several factors can enhance the prospects. Probably the most important factor is initial appearance. The interviewer needs to dress professionally and in a manner that will be comfortable to the respondent. In some contexts, a business suit and briefcase may be appropriate; in others, it may intimidate. The way the interviewers appear initially to the respondent has to communicate some simple messages; that they're trustworthy, honest, and nonthreatening. Cultivating a manner of professional confidence, the sense that the respondent has nothing to worry about because the interviewers know what they're doing, is a difficult skill to teach interviewers and an indispensable skill for achieving initial entry.
- *Doorstep technique.* If the interviewer is standing on the doorstep and someone has opened the door, even if only halfway, the interviewer needs to smile and briefly state why he or she is there. Have your interviewers suggest what they would like the respondent to do. Not ask. Suggest. Instead of saying, "May I come in to do an interview," have them try a more imperative approach like, "I'd like to take a few minutes of your time to interview you for a very important study."

- *Introduction.* If interviewers get this far without having doors slammed in their faces, chances are they will be able to get an interview. Without waiting for the respondent to ask questions, they should introduce themselves. Be sure your interviewers have this part of the process memorized so they can deliver the essential information in twenty to thirty seconds at most. They should state their name and the name of the organization they represent, as well as show their identification badge and the letter that introduces them. You want them to have as legitimate an appearance as possible. If they have a three-ring binder or clipboard with the logo of your organization, they should have it out and visible. They should assume that the respondent will be interested in participating in an important study.

- *Explaining the study.* At this point, the interviewers have been invited to come in. (After all, they're standing there in the cold, holding an assortment of materials, clearly displaying their credentials, and offering the respondent the chance to participate in an interview; to many respondents, it's a rare and exciting event. They are seldom asked their views about anything, and yet they know that important decisions are made all the time based on input from others.) When the respondent has continued to listen long enough, the interviewer needs to explain the study. There are three rules to this critical explanation: 1) Keep it short; 2) Keep it short; and 3) Keep it short! The respondent doesn't have to or want to know all of the nuances of this study, how it came about, how you convinced your thesis committee to buy into it, and so on. Provide the interviewers with a one- or two-sentence description of the study and have them memorize it. No big words. No jargon. No detail. There will be more than enough time for that later. (Interviewers should bring some written materials to leave at the end for that purpose.) Provide a 25-words-or-less description. What the interviewers *should* spend some time on is disclosing fully the purpose of the study, and assuring the respondent that they are interviewing them confidentially and that their participation is voluntary.

- *Asking the questions.* The interviewer has gotten in and established an initial rapport with the respondent. It may be that the respondent was in the middle of doing something when the interviewer arrived and needs a few minutes to finish the phone call or send the kids off to do homework. Then it's time to begin the interview itself. Here are some hints you can give your interviewers:

- *Use questionnaire carefully, but informally.* The interview questionnaire is the interviewer's friend. It was developed with a lot of care and thoughtfulness. While interviewers have to be ready to adapt to the needs of the setting, their first instinct should always be to trust the instrument that was designed. However, they also need to establish a rapport with the respondent. If they bury their faces in the instrument and read the questions, they'll appear unprofessional and disinterested. Reassure them that even though they may be nervous, the respondent is probably even more nervous. Encourage interviewers to memorize the first few questions, so that they need to refer to the instrument only occasionally, using eye contact and a confident manner to set the tone for the interview and help the respondent get comfortable.

- *Ask questions exactly as written.* Sometimes interviewers will think that they could improve on the tone of a question by altering a few words to make it simpler or more friendly. Urge them not to. They should ask the questions as they appear on the instrument. During the training and rehearsals, allow the interviewers to raise any issues they have with the questions. It is important that the interview be as standardized as possible across respondents. (This is true except in certain types of exploratory or interpretivist research where standardization may not be the focus.)

- *Follow the order given*. When interviewers know an interview well, they may see a respondent bring up a topic that they know will come up later in the interview. They may be tempted to jump to that section. Urge them not to. This can cause them to lose their place or omit questions that build a foundation for later questions.
- *Ask every question*. Sometimes interviewers will be tempted to omit a question because they thought they already heard what the respondent will say. Urge them not to assume. For example, let's say you were conducting an interview with college-age women about the topic of date rape. In an earlier question, the respondent mentioned that she knew of a woman on her dormitory floor who had been raped on a date within the past year. A few questions later, the interviewer is supposed to ask, "Do you know of anyone personally who was raped on a date?" Interviewers might figure they already know that the answer is yes and decide to skip the question. Encourage them to say something like, "I know you may have already mentioned this, but do you know of anyone personally who was raped on a date?" At this point, the respondent may say, "Well, in addition to the woman who lived down the hall in my dorm, I know of a friend from high school who experienced date rape." If the interviewer hadn't asked the question, this detail would have remained undiscovered.
- *Don't finish sentences*. Silence is one of the most effective devices for encouraging respondents to talk. If interviewers finish their sentences for them, they imply that what they had to say is transparent or obvious, or that they don't want to give them the time to express themselves in their own language.

7.5e Obtaining Adequate Responses—The Probe

When the respondent gives a brief, cursory answer, your interviewer needs to probe the respondent to elicit a more thoughtful, thorough response. Teach the following probing techniques:

- *The silent probe*. The most effective way to encourage someone to elaborate is to do nothing at all—just pause and wait. This is referred to as the silent probe. It works (at least in certain cultures) because the respondent is uncomfortable with pauses or silence. It suggests to the respondents that the interviewer is waiting, listening for what they will say next.
- *Overt encouragement*. At times, interviewers can encourage the respondent directly. They should try to do so in a way that does not imply approval or disapproval of what the respondent said (that could bias their subsequent results). Overt encouragement could be as simple as saying "uh-huh" or "okay" after the respondent completes a thought.
- *Elaboration*. Interviewers can encourage more information by asking for elaboration. For instance, it is appropriate to ask questions like, "Would you like to elaborate on that?" or "Is there anything else you would like to add?"
- *Ask for clarification*. Sometimes, interviewers can elicit greater detail by asking the respondent to clarify something that was said earlier by saying something like, "A minute ago you were talking about the experience you had in high school. Could you tell me more about that?"
- *Repetition*. This is the old psychotherapist trick. You say something without really saying anything new. For instance, the respondent just described a traumatic childhood experience. The interviewer might say "What I'm hearing

you say is that you found that experience very traumatic" and then pause. The respondent is likely to say something like, "Well, yes, and it affected the rest of my family as well. In fact, my younger sister…."

7.5f Recording the Response

Although we have the capability to record a respondent in audio and/or video, most interview methodologists don't think it's a good idea. Respondents are often uncomfortable when they know their remarks will be recorded word-for-word. They may strain to say things only in a socially acceptable way. Although you would get a more detailed and accurate record, it is likely to be distorted by the process of obtaining it. This may be more of a problem in some situations than in others. It is increasingly common to be told that your conversation may be recorded during a phone interview; and most focus-group methodologies use unobtrusive recording equipment to capture what's being said. However, in general, personal interviews are still best when recorded by the interviewer using pen and paper. Here, we assume the paper-and-pen approach (or the electronic equivalent with a tablet or notebook computer).

- *Record responses immediately.* The interviewers should record responses as they are being stated. This appropriately conveys the idea that they are interested enough in what the respondent is saying to write it down. The interviewers don't have to write down every single word; but you may want them to record certain key phrases or quotes verbatim. Implement a system for distinguishing what the respondent says verbatim from what interviewers are characterizing (use quotation marks, for instance, for verbatim material).
- *Include all probes.* Have your interviewers indicate every single probe that you use. Develop shorthand for different standard probes. Use a clear form for writing them in (for example, place probes in the left margin). Use abbreviations where possible; abbreviations will help interviewers capture more of the discussion. Develop a standardized system (e.g., R = respondent; DK = don't know). If interviewers create an abbreviation on the fly, have them indicate its origin. For instance, if your interviewer decides to abbreviate Spouse with an S, have them make a notation in the right margin saying S = Spouse.

7.5g Concluding the Interview

To bring the interview to closure, have your interviewers remember the following:

- *Thank the respondent.* Don't forget to do this. Even if the respondents were troublesome or uninformative, it is important to be polite and thank them for their time.
- *Tell them when you expect to send results.* You owe it to your respondents to show them what they contributed to and what you learned. Now, they may not want your entire 300-page dissertation. It's common practice to prepare a short, readable, jargon-free summary of interviews to send to the respondents.
- *Don't be brusque or hasty.* Interviewers need to allow for a few minutes of winding-down conversation. The respondent may be interested in how the results will be used. While the interviewers are putting away their materials and packing up to go, have them engage the respondent. Some respondents may want to keep on talking long after the interview is over. Provide your interviewers with a way to cut off the conversation politely and make their exit. For instance, you might have your interviewers say, "I would love to stay to discuss this more with you but, unfortunately, I have another interview appointment I must keep."

- *Immediately after leaving, have the interviewer write down any notes about how the interview went.* Sometimes interviewers will have observations about the interview that they didn't want to write down while they were with the respondent. (Perhaps they noticed the respondent become upset by a question, or detected hostility in a response.) Immediately after the interview, have them go over their notes and make any other comments and observations; but to be sure to distinguish these from the notes made during the interview (by using a different color pen, for instance).

SUMMARY

We covered a lot here. You've learned about the different types of surveys: questionnaires and interviews and how to choose between them. You learned how to construct a questionnaire and address issues of question content, response formats, and question wording and placement. You learned how to train interviewers and the basic steps involved in conducting an interview. Based on this chapter, you should feel pretty confident taking a crack at developing your own survey. There's no substitute for getting experience at designing and implementing a survey project.

Key Terms

dichotomous response format p. 182
double-barreled question p. 185
dual-media surveys p. 174
electronic surveys or e-surveys p. 174
email surveys p. 174
filter or contingency question p. 184
focus group p. 175
group-administered questionnaire p. 173

group interview p. 175
household drop-off survey p. 173
interval-level response format p. 183
Likert-type response scale p. 183
mail survey p. 173
multi-option or multiple-response variable p. 188
nominal response format p. 183
ordinal response format p. 183

personal interview p. 174
point-of-experience survey p. 173
response brackets p. 187
response format p. 187
single-option variable p. 189
structured response format p. 187
survey p. 172
telephone interview p. 175
unstructured response format p. 189
web surveys p. 174

Suggested Websites

The Survey Research Center

http://www.src.isr.umich.edu/
The Survey Research Center at the University of Michigan has been a leader in the field for more than sixty years. The website is rich with resources, including data, publications, and the famous summer training institute.

Gallup

http://www.gallup.com/home.aspx

This company has long been known for political opinion research, and now it maintains a worldwide program of research that allows them to say that they "know more about the attitudes and behaviors of the world's constituents, employees, and customers than any other organization" (from the website, accessed November 20, 2012).

Review Questions

1. Which of the following is *not* a type of survey technique?

a. mail survey
b. group-administered questionnaire
c. telephone interview
d. focus group
(Reference: 7.2)

2. To determine whether your respondent is qualified to answer your survey questions, you might

a. ask simple questions to see if the respondent understands
b. ask open-ended questions to see what the respondent tells you
c. locate data to determine the respondent's knowledge base
d. ask a filter or contingency question
(Reference: 7.3c)

3. Researchers and respondents can both introduce bias and prejudices into the survey process. The bias that makes us want to look good is called

a. social desirability
b. social bias
c. self-esteem
d. faking bad
(Reference: 7.3e)

4. When a survey question offers two possible responses, it is considered _____

a. double-barreled
b. dichotomous
c. bipolar
d. filter or contingency
(Reference: 7.4a)

5. What kind of survey question helps you determine what the next question should be based on the response?

a. double-barreled
b. dichotomous
c. bipolar
d. filter or contingency
(Reference: 7.4a)

6. If a survey question asks about two things in the same item (e.g., "How much do you like dogs and cats?"), it is considered

a. a forced choice question
b. a double-barreled question
c. a loaded question
d. a question that is too specific
(Reference: 7.4b)

7. How can you increase the likelihood that the respondent will respond truthfully to a question?

a. Pose the question in terms of a hypothetical respondent
b. Specify the responses in as much detail as possible
c. Reword the question to appear biased or loaded
d. Instruct the respondent that only truthful responses are acceptable
(Reference: 7.4b)

8. What survey response format would you choose if you wanted to know the types of small electronic appliances the respondent had purchased in the last year?

a. fill-in-the-blank
b. check the answer
c. true or false
d. delete the answer
(Reference: 7.4c)

9. A survey asks respondents to "check all that apply" regarding a list of common household items they have used. Those recorded answers would be considered

a. a single-option variable
b. multioption or multiple-response variables
c. dichotomous variables
d. checklist variables
(Reference: 7.4c)

10. When we ask respondents to state their opinion in written form in a survey, we are asking for what type of response format?

a. double-barreled
b. structured
c. single-option
d. unstructured
(Reference: 7.4c)

11. A dichotomous response format includes two possible options for response—for example, "True" and "False"

a. True
b. False
(Reference: 7.4c)

12. The open-ended comments cards found in many businesses and restaurants are good examples of a structured-response format

a. True
b. False
(Reference: 7.4c)

13. The most important items should always be placed last in a survey so that respondents are sufficiently "warmed up" to the topic

a. True
b. False
(Reference: 7.4e)

14. The "Golden Rule" as applied to survey research means that you should always offer compensation to your respondents

a. True
b. False
(Reference: 7.4f)

15. Most methodologists agree that audio- or videotaping your respondents is not advisable, because most people will be at least a little uncomfortable with being taped and may change their response as a result

a. True
b. False

(Reference: 7.5f)

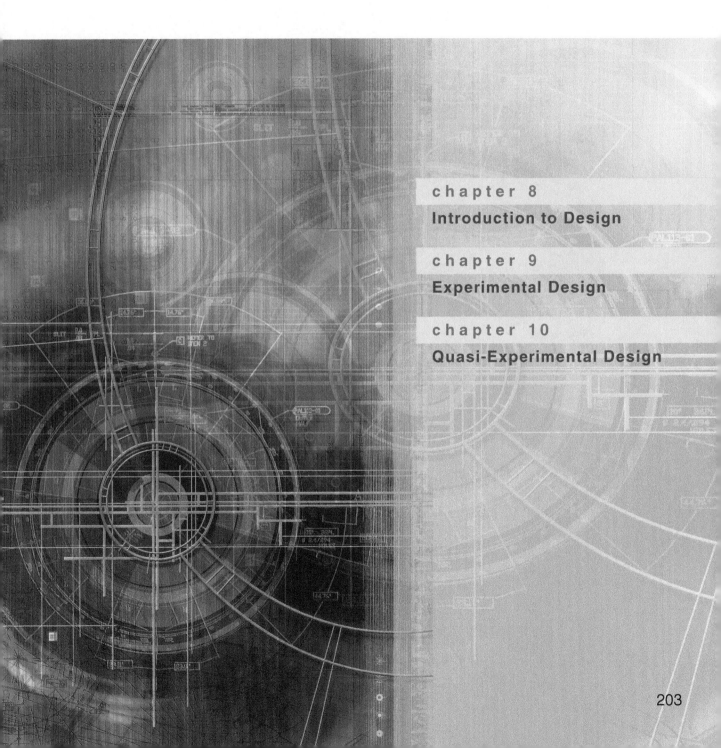

PART 4 Design

203

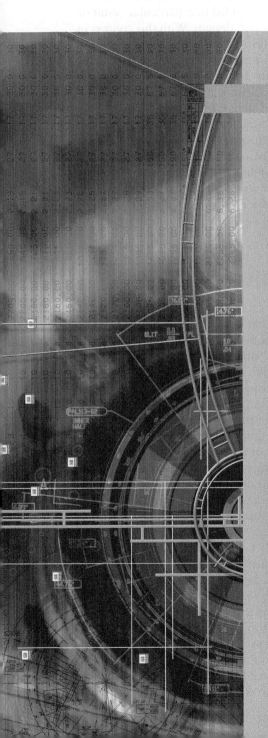

8

Introduction to Design

8.1 Foundations of Design

In many aspects, the research design provides the glue that holds various elements in a research project together. The design is used to structure the research, to show how all of the major parts of the project—the sample, measures, treatments or programs, and methods of assignment—work together to address the central hypothesis of interest. In this chapter, we begin by showing that issues of cause-and-effect (whether a program or intervention led to a particular result or outcome) are closely linked to the choice of research designs. With this context in mind, we discuss internal validity as the first major section of this chapter because it is mainly concerned with the accuracy of causal inferences. We then describe major threats to the internal validity of a study and also present how research designs may be used to address them. Next, we'll show the basic components of research designs and the classification of chief design types. You'll see that an important distinction is made between the experimental designs that use random assignment to groups/programs and the quasi-experimental designs that don't use random assignment. Finally, we will also describe a few simple ways to expand on the basic versions of research designs. Understanding the linkages between different design types, as well as the strengths and weaknesses of each, will help you in selecting the most appropriate research design for your project.

8.2 Research Design and Causality

Many social science research studies address causal questions. Okay, we have to interrupt this right off the bat and tell you that one of our pet peeves is when students misspell causal as casual. We know both words have exactly the same letters. And, we know it's easy to switch the "a" and the "u" when you're typing. And, yes, we know that your spell checker didn't tell you it was spelled wrong. But it's a bit disconcerting, to say the least, to see in the middle of a research paper that you are investigating a *casual* hypothesis! Anyway, getting back on topic, a *cause* is something that leads to some outcome or event. A cause makes something happen. If we want to understand something we are observing, we often ask ourselves what caused it to be the way it is or what caused it to take place.

For example, following the theater shootings during the screening of the Batman movie, *The Dark Knight Rises*, in Aurora, Colorado, it was alleged by many that violence in movies leads to crime in real life. That is, movie violence *causes* real violence. Now, how would you evaluate this connection? What evidence would you need to justify your argument? One option would be to interview a sample of the audience right after they finish watching a particularly violent movie. After the movie, you might ask them how "aggressive" they feel on a scale of 1 to 10. Even if a large percentage of the sample provided a high score on your aggressiveness scale, it may not necessarily mean that the movie was the cause behind their heightened aggressive attitude. It is possible that individuals who generally are a little more aggressive tend to watch such violent movies in the first place, and that's primarily why they score high on aggressiveness scales when you interview them afterwards. But even if they score high on the aggressive attitude scale, it is obvious that relatively few people actually commit violent acts after seeing a film. What if you were to interview individuals before the movie started and then again, after the movie ended? You can then compare their pre- and post- answers to reach some conclusions on the influence of movie violence on their attitudes on aggressiveness. This method might be better than the first one because it allows you to account for individuals' initial attitudes. But it doesn't solve the problem that more violent people might elect to see such a movie in the first place (**Figure 8.1**). This example hints at the complexity of definitively answering a causal research question. The rest of the chapter will provide more of the details on how this can be done through the research design choices you make.

Figure 8.1 What do you think caused the Aurora theater shooting?

In essence, there are many different ways in which you can attempt to answer questions related to causal inference in your research—however, the design you end up employing has significant implications for the credibility and validity of your conclusions. A thoughtful investigator will consider the possible alternative interpretations of results before even conducting the study. This chapter will help you understand how to design a study with strong internal validity, meaning that you have a good chance of producing a credible answer to your causal question.

8.2a Establishing Cause and Effect in Research Design

How do you establish a cause-effect (causal) relationship? More specifically, what features need to be considered in your research design in order to evaluate how well it can support causal conclusions? Generally, you must meet three criteria before you can say that you have evidence for a causal relationship:

1. **Temporal precedence**
2. **Covariation of the cause and effect**
3. **No plausible alternative explanations**

Temporal Precedence

Temporal precedence is basically a fancy term for time order. To establish temporal precedence, you have to show that the cause happened *before* the effect. For example, if the Aurora theater shootings happened before the release of any of the three Batman movies, we probably won't be asking this particular causal question at all. Sounds easy, huh? Of course, you say, the cause has to happen before the effect! Did you ever hear of an effect happening before its cause? Consider a classic example from economics where temporal precedence may be harder to disentangle: does inflation cause unemployment? It certainly seems plausible that as inflation rises, more employers find that to meet costs they have to lay off employees. So it is possible that inflation could, at least partially, be a cause for unemployment. However, both inflation and employment rates are occurring together on an ongoing basis. Is it possible that fluctuations in employment can affect

temporal precedence The criterion for establishing a causal relationship that holds that the cause must occur before the effect.

Figure 8.2 The difficulty in establishing temporal precedence in a causal relationship.

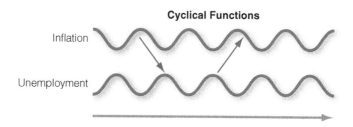

inflation? If employment in the workforce increases (lower unemployment) there is likely to be more demand for goods, which would tend to drive up the prices (that is, inflate them), at least until supply can catch up. It is then also possible for employment to influence inflation. So which is the cause and which the effect, inflation or unemployment? What comes first, the chicken or the egg? It turns out that this kind of cyclical situation involves ongoing processes that interact where both may cause and, in turn, be affected by, the other (see **Figure 8.2**). It is hard to establish a single causal relationship in this situation because it's likely that the variables influence each other causally.

Covariation of the Cause and Effect

What does this mean? Before you can show that you have a causal relationship you have to show that you have a relationship at all. For instance, consider the logical argument:

If X then Y

If not X then not Y

If you observe that whenever X is present, Y is also present, and whenever X is absent, Y is also absent, you have demonstrated that there is an associational relationship between X and Y. We don't know about you, but sometimes we find it's not easy to think about X's and Y's. Let's put this same logic in terms of an example:

If Batman movie then violence

If no Batman movie then no violence

Or, in terms of a treatment and its effect: whenever you administer a program, you observe the outcome, but when you don't give the program, you don't observe the outcome. This provides evidence that the program (independent variable) and outcome (dependent variable) are related. Notice, however, that this logical argument doesn't provide evidence that the program *caused* the outcome; perhaps some other factor that is always present with the program caused the outcome rather than the program itself. The relationships described so far are simple relationships—that is, they simply link the observed effect to its plausible cause. Sometimes you want to know whether different amounts of the program lead to different amounts of the outcome—a continuous relationship:

If more of the program then more of the outcome

If less of the program then less of the outcome

No Plausible Alternative Explanations

As noted in the previous section, just because you show there's a relationship doesn't mean that it's a causal one. Suppose you measure the association between children's shoe size and their reading ability. In almost all cases you will find a positive relationship. But it's not the case that increasing shoe size causes an increase in reading ability or vice-versa! It's just that the child is getting older, and that influences both the shoe size as well as his/her reading ability. This should remind you of the popular adage, "Correlation does not mean causation!"

It's always possible that some other variable or factor is causing the outcome. This is sometimes referred to as the third-variable or missing-variable problem, and it's at the heart of internal validity, as we will see later in the chapter. What are some of the possible alternative explanations? Later in this chapter, we'll discuss the threats to internal validity, and you'll see that each threat describes a type of alternative explanation.

Let's look at a common example of some of the issues that come up when we look at a causal relationship in applied social research. Assume that you provide a smoking cessation program to a group of people. You measure their smoking behavior before you begin the program (to establish a baseline), you give the group the program, and then you measure the group's behavior afterwards in a posttest. You see a marked decrease in their smoking habits, which you would like to infer is caused by your program. What are the plausible alternative explanations for the results?

One alternative explanation is that it's not your program that caused the decrease but some other specific, say historical, event. For instance, your anti-smoking campaign did not cause the reduction in smoking; but rather the U.S. Surgeon General's latest report was issued between the time you gave your pre- and posttests, and that caused the effect. How do you rule this out with your research design? One of the simplest ways would be to incorporate the use of a **control group**—a group comparable in every way possible to your program group except that they didn't receive your program (**Figure 8.3**). However, both groups experienced the Surgeon General's latest report. If you find that your program group shows a reduction in smoking greater than the control group, then this can't be due to the Surgeon General's report because both groups experienced that. You have effectively "ruled out" the Surgeon General's report as a plausible alternative explanation for the outcome, thereby strengthening the credibility of your causal conclusions.

In most applied social research that involves interventions or evaluating programs, temporal precedence is not a difficult criterion to meet because you administer the program before you measure effects. Establishing covariation or association is also relatively simple, because you have some control over the program and can set things up so you have some people who get it and some who don't (and then you can measure what happens if there is X and conversely, if there is no X). Typically the most difficult criterion to meet is the third—ruling out alternative explanations for the observed effect. Research designs give us a mechanism that helps us rule out some of the most important alternative explanations. That is why research design is so important when it comes to causality and also why it is intimately linked to the idea of internal validity.

8.2b Internal Validity

Internal validity is the approximate truth about inferences regarding cause-effect or causal relationships. Now, that was a mouthful! What does it really mean? Establishing internal validity is just a technical way of saying that you have ruled

McCracken, Theresa/CartoonStock

Figure 8.3 Control groups are important in research designs for helping address alternative explanations for a causal relationship. Okay, so researchers do have a sense of humor! No, there really aren't "out-of-control" groups in research designs, or at least we hope not.

control group In a comparative research design, like an experimental or quasi-experimental design, the control or comparison group is compared or contrasted with a group that receives the program or intervention of interest.

In *this* study...

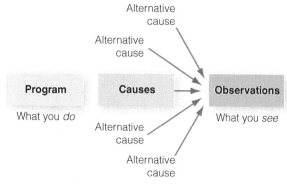

Figure 8.4 A schematic for understanding internal validity.

out plausible alternative explanations and have thus shown that indeed a causal relationship likely exists between your program and the intended outcome. The key question in internal validity is whether observed changes can be attributed to your program or intervention (the cause) and not to other possible causes (described as alternative explanations for the outcome) (see **Figure 8.4**). And, how do you go about establishing internal validity? One of the major ways is through the research design you apply. That is, a specific research design can address particular issues and help you eliminate the alternative explanations for your results, thereby enhancing the internal validity of your causal conclusions.

So, you can't really understand why certain designs might be better than others until you understand how the designs try to address particular internal validity concerns. If we only taught you the different standard research designs, you could probably learn to apply them. But you would never really understand them, and you certainly would not know how to adapt them to address the specific concerns in your study. If you learn about internal validity and how threats to it shape research design, you'll not only be able to use a few standard designs, but you'll also better understand how to tailor designs to your specific needs.

One of the things that's most important to grasp about internal validity is that it is only relevant to the specific study in question. That is, you can think of internal validity as a zero-generalizability issue. Unlike external validity, internal validity doesn't care about whether the cause-effect relationship would be found in any other context outside the current study. All that internal validity means is that you have sufficient evidence that what you did in your immediate study (the program) caused what you observed (the outcome) to happen. Additionally, internal validity is also not concerned with whether the program you implemented was what you wanted to implement or whether what you observed was what you wanted to observe; those are construct validity concerns (see Chapter 5, "Introduction to Measurement").

In fact, it is possible to have internal validity in a study and not have construct or external validity. For instance, imagine a study in which you are looking at the effects of a new computerized tutoring program on math performance in first-grade students. Imagine that the tutoring is unique in that it has a heavy computer-game component, and you think that will really improve math performance. Finally, imagine that you were wrong. (Hard, isn't it?) Let's say that it turns out that math performance did improve, and that it was because of something you did in the study, but that it had nothing to do with the computer program. For example, perhaps what caused the improvement was the individual attention that the adult who introduced the program to the child gave them; the computer program didn't make any difference. This study would have internal validity because *something* you did affected *something* that you observed. (You did cause *something* to happen.) The study would not have construct validity because the label "computer-math program" does not accurately describe the actual cause or the treatment. A more accurate label for the cause that actually influenced the effect might be "personal adult attention."

So, the key issue in internal validity is a causal one. Now, we'll discuss the different **threats to internal validity**—the kinds of arguments your critics will raise

threats to internal validity Any factor that leads you to draw an incorrect conclusion that your treatment or program causes the outcome.

when you try to conclude that your program caused the outcome. For convenience, we divide the threats to internal validity into three categories. The first involves the **single-group threats**—challenges that may be raised when you are only studying a single group that receives your program. The second consists of the **multiple-group threats**—challenges that are likely to be raised when you have several groups in your study (typically this involves a program and a control/comparison group). Finally, we'll discuss the **social threats to internal validity**—threats that arise because social research is conducted in real-world human contexts where people will react to not only what affects them, but also to what is happening to others around them.

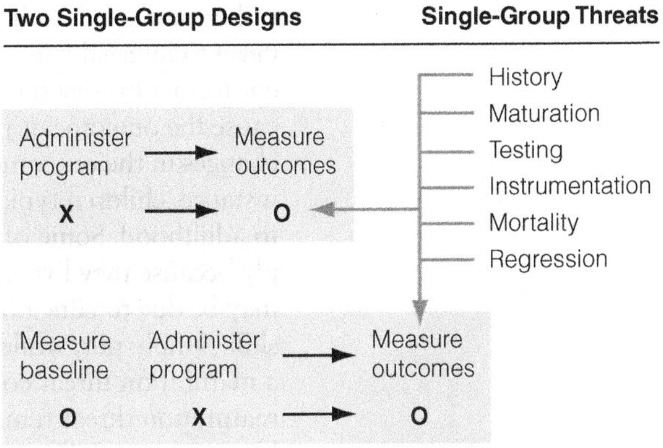

Figure 8.5 Single-group threats to internal validity.

Single-Group Threats

What is meant by a single-group threat? Let's consider two single-group designs and then consider the threats that are most relevant with respect to internal validity. Consider the top design in **Figure 8.5**. Here, a group of people receives your program and afterwards is given a posttest. This is known as a **posttest-only single-group design**. In the bottom part of the figure, you see a **pretest-posttest, single-group design**. In this case, the participants are first tested on a pretest or baseline measure, then provided the program or treatment, and finally tested again with a posttest.

To help make this a bit more concrete, let's imagine that you are studying the effects of a program on an outcome measure. In the post-only design (the first case), you would give the program and then give a posttest. You might choose not to give them a baseline measure because you have reason to believe that, prior to the program, they would have no basis for performance on the outcome. It wouldn't make sense to pretest them if they have no idea what you're testing them about. In the pre-post design (the second case), you are not willing to assume that your group members have no prior knowledge. You measure the baseline to determine where the participants start out on the outcome. In fact, your pretest determines the extent of prior knowledge, and it thereby allows you to separate its effect from that of the program in influencing your final outcome. You might hypothesize that the change or gain from pretest to posttest is due to your program.

With this scenario in mind, consider what would happen if you observe a certain level of posttest performance or a change or gain from pretest to posttest. You want to conclude that the observed outcome is due to your program. How could you be wrong? Here are some of the threats to internal validity that your critics might raise:

- *History threat:* When some event other than your program occurs at the same time and affects the outcome, we say there is a **history threat** to internal validity. It's not your program that caused the outcome; it's something else, some other event that occurred between the pretest and the posttest. We refer to this other event as an "historical" one, because it is something that happens during the course of your study; it, like your program, is an event in history, so to speak. In the example above, the Surgeon General's announcement occurring at the same time as the implementation of your antismoking program is a type of history threat.
- *Maturation threat:* It is possible to see the desired outcome level or change in outcome over time due to normal maturation or internal growth in the outcome,

single-group threats A threat to internal validity that occurs in a study that uses only a single program or treatment group and no comparison or control.

multiple-group threats An internal validity threat that occurs in studies that use multiple groups, for instance, a program and a comparison group.

social threats to internal validity Threats to internal validity that arise because social research is conducted in real-world human contexts where people will react to not only what affects them, but also to what is happening to others around them.

posttest-only single-group design A design that has an intervention and a posttest where measurement of outcomes is only done within a single group of program recipients.

pretest-posttest Any research design that uses measurement both before and after an intervention or program.

single-group design Any research design that involves only a single group in measuring outcomes.

history threat A threat to internal validity that occurs when some historical event affects your study outcome.

rather than as a result of your program. In such a case, we say that a **maturation threat** to internal validity is present. How is this maturation explanation different from a history threat? In general, if a specific event or chain of events could cause the outcome, it is a history threat. A maturation threat consists of all of the changes in the outcome that would occur naturally with the passage of time. For instance, children typically improve in language skills as they move from infancy to adulthood. Some of this improvement is natural maturation that occurs simply because they live in a context with others who also use language. Some of it may be due to education or to programs designed to provide them with greater skill. Simply put, while a history threat is due to an external event or occurrence, a maturation threat comes into play due to internal changes in a person. The maturation threat reminds us that we cannot attribute all change in an outcome like language skill to a program or intervention we might be studying. Even without any intervention, kids would improve in language naturally.

- *Testing threat:* This threat only occurs in a pre-post design. What if the pretest made some of the respondents more aware of what you are measuring? Perhaps it sensitized or primed them for the program, so that when they received it they were ready for it in a way that they would not have been without the pretest. This is what is meant by a **testing threat** to internal validity; it is taking the pretest, not participating in the program, that affects how the respondents do on the posttest. In other words, here the pretest or the baseline testing experience becomes the treatment rather than the actual program.

- *Instrumentation threat:* Like the testing threat, the **instrumentation threat** operates only in a pretest-posttest situation. What if the change from pretest to posttest is due not to your program but rather to a change in the way you measured the outcome? For example, as compared to the pretest, you might use a different measurement instrument in the posttest—say, a computer-based survey as compared to the paper-and-pencil survey used at baseline. Perhaps part or all of any pre-post gain is attributable to the change in instrument, rather than to your program. It is possible that a part of your sample is uncomfortable with using computers, and that is reflected in their answers on the posttest. Instrumentation threats are also likely when the instrument is essentially a human observer. The observers may get tired over time or bored with taking down the responses. Conversely, they might get better at making the observations as they practice more. In either event, the change in instrumentation, not the program, leads to the outcome.

- *Mortality threat:* A **mortality threat** to internal validity doesn't mean that people in your study are dying (although if they are, it would certainly be a mortality threat!). Mortality is used metaphorically here. It refers to people dropping out of your study. In a sense, the dropouts "die" in terms of your data collection. What's wrong with mortality in a study? Let's assume that in your program you have a considerable number of people who drop out between pretest and posttest. Assume also that the participants who are dropping out had lower pretest scores than those who remained. If you look at the average gain from pretest to posttest, using all of the scores available to you on each occasion, you would include these low-pretest, subsequent dropouts in the pretest and not in the posttest. You'd be dropping out the potential low scorers from the posttest, or you'd be artificially inflating the posttest average over what it would have been if no respondents had dropped out. Thus, in cases of a mortality threat, the differences between the treatment and control group at the end of the study may be due to differences in those who remained in each group, rather than due to the effects of the treatment. For this reason, mortality threats are also often known as "selective attrition."

maturation threat A threat to internal validity that occurs as a result of natural maturation between pre- and post-measurement.

testing threat A threat to internal validity that occurs when taking the pretest affects how participants do on the posttest.

instrumentation threat A threat to internal validity that arises when the instruments (or observers) used on the posttest and the pretest differ.

mortality threat A threat to internal validity that occurs because a significant number of participants drop out.

You won't necessarily solve this problem by comparing pre-post averages for only those who stayed in the study, either. This subsample would certainly not be representative of the original sample, much less of the population. Furthermore, you know that because of regression threats (see the following section) these participants may appear to actually do worse on the posttest, simply as an artifact of the nonrandom dropout or mortality in your study. When mortality is a threat, the researcher can often estimate the degree of the threat by comparing the dropout group against the non-dropout group on pretest measures. If there are no major differences, it may be more reasonable to assume that mortality was happening across the entire sample and is not biasing results greatly. However, if the pretest differences are large, you must be concerned about the potential biasing effects of mortality. Note, in order to make this comparison, you need to first know which groups were selectively dropping off and why.

- *Regression threat:* A **regression threat** to internal validity, also known as a **regression artifact** or **regression to the mean**, is a statistical phenomenon that falsely makes it appear that your group changed to be more like the overall population between the pretest and posttest. If your group was above the population average on the pretest, they will look like they lost ground on the posttest. If they were below the population average on the pretest, they will falsely appear to improve. Regression occurs whenever you have a nonrandom sample from a population and two measures that are imperfectly correlated. Okay, for most of you, that explanation was probably gibberish. We'll try again.

  Assume that your two measures are a pretest and posttest. You can certainly bet these aren't perfectly correlated with each other—they would only be perfectly correlated if the highest scorer on the pretest was the highest on the posttest, the next-highest pretest scorer was second-highest on the posttest, and so on down the line. Furthermore, assume that your sample consists of pretest scorers who, on average, score lower than the population average. The regression threat means that the pretest average for the group in your study will appear to increase or improve (relative to the overall population) even if you don't do anything to them, even if you never give them a treatment. Regression is a confusing threat to understand at first. We like to think about it as the *you can only go up (or down) from here* phenomenon. For example, if you include in your program only the participants who constituted the lowest 10 percent on the pretest, what are the chances that they would constitute exactly the lowest 10 percent on the posttest? Virtually none. That could only happen in the unlikely event that the pretest and posttest are perfectly correlated. While most of them would score low on the posttest, it is very unlikely that the lowest 10 percent on the pretest would be exactly the lowest 10 percent on the posttest. This purely statistical phenomenon is what we mean by a regression threat. The key idea is that there will always be some participants who perform very well (or poorly) on the pretest due to chance. Because of this chance factor, it is unlikely that they will perform just as well (or just as poorly) on the posttest. So, it might seem that they are getting worse (or better) in the posttest as compared to the baseline, suggesting that your treatment had an effect. But, this is just regression to the mean.

  How do you deal with these single-group threats to internal validity? Although you can rule out threats in several ways, one of the most common approaches to ruling them out is through your research design. For instance, instead of doing a single-group study, you could incorporate a second group, typically called a control group. In such a study you would have two groups: one receives your program and the other one doesn't. In fact, the only difference between these groups should be

regression threat or **regression artifact** or **regression to the mean** A statistical phenomenon that causes a group's average performance on one measure to regress toward or appear closer to the mean of that measure, more than anticipated or predicted. Regression occurs whenever you have a nonrandom sample from a population and two measures that are imperfectly correlated. A regression threat will bias your estimate of the group's posttest performance and can lead to incorrect causal inferences.

the presence or absence of the program. When that's the case, the control group would experience all the same history and maturation threats, have the same testing and instrumentation issues, and have similar rates of mortality and regression to the mean. In other words, a good control group is one of the most effective ways to rule out all of the single-group threats to internal validity. Of course, when you add a control group, you no longer have a single-group design, and you still have to deal with two major types of threats to internal validity: the multiple-group threats to internal validity and the social threats to internal validity.

Multiple-Group Threats

A multiple-group design typically involves at least two groups and before-after measurements. Most often, one group receives the program or treatment while the other does not. The group that does not receive the treatment is called the control or the comparison group. However, sometimes one group gets the program and the other gets either no program or another program you would like to compare. In this case, you would be comparing two programs for their relative outcomes. Typically, you would construct a multiple-group design so that you could compare the groups directly. In such designs, the key internal validity issue is the degree to which the groups are comparable before the study. If they are comparable, and the only difference between them is the program, posttest differences can be attributed to the program; but that's a big *if*. If the groups aren't comparable to begin with, you won't know how much of the outcome to attribute to your program, versus the initial differences between groups.

There really is only one multiple-group threat to internal validity: that the groups were not comparable before the study. This threat is called a **selection bias** or **selection threat**. A selection threat is *any* factor that leads to differences between the groups at the start of the study. Whenever you suspect that outcomes differ between groups, not because of your program but because of prior group differences, you are suspecting a selection bias. Although the term selection bias is used as the general category for all prior differences, when you know specifically what the group difference is, you usually hyphenate it with the selection term. The multiple-group selection threats directly parallel the single-group threats. For instance, whereas history is a single-group threat, selection-history is its multiple-group analogue.

Here are the major multiple-group threats to internal validity:

- *Selection-History threat:* A **selection-history threat** is any other event that occurs between pretest and posttest that influences the treatment and control groups differently. Because this is a selection threat, the groups by definition differ in some way at the beginning of the study. Because it's a history threat, the way the groups differ is with respect to their reactions to certain historical events that may occur between the pretest and the posttest. Consider the example of the antismoking program you are implementing in order to increase smoking cessation rates. Suppose that the treatment group is low-income and the control group is higher-income. Further, imagine that the Surgeon General's report is released during program implementation. Because low-income individuals are also less likely to have health insurance, a large percentage of them may quit smoking after hearing about harmful health consequences from the Surgeon General's report. On the other hand, higher-income people are more likely to have health insurance, and thus may be more willing to take a chance with their health. So, we might see the treatment group (low-income folks) quitting, while the control group (high-income individuals) may not change their behavior at all. You might think that your program is making a difference, when all that's happening is

selection threat or **selection bias** Any factor other than the program that leads to pretest differences between groups.

selection-history threat A threat to internal validity that results from any other event that occurs between pretest and posttest that the groups experience differently.

two groups experiencing a relevant event differentially between the pretest and posttest—all of this mainly because the two groups were very different to begin with! **Figure 8.6** shows a poster from the Great American Smokeout, a national campaign designed to encourage the population to stop smoking on a single day. What are some of the validity threats that researchers evaluating the program should try to take into account?

Figure 8.6 Can you think of ways to determine if the Great American Smokeout really helps people quit smoking?

- *Selection-Maturation threat:* A **selection-maturation threat** results from differential rates of normal growth between the pretest and posttest for the groups. It's important to distinguish between history and maturation threats. In general, history refers to a discrete event or series of events, whereas maturation implies the normal, ongoing developmental process that takes place. In any case, if the groups are maturing at different rates with respect to the outcome, you cannot assume that posttest differences are due to your program; they may be selection-maturation effects.

- *Selection-Testing threat:* A **selection-testing threat** occurs when a *differential* effect of taking the pretest exists between groups on the posttest. Perhaps the test primed the respondents in each group differently or they may have learned differentially from the pretest. In these cases, an observed posttest difference can't be attributed to the program. It could be the result of selection testing.

- *Selection-Instrumentation threat:* A **selection-instrumentation threat** refers to any differential change in the test used for each group from pretest to posttest. In other words, the test changes differently for the two groups. Perhaps the instrument is based on observers who rate outcomes for the two groups. What if the program group observers, for example, become better at doing the observations while, over time, the comparison group observers become fatigued and bored. Differences on the posttest could easily be due to this differential instrumentation—selection-instrumentation—and not due to the program.

- *Selection-Mortality threat:* A **selection-mortality threat** arises when there is differential nonrandom dropout between pretest and posttest. Different types of participants might drop out of each group, or more may drop out of one than the other. Posttest differences might then be due to the different types of dropouts—the selection-mortality—and not to the program.

- *Selection-Regression threat:* Finally, a **selection-regression threat** occurs when there are different rates of regression to the mean in the two groups. This might happen if one group was more extreme on the pretest than the other. It may be that the program group had a disproportionate number of low pretest scorers. Its mean regresses a greater distance toward the overall population posttest mean, and its group members appear to gain more than their comparison-group counterparts. This is not a real program gain; it's a selection-regression artifact.

When you move from a single-group to a multiple-group study, what advantages do you gain from the rather significant investment in a second group? If the second group is a control group and is comparable to the program group, you can rule out entirely the single-group threats to internal validity because those threats will all be reflected in the comparison group and cannot explain why posttest group differences would occur. But the key is that the groups must be comparable. How can you possibly hope to create two groups that are truly comparable? The best way to do that is to randomly assign persons in your sample into the two groups—that is, you conduct a randomized or true experiment (see the discussion of experimental designs in Chapter 9).

selection-maturation threat A threat to internal validity that arises from any differential rates of normal growth between pretest and posttest for the groups.

selection-testing threat A threat to internal validity that occurs when a differential effect of taking the pretest exists between groups on the posttest.

selection-instrumentation threat A threat to internal validity that results from differential changes in the test used for each group from pretest to posttest.

selection-mortality threat A threat to internal validity that arises when there is differential nonrandom dropout between pretest and posttest.

selection-regression threat A threat to internal validity that occurs when there are different rates of regression to the mean in the two groups between pretest and posttest.

However, in many applied research settings you can't randomly assign, either because of logistical or ethical factors. In those cases, you typically try to assign two groups nonrandomly so that they are as equivalent as you can make them. You might, for instance, have one preexisting group assigned to the program and another to the comparison group. In this case, you would hope the two are equivalent, and you may even have reasons to believe that they are. Nonetheless, they may not be equivalent. Therefore, you have to take extra care to look for preexisting differences and adjust for them in the analysis, because you did not use a procedure like random assignment to at least ensure that they are probabilistically equivalent. If you measure the groups on a pretest, you can examine whether they appear to be similar on key measures before the study begins and make some judgment about the plausibility that a selection bias or preexisting difference exists. There are also ways to adjust statistically for preexisting differences between groups if they are present, although these procedures are notoriously assumption-laden and fairly complex. Research designs that look like randomized or true experiments (they have multiple groups and pre-post measurement) but use nonrandom assignment to choose the groups are called **quasi-experimental designs** (see the discussion of quasi-experimental designs in Chapter 10, "Quasi-Experimental Design").

Even if you move to a multiple-group design and have confidence that your groups are comparable, you cannot assume that you have assured strong internal validity. There would still remain a number of social threats to internal validity that arise from the human interaction in applied social research, and you will need to address them.

Social Interaction Threats

Applied social research is a human activity. The results of such research are affected by the human interactions involved. The social threats to internal validity refer to the social pressures in the research context that can lead to posttest differences not directly caused by the treatment itself. Most of these threats occur because the various groups (for example, program and comparison) or key people involved in carrying out the research are aware of each other's existence and of the role they play in the research project, or are in contact with one another. Many of these threats can be minimized by *isolating the two groups from each other,* but this leads to other problems. For example, it's often hard within organizational or institutional constraints to both randomly assign and then subsequently isolate the groups from each other; this is likely to reduce generalizability or external validity (see external validity in Chapter 4). Here are the major social interaction threats to internal validity:

- *Diffusion or imitation of treatment:* **Diffusion or imitation of treatment** occurs when a comparison group learns about the program either directly or indirectly from program group participants. Participants from different groups within the same organization might share experiences when they meet casually. Or, comparison group participants, seeing what the program group is getting, might set up their own experience to try to imitate that of the program group. In either case, if the diffusion or imitation affects the posttest performance of the comparison group, it can jeopardize your ability to assess whether your program is causing the outcome. For instance, in a school context, children from different groups within the same school might share experiences during the lunch hour. Or, comparison group students, seeing what the program group is getting, might set up their own experience to try to imitate that of the program group. Notice that this threat to validity tends to equalize the outcomes between groups, reducing the chance of seeing a program effect even if there is one.

quasi-experimental design Research designs that have several of the key features of randomized experimental designs, such as pre-post measurement and treatment-control group comparisons, but lack random assignment groups.

diffusion or imitation of treatment A social threat to internal validity that occurs because a comparison group learns about the program either directly or indirectly from program group participants.

- *Compensatory rivalry:* In the **compensatory rivalry** case, the comparison group knows what the program group is getting and develops a competitive attitude with the program group. The participants in the comparison group might see the program the other group is getting and feel jealous. This could lead them to compete with the program group just to show how well they can do. Sometimes, in contexts like these, the participants are even encouraged by well-meaning administrators to compete with each other. (Although this might make organizational sense as a motivation for the participants in both groups to work harder, it works against the ability of researchers to see the effects of their program.) If the rivalry between groups affects posttest performance, it could make it more difficult to detect the effects of the program. As with diffusion and imitation, this threat generally equalizes the posttest performance across groups, increasing the chance that you won't see a program effect, even if the program is effective.

- *Resentful demoralization:* **Resentful demoralization** is almost the opposite of compensatory rivalry. Here, participants in the comparison group know what the program group is getting and, instead of developing a rivalry, the group members become discouraged or angry and give up. Or, if the program group is assigned to an especially difficult or uncomfortable condition, they can rebel in the form of resentful demoralization. Unlike the previous two threats, this one is likely to exaggerate posttest differences between groups, making your program look even more effective than it actually is.

- *Compensatory equalization of treatment:* **Compensatory equalization of treatment** is the only threat of the four that primarily involves the people who help manage the research context rather than the participants themselves. When program and comparison group participants are aware of each other's conditions, they might wish they were in the other group (depending on the perceived desirability of the program, it could work either way). They might pressure the administrators to have them reassigned to the other group. The administrators may begin to feel that the allocation of goods to the groups is not fair and may compensate one group for the perceived advantage of the other. If the program is a desirable one, you can bet that the participants assigned to the comparison group will pressure the decision makers to equalize the situation. Perhaps these decision makers will give the comparison group something to compensate for their not getting the desirable program. If these compensating programs equalize the groups on posttest performance, they will tend to work against your detecting an effective program even when it does work.

As long as people engage in applied social research, you have to deal with the realities of human interaction and its effect on the research process. The threats described here can often be minimized by constructing multiple groups that are unaware of each other (for example, a program group from one organization or department and a comparison group from another) or by training administrators in the importance of preserving group membership and not instituting equalizing programs. However, researchers will never be able to eliminate entirely the possibility that human interactions are making it more difficult to assess cause-effect relationships.

Other Ways to Rule out Threats to Internal Validity

Before we get to constructing designs themselves, it would help to think about methods *other than* the research design that may be used to rule out internal validity threats. Previously, we discussed that good research designs minimize the plausible alternative explanations for the hypothesized cause-effect relationship.

compensatory rivalry A social threat to internal validity that occurs when one group knows the program another group is getting and, because of that, develops a competitive attitude with the other group. Often it is the comparison group knowing that the program group is receiving a desirable program (e.g., new computers) that generates the rivalry.

resentful demoralization A social threat to internal validity that occurs when the comparison group knows what the program group is getting, and instead of developing a rivalry, control group members become discouraged or angry and give up.

compensatory equalization of treatment A social threat to internal validity that occurs when the control group is given a program or treatment (usually by a well-meaning third party) designed to make up for or "compensate" for the treatment the program group gets. By equalizing the group's experiences, this threat diminishes the researcher's ability to detect the program effect.

But research design is not the only way you can rule out threats. Here, we present four alternative ways to minimize any threats to any type of validity:

- *By argument*—The most straightforward way to rule out a potential threat to validity is simply to make an argument that the threat in question is not a reasonable one. Such an argument may be made either *a priori* or *a posteriori*. (That's before the fact or after the fact, for those of you who never studied dead languages.) The former is usually more convincing than the latter. For example, depending on the situation, you might argue that an instrumentation threat is not likely because the same test is used for pre- and posttest measurement and did not involve observers who might improve or change over time. In most cases, ruling out a potential threat to validity by argument alone is weaker than using other approaches (see below). As a result, the most plausible threats in a study should not, except in unusual cases, be ruled out by argument alone.

- *By measurement or observation*—In some cases it is possible to rule out a threat by measuring it and demonstrating that either it does not occur at all or occurs so minimally as to not be a strong alternative explanation for the cause-effect relationship. Consider, for example, a study of the effects of an advertising campaign on subsequent sales of a particular product. In such a study, history (meaning the occurrence of events other than the advertising campaign that might lead to an increased desire to purchase the product) would be a plausible alternative explanation. For example, a change in the local economy, the removal of a competing product from the market, or similar events could cause an increase in product sales. You can attempt to minimize such threats by measuring local economic indicators and the availability and sales of competing products. If there are no changes in these measures coincident with the onset of the advertising campaign, these threats would be considerably minimized. Similarly, if you are studying the effects of special mathematics training on math achievement scores of children, it might be useful to observe everyday classroom behavior to verify that students were not receiving any other math training in addition to what was provided in the study.

- *By analysis*—Statistical analysis offers you several ways to rule out alternative explanations. For instance, you could study the plausibility of an attrition or mortality threat by conducting a two-way factorial experimental design (see Chapter 9 on "Experimental Design"). One factor in this study might be the original treatment group designations (for example, program vs. comparison group), while the other factor would be attrition (for example, dropout vs. non-dropout group). The dependent measure could be the pretest or other available pre-program measures. A main effect on the attrition factor would be indicative of a threat to external validity or generalizability; whereas an interaction between group and attrition factors would point to a possible threat to internal validity. Where both effects occur, it is reasonable to infer that there is a threat to both internal and external validity. The plausibility of alternative explanations might also be minimized using covariance analysis (see the discussion on covariance analysis in Chapter 12). For example, in a study of the effects of workfare programs on social welfare caseloads, one plausible alternative explanation might be the status of local economic conditions. Here, it might be possible to construct a measure of economic conditions and include that measure as a covariate in the statistical analysis, in order to adjust for or remove this factor from the outcome scores. You must be careful when using covariance adjustments of this type; perfect covariates do not exist in most social research, and the use of imperfect covariates does not completely adjust for potential alternative explanations. Nevertheless, demonstrating that treatment effects occur, even after adjusting on a number of good covariates, strengthens causal assertions.

- *By preventive action*—When you anticipate potential threats, you can often rule them out by taking some type of preventive action. For example, if the program is a desirable one, it is likely that the comparison group would feel jealous or demoralized. You can take several actions to minimize the effects of these attitudes, including offering the program to the comparison group upon completion of the study or using program and comparison groups that have little opportunity for contact and communication. In addition, you can use auditing methods and quality control to track potential experimental dropouts or to ensure the standardization of measurement.

These four methods for reducing the threats to internal validity should not be considered mutually exclusive. They may also be used in addition to the research design itself to strengthen the case for internal validity. For example, the inclusion of measurements designed to minimize threats to validity will obviously be related to the design structure and is likely to be a factor in the analysis. A good research plan should, wherever possible, make use of multiple methods for reducing threats. In general, reducing a particular threat by design or preventive action is stronger than by using one of the other three approaches. Choosing which strategy to use for any particular threat is complex and depends at least on the cost of the strategy and on the potential seriousness of the threat.

8.3 Developing a Research Design

Here is where the rubber meets the road. Research design can be thought of as the *structure* of research; the research design tells you how all the elements in a research project fit together. First, we'll take a look at the different elements or pieces in a design and then show you how you might think about putting them together to create a tailored design to address your own research context. We will also describe the notations that researchers often use to describe a design. These notations enable them to summarize a complex design structure efficiently.

Every research design should include the following elements:

- *Time*—A causal relationship, by its very nature, implies that some time has elapsed between the occurrence of the cause and the consequent effect. Although for some phenomena, the elapsed time is measured in microseconds and is therefore unnoticeable to a casual observer, you normally assume that the cause and effect in social science arenas do not occur simultaneously. In design notation, you indicate the passage of time horizontally. You place the symbol used to indicate the presumed cause to the left of the symbol that indicates measurement of the effect. Thus, as you read from left to right in design notation, you are reading across time. Complex designs might involve a lengthy sequence of observations and programs or treatments across time.
- *Treatments or programs:* These are symbolized with an X in design notation. The X can refer to a simple intervention (such as a one-time surgical technique) or to a complex hodgepodge program (such as an employment-training program). Usually, a control or a comparison group (that does not receive a treatment) has no symbol for the treatment (although some notational systems use X+ and X− to indicate the treatment and control, respectively). As with

observations, you can use subscripts to distinguish different programs or program variations. For example, when multiple programs or treatments are being studied using the same design, you keep the programs distinct by using subscripts such as X_1 or X_2.

- **Observations or measures:** These are symbolized by an O in design notation. An O can refer to a single measure (a measure of body weight), a single instrument with multiple items (a 10-item, self-esteem scale), a complex multipart instrument (a survey), or a whole battery of tests or measures given out on one occasion. If you need to distinguish among specific measures, you can use subscripts with the O, as in O_1, O_2, and so on. If the same measurement or observation is taken at every point in time in a design, a simple O is sufficient.

- **Groups or individuals**—This describes the individuals who participate in various conditions. Typically, there will be one or more program and comparison groups. Each group in a design is given its own line in the design structure. For instance, if the design notation has three lines, the design contains three groups. Furthermore, the manner in which groups are assigned to the conditions can be indicated by an appropriate symbol at the beginning of each line. In these cases, R represents a randomly assigned group, N depicts a nonrandomly assigned group (a nonequivalent group or cohort), and C indicates that the group was assigned using a cutoff score on a measurement.

It's always easier to explain design notation through examples than it is to describe it in words. **Figure 8.7** shows the design notation for a pretest-posttest (or before-after) treatment versus comparison group randomized experimental design. Let's go through each of the parts of the design. There are two lines in the notation, so you should realize that the study has two groups. There are four Os in the notation: two on each line and two for each group. When the Os are stacked vertically on top of each other, it means they are collected at the same time. In the notation, the two Os taken before (to the left of) the treatment are the pretest. The two Os taken after the treatment is given are the posttest. The R at the beginning of each line signifies that the two groups are randomly assigned (making it an experimental design as described in Chapter 9).

The design is a treatment-versus-comparison-group one, because the top line (treatment group) has an X, whereas the bottom line (control group) does not. You

Figure 8.7 A detailed example of design notation.

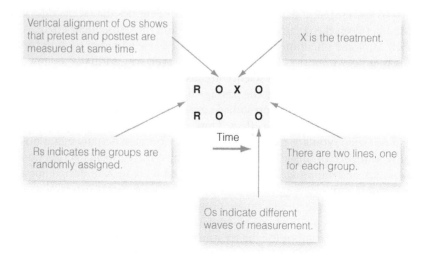

should be able to see why many students call this type of notation the tic-tac-toe method of design notation; there are lots of Xs and Os! Sometimes you have to use more than simply the Os or Xs. **Figure 8.8** shows the identical research design with some subscripting of the Os. What does this mean? Because all of the Os have a subscript of 1, some measure or set of measures was collected for both groups on both occasions. But the design also has two Os with a subscript of 2, both taken at the posttest. This means that some additional measure or set of measures was collected *only* at the posttest.

With this simple set of rules for describing a research design in notational form, you can concisely explain complex design structures. Additionally, using a notation helps to show common design substructures across different designs that you might not recognize as easily without the notation. For example, three of the designs we'll show you are named the analysis of covariance randomized experimental design (see Chapter 9), the nonequivalent groups design, and the regression discontinuity design (see Chapter 10). From their titles and even their written descriptions, it is not immediately apparent what these designs have in common. But, one glance at their design notation immediately shows that all three are similar in that they have two groups and before-and-after measurement, and that the only key difference between the designs is in their group assignment method.

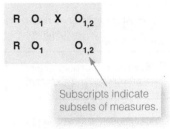

$$R \quad O_1 \quad X \quad O_{1,2}$$
$$R \quad O_1 \qquad O_{1,2}$$

Subscripts indicate subsets of measures.

Figure 8.8 An example of a design notation that includes subscripts.

8.4 Types of Designs

What are the different major types of research designs? You can classify designs into a simple threefold classification by asking some key questions, as shown in **Figure 8.9**.

First, does the design use random assignment to groups? (Don't forget that random *assignment* is not the same thing as random selection of a sample from a population!) If random assignment is used, the design is a randomized experiment or *true* experiment. If random assignment is not used, ask a second question: Does the design use *either* multiple groups or multiple waves of measurement? If the answer is yes, label it a quasi-experimental design. If no, call it a nonexperimental design.

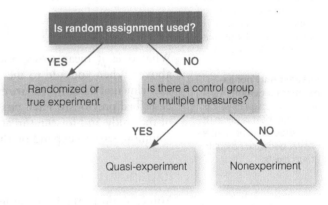

Figure 8.9 Basic questions that distinguish the major types of designs.

This threefold classification is especially useful for describing the design with respect to internal validity. A randomized experiment generally is the strongest of the three designs when your interest is in establishing a cause-effect relationship. A nonexperiment is generally the weakest in this respect. We have to hasten to add here that we don't mean that a nonexperiment is the weakest of the three designs *overall*, but only with respect to internal validity or causal assessment. In fact, the simplest form of nonexperiment is a one-shot survey design that consists of nothing but a single observation O. This is probably one of the most common forms of research and, for some research questions—especially descriptive ones—is clearly a strong design. When we say that the nonexperiment is the weakest with respect to internal validity, all we mean is that it isn't a particularly good method for assessing cause-effect relationships that you think might exist between a program and its outcomes.

Posttest-only randomized experiment	R	X	O
	R		O
Pretest-posttest nonequivalent groups quasi-experiment	N	O	X O
	N	O	O
Posttest-only nonexperiment		X	O

Figure 8.10 Notational examples of each of the three major classes of research design.

posttest-only randomized experiment An experiment in which the groups are randomly assigned and receive only a posttest.

pre-post nonequivalent groups quasi-experiment A two-group quasi-experimental design structured like a randomized experiment, but lacking random assignment to group.

posttest-only nonexperimental design A research design in which only a posttest is given. It is referred to as nonexperimental because no control group exists.

Figure 8.11 A double-pretest, single-group design created by expanding across time.

O X O   O X O

Figure 8.12 An add-remove design formed by expanding program and observation elements over time.

To illustrate the different types of designs, consider one of each in design notation as shown in **Figure 8.10**. The first design is a **posttest-only randomized experiment**. You can tell it's a randomized experiment because it has an R at the beginning of each line, indicating random assignment (see Chapter 9 for a detailed discussion). The second design is a **pre-post nonequivalent groups quasi-experiment** (see Chapter 10 for a detailed discussion). You know it's not a randomized experiment because random assignment wasn't used. Additionally you know it's not a nonexperiment because both multiple groups and multiple waves of measurement exist. That means it must be a quasi-experiment. You add the label *nonequivalent* because in this design you do not explicitly control the assignment and the groups may be nonequivalent or not similar to each other. Finally, you see a **posttest-only nonexperimental design**. You might use this design if you want to study the effects of a natural disaster like a flood or tornado and you want to do so by interviewing survivors. Notice that in this design, you don't have a comparison group (for example, you didn't interview in a town down the road that didn't have the tornado to see what differences the tornado caused) and you don't have multiple waves of measurement (a pre-tornado level of how people in the ravaged town were doing before the disaster). Does it make sense to do the nonexperimental study? Of course! You could gain valuable information from well-conducted post-disaster interviews. However, you may have a hard time establishing which of the things you observed are due to the disaster rather than to other factors like the peculiarities of the town or pre-disaster characteristics.

8.4a Expanding on Basic Designs

What does it mean to expand on the basic versions of the designs described above? When you add to the most basic design (the X → O posttest-only nonexperimental design), you are essentially expanding one of the four basic elements described previously. Each possible expansion has implications both for the cost of the study and for the threats that might be ruled out. Here are the four most common ways to expand on this simple design.

Expanding Across Time

You can add to the basic design by including additional observations either before or after the program, or by adding or removing the program or different programs. For example, you might add one or more pre-program measurements and achieve the design shown in **Figure 8.11**.

The addition of such pretests provides a baseline that, for instance, helps to assess the potential of a maturation or testing threat. Similarly, you could add additional post-program measures, which would be useful for determining whether an immediate program effect decays over time, or whether there is a lag in time between the initiation of the program and the occurrence of an effect. You might also add and remove the program over time, as shown in **Figure 8.12**.

Expanding Across Programs

You have just seen that you can expand the program by adding it or removing it across time. Another way to expand the program would be to divide it into different levels of treatment. For example, in a study of the effect of a novel drug on subsequent behavior, you might use more than one dosage of the drug (see the design notation in **Figure 8.13**).

This is a common strategy in a sensitivity or parametric study where the primary focus is on understanding the effects obtained at various program levels. In a similar manner, you might expand the program by varying specific components of it across groups, which might be useful if you wanted to study different modes of the delivery of the program, different sets of program materials, and the like. Finally, you can expand the program by using theoretically polarized or opposite treatments. A comparison group is one example of such a polarization. Another might involve use of a second program that you expect will have an opposite effect on the outcome measures. A strategy of this sort provides evidence that the outcome measure is sensitive enough to differentiate between different programs.

Figure 8.13 A two-treatment design formed by expanding across programs.

Expanding Across Observations

At any point in time in a research design, it is usually desirable to collect multiple or redundant measurements. For example, you might add a number of measures that are similar to each other to determine whether their results converge. Or, you might want to add measurements that theoretically should not be affected by the program in question to demonstrate that the program discriminates between effects. Strategies of this type are useful for achieving convergent and discriminant validity of measures, as discussed in the Chapter 5, "Introduction to Measurement.". Another way to expand the observations is by proxy measurements (see the discussion of proxy pretest designs in Chapter 10, "Quasi-Experimental Design"). Finally, you might also expand the observations through the use of "recollected" or "retrospective" measures. Again, if you were conducting a study and had neglected to administer a pretest or desired information in addition to the pretest information, you might ask participants to recall how they felt or behaved prior to the study and use this information as an additional measure. Different measurement approaches obviously yield data of different quality. What is advocated here is the use of multiple measurements rather than reliance on only a single strategy.

Figure 8.14 The basic pre-post randomized experimental design.

Expanding Across Groups

Often, it will be to your advantage to add additional groups to a design to rule out specific threats to validity. For example, consider the pre-post two-group randomized experimental design in **Figure 8.14**.

If this design were implemented within a single institution where members of the two groups were in contact with each other, one might expect intergroup communication, group rivalry, or demoralization of a group denied a desirable treatment or given an undesirable one, to pose threats to the validity of the causal inference. These are all social desirability threats. In such a case, you might add an additional group from a similar but different institution that consists of persons unaware of the original two groups (see **Figure 8.15**).

In a similar manner, whenever you use nonequivalent groups, it is usually advantageous to have multiple replications of each group. The use of multiple nonequivalent groups helps minimize the potential of a particular selection bias affecting the results. In some cases, it may be desirable to include the "norm group" as an additional group in the design. Norming group averages are available for most standardized achievement tests, for example, and might comprise an additional nonequivalent control group. You can also use cohort groups in a number of ways. For example, you might use a single measure of a cohort group to help rule out a testing threat (see **Figure 8.16**).

Figure 8.15 A randomized experiment expanded with a nonequivalent control group.

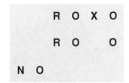

Figure 8.16 A randomized experiment expanded with a nonequivalent group to help rule out a testing threat.

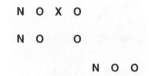

Figure 8.17 A nonequivalent group design expanded with an additional nonequivalent group.

In this design, the randomized groups might be sixth graders from the same school year, and the cohort might be the entire sixth grade from the previous academic year. This cohort group did not take the pretest, and if it is similar to the randomly selected control group, it would provide evidence for or against the notion that taking the pretest had an effect on posttest scores. You might also use pre-post cohort groups (see **Figure 8.17**). Here, the treatment group consists of sixth graders, the first comparison group of seventh graders in the same year, and the second comparison group consists of the following year's sixth graders (the fifth graders during the study year). As you will see below and in the chapters that follow, strategies of this sort are particularly useful in nonequivalent designs where selection bias is a potential problem and where routinely collected institutional data is available (**Figure 8.18**).

Figure 8.18 What are some of the ways that different classes of students might not be equivalent for comparison in a study?

SUMMARY

Research design helps you put together all of the disparate pieces of your research project: the participants or sample, the measures, and the data analysis. This chapter showed that research design is intimately connected with the topic of internal validity, because the type of research design you construct determines whether you can address causal questions, such as whether your treatment or program made a difference on outcome measures. There are three major types of problems—threats to validity—that occur when trying to assure internal validity. Single-group threats occur when you have only a single program group in your study. Researchers typically try to avoid single-group threats by using a comparison group, but this leads to multiple-group threats (selection threats) when the groups are not comparable. Since all social research involves human interaction, you must also be concerned about social threats to internal validity that can make your groups perform differently but that are unrelated to the treatment or program. Research designs can get somewhat complicated. To keep them straight and describe them succinctly, researchers use design notation that describes the design in abstract form.

Key Terms

compensatory equalization of treatment p. 217
compensatory rivalry p. 217
control group p. 209
diffusion or imitation of treatment p. 216
history threat p. 211
instrumentation threat p. 212
maturation threat p. 212
mortality threat p. 212
multiple-group threats p. 211
posttest-only nonexperimental design p. 222
posttest-only randomized experiment p. 222

posttest-only single-group design p. 211
pre-post nonequivalent groups quasi-experiment p. 222
pretest-posttest p. 211
quasi-experimental design p. 216
regression artifact p. 213
regression threat p. 213
regression to the mean p. 213
resentful demoralization p. 217
selection bias p. 214
selection threat p. 214
selection-history threat p. 214
selection-instrumentation threat p. 215

selection-maturation threat p. 215
selection-mortality threat p. 215
selection-regression threat p. 215
selection-testing threat p. 215
single-group design p. 211
single-group threat p. 211
social threats to internal validity p. 211
temporal precedence p. 207
testing threat p. 212
threats to internal validity p. 210

Suggested Websites

Centre for Evidence-Based Medicine

www.cebm.net.

The Centre for Evidence-Based Medicine at the University of Oxford promotes evidence-based health care. Here you can see the application of design principles in health research and health policy. Be sure to check out the "EBM tools" section, where you'll find support for all steps in the research process, including the design considerations in this chapter.

The Claremont Graduate University Debate on Causality

http://ccdl.libraries.claremont.edu/cdm/singleitem/collection/lap/id/68

The Claremont Graduate University debate on Causality in Program Evaluation and Applied Research. The participants were Mark Lipsey and Michael Scriven and the debate was moderated by Stewart Donaldson. They discussed their perspectives on quantitative and qualitative design approaches to questions of causality.

Review Questions

1. What term reflects the accuracy of any cause-effect relationship you might infer as a result of a study?

a. generalizability
b. conclusion validity
c. internal validity
d. construct validity
(Reference: 8.1)

2. Which of the following is a condition that must be met in order to establish evidence for a cause-effect relationship between a treatment and an outcome?

a. Changes in the presumed cause must not be related to changes in the presumed effect.
b. The presumed cause must occur after the effect is measured.

c. There are no plausible explanations for the presumed effect other than your hypothesized cause.
d. You are able to demonstrate that you have both discrimination and convergence of the presumed cause and effect.
(Reference: 8.2a)

3. Which of the following is *not* a way to minimize threats to the validity of research?

a. Establish a convincing argument against alternative explanations.
b. Establish a working relationship with individuals who share your hypothesis.
c. Rule out alternative explanations using statistical analyses.
d. Rule out alternative explanations by adding preventive features to the research design.
(Reference: 8.2b)

4. If the results of a study determine that an intervention worked, though not for the reason anticipated, the study would have ____ validity but not ____ validity.

a. internal, construct
b. construct, internal
c. construct, causal
d. conclusion, construct
(Reference: 8.2b)

5. Regression toward the mean is a phenomenon that affects a(n) ___ score.

a. individual
b. group
c. statistical
d. variable
(Reference: 8.2b)

6. Which of the following strategies would help rule out social threats to internal validity?

a. expanding across time
b. expanding across programs
c. expanding across groups
d. expanding across observations
(Reference: 8.2b)

7. Which of the following strategies would help achieve convergent and discriminant validity of measures?

a. expanding across time
b. expanding across programs

c. expanding across groups
d. expanding across observations
(Reference: 8.4a)

8. When we study whether differences in attrition are related to changes observed in a treatment and comparison group, we are attempting to rule out a _____threat to internal validity.

a. social
b. maturation
c. mortality
d. testing
(Reference: 8.2b1)

9. In simplest terms, a good research design does which of the following?

a. guarantees statistical significance
b. minimizes the need for further studies
c. minimizes the plausible alternative explanations for the hypothesized cause-effect relationship
d. includes multiple measures of every theoretically related construct
(Reference: 8.3)

10. Redundancy in research design is a good thing because

a. if you can include more measures, groups, and so on, you will need a larger budget, and larger grants are helpful in the researcher's career development.
b. including multiple measures, for example, can minimize bias due to a particular form of measurement.
c. as in Murphy's Law, what can go wrong probably will go wrong, and redundancy in design is the best way to prepare for this.
d. Actually, redundancy is a bad thing in research design.
(Reference: 8.4a)

11. Generally speaking, the best strategies to use in attempting to control threats to validity are to try preventing them in the first place or by design.

a. True
b. False
(Reference: 8.2b)

12. The purpose of overexpanding the components of the basic factorial design is to make a study more publishable.

a. True
b. False
(Reference: 8.4a)

13. Covariation and causation mean the same thing.

a. True
b. False
(Reference: 8.2a)

14. Controlling threats to validity on the basis of a good argument done in an *a priori* manner is usually sufficient to account for nearly all threats to validity.

a. True
b. False
(Reference: 8.2)

15. It is necessary to have at least two time points in research that attempts to study a causal relationship.

a. True
b. False
(Reference: 8.2a)

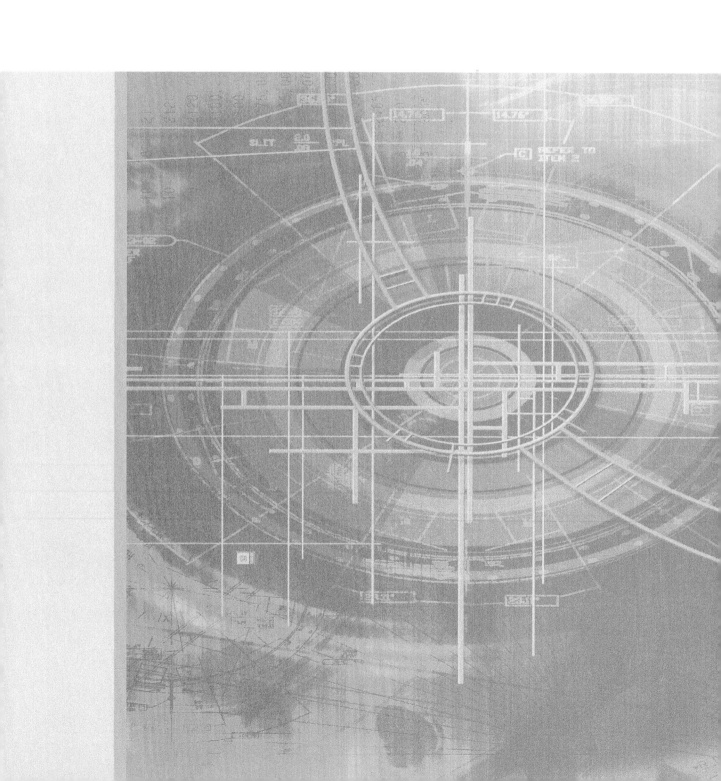

9

Experimental Design

9.1 Foundations of Experimental Design

As a college student, you are probably familiar with being invited to participate in an "experiment." In many cases, participation also gets you prizes, cash, or gift certificates. What is so valuable about experiments that researchers are willing to offer generous compensation in return for your participation? The answer to this question gets at the heart of this chapter. Experimental designs are touted by many as the most "rigorous" of all research designs. In fact, they are often labeled as the "gold standard" against which all other designs are judged. From a theoretical perspective, they probably are the strongest of designs, but we have to be careful to understand what is meant by that. If you can implement an experimental design well (and that is a big *if* indeed), the experiment is probably the strongest design with respect to internal validity (see Chapter 8, "Design"). We hope to show you why that is the case. However, we also want to convey some of the challenges involved in carrying out an experiment and some of the weaknesses of this approach. It is one of the strongest tools we have when we want to establish a cause-and-effect relationship between a program or intervention and an outcome. Nevertheless, like everything else in research, it's not perfect!

This chapter begins by tracing the historical roots of experimental design. It goes on to discuss the characteristics of the design that make it strong in internal validity. We show that a key distinguishing feature of experimental design—random assignment to group—depends on the idea of probabilistic equivalence. We then try to head off one of the biggest sources of confusion to most students—the distinction between *random selection* and *random assignment*. After this, we present a way to classify the different experimental designs based on whether they are "signal enhancers" or "noise reducers." Each type of design is presented in turn. We conclude the chapter by emphasizing the limitations of the experimental design.

9.2 Introduction: The Origins of Experimental Design

Experimental design is sometimes considered to have its roots, so to speak, in the agricultural experiments of the early twentieth century. The famous statistician, Sir Ronald Fisher, is popularly credited with its invention (Armitage, 2003) (Figure 9.1). So, how did the experimental design come about? Around 1915, the Rothamsted experimental station in England (a major center for agricultural research) was involved in an important endeavor—scientists were trying to study which combinations of newly developed artificial fertilizer treatments (Okay, no crude jokes here, please!) improved agricultural yield. However, even after years (90 years, to be precise) of data collection, researchers were having difficulty distinguishing the effects of fertilizers from those of other factors, such as soil quality, types of drainage, and weather conditions, in causing changes to crop productivity (Salsburg, 2001). That is, there was no way to pull apart the effects of all these different factors, including those of the fertilizers, on agricultural yield—a problem that Fisher later termed "confounding."

Figure 9.1 Sir Ronald Fisher, credited with developing the randomized experimental design.

A. Barrington Brown/Science Source

random assignment The process of assigning your sample into two or more subgroups by chance. Procedures for random assignment can vary from flipping a coin to using a table of random numbers to using the random-number capability built into a computer.

The breakthrough came in 1919 with the appointment of Fisher at the experimental station. Fisher recommended a simple but revolutionary solution—he suggested that the researchers divide the large farm into smaller plots and then randomly assign different fertilizers to different areas (Salsburg, 2001). He demonstrated that *randomness* had the ability to untangle the effect of a particular intervention from all of the other potential confounding factors. In effect, the intervention or the treatment was forced to be "independent" of all other factors by **random assignment** to conditions.

Even though Fisher is officially credited with inventing the randomized experimental design, there is evidence that significant developments in this area had already begun in the field of educational research. As early as 1901, American psychologists Edward L. Thorndike (**Figure 9.2**) and Robert S. Woodworth had identified the need for employing a control group in conducting experiments to study outcomes like "learning" and "intelligence." In 1923 (two years before Fisher published his first book), William McCall, who was a student of Thorndike's at Columbia University, detailed the "rotational experiment" (known now as the Latin-square design). This method was a close cousin of Fisher's randomized design—it involved a systematic rotation of educational programs, where different programs were provided to different children and were later switched in order to tease out the effect of the intervention (Campbell & Stanley, 1963). This suggests that the idea of a randomized experiment can be traced back to at least two different disciplines.

In recent years, randomized experiments have gained great prominence in many fields, but perhaps in none as much as in the field of medicine. Its use there was initiated by Sir Austin Bradford Hill, an English epidemiologist who, in 1948, conducted

Science Source

Figure 9.2 E. L. Thorndike did foundational work on randomized experimental designs.

**"There's a flaw in your experimental design.
All the mice are scorpios."**

Aaron Bacall/The New Yorker Collection/Cartoon Bank

Figure 9.3 The complexities of experimental design.

the first "clinical trial" by using streptomycin in the treatment of pulmonary tuberculosis. The streptomycin trial was followed by numerous other drug trials in England. Shortly thereafter, in 1955, the foundational randomized clinical trial was organized by Jonas Salk in Pittsburgh to test the polio vaccine. This was a massive effort that was unparalleled in scale—around 1.8 million children were enrolled through the logistical efforts of thousands of volunteers. The trial demonstrated successful results that firmly established the vaccine as an effective public health intervention against polio. Since then, randomized clinical trials have evolved as a standard procedure in the evaluation of all new treatments in medicine.

9.2a Distinguishing Features of Experimental Design

With this history in mind, we can begin to examine the theoretical strengths of the method. As mentioned earlier, experimental designs are usually considered the strongest of all designs in terms of internal validity (see the discussion on internal validity in Chapter 8, "Design"). Why? Recall that internal validity is at the center of all causal or cause-effect conclusions or inferences. When you want to determine whether some program or treatment causes some outcome or outcomes to occur, you are interested in having strong internal validity (**Figure 9.3**). Essentially, you want to assess the proposition:

If X, then Y.

Or, in everyday terms:

If a particular program is given, then a certain outcome occurs.

Unfortunately, it's not enough to show that when the program or treatment occurs, the expected outcome also happens. That is because there are almost always many other reasons besides the program that might account for why you observed the outcome. For example, as described in the case of the Rothamsted experimental station, factors other than a fertilizer intervention could have led to changes in agricultural productivity. To show that there is a causal relationship, you have to simultaneously address the two propositions:

If X, then Y
and
If not X, then not Y.

Or, once again more informally:

If the program is given, then the outcome occurs
and
If the program is **not** given, then the outcome does **not** occur.

If you are able to provide evidence for both of these propositions, you've in effect isolated the program from all of the other potential causes of the outcome. Going back to the Rothamsted example, establishing both propositions would mean that you have teased out the effect of fertilizers from that of other factors such as soil quality and amount of rainfall. In other words, you've shown that when the program is present, the outcome occurs (crops grow better) and when it's not present, the outcome doesn't occur (they grow at normal levels). Simply stated, you've provided evidence that the implemented program has *caused* the observed outcome to occur, assuming there are no other plausible threats to internal validity (as described in Chapter 8).

There is another way to think about causation. Imagine a fork in the road. Down one path, you implement the program and observe the outcome. Down the other path, you don't implement the program and the usual non-program outcome occurs. But, can you take both paths in the road in the same study? How can you be in two places at once? Ideally, what you want is to have the same conditions—the same people, context, time, and so on—and see whether, when the program or treatment is given, you get the outcome, and when the program is not given, you don't. Obviously, you can never achieve this hypothetical situation because if you give the program to a group of people, you can't simultaneously "not give it"! So, how do you solve this apparent dilemma?

As discussed in the previous section, Fisher and others answered this problem by thinking about it a little differently. What if you could create two groups or contexts that are as similar as you can possibly make them? If you could be confident that the two situations are as comparable as possible, you could administer your program in one (and see whether the outcome occurs) and not administer the program in the other (and see whether the outcome doesn't occur). In other words, setting up comparable groups affords us the luxury of taking both forks in the road simultaneously. You can have your cake and eat it, too, so to speak!

Because random assignment facilitates the creation of two comparable groups, experimental designs are said to be the strongest in terms of establishing internal validity. In the simplest type of experiment, you create two groups that are equivalent to each other. One group (the **program** or **treatment group**) gets the program and the other group (the **comparison group**) does not. In all other respects, the groups are treated in the same manner. On average, they have similar people who live in similar contexts, have similar backgrounds, and so on. Now, if you observe differences in outcomes between these two groups, the differences must be due to the only thing that differs between them—that is, one group received the program and the other didn't.

Okay, so the next question is, how do experiments create two equivalent groups? The approach used in experimental design is to assign people randomly from a common pool of people into the two groups. The experiment relies on this idea of random assignment to groups as the basis for obtaining two similar groups. If you remember, we discussed this method of random assignment above as Fisher's big idea.

Note that, even with random assignment, you never expect the groups you create to be exactly the same on any given measure. That is, even if randomly assigned, it is impossible to have two groups that are 100 percent alike because, after all, they are made up of different people. So, where does this leave us? The answer is that you rely on the idea of probability and assume that the two groups are **probabilistically equivalent** or equivalent within known probabilistic ranges.

What do we mean by the term probabilistic equivalence, and why is it important to experimental design? Well, to begin with, we certainly *don't* mean that two groups are equal to each other. When you deal with human beings, it is impossible to say that any two individuals or groups are equal or equivalent. Clearly, the important term in the phrase is *probabilistic*. This means that the type of equivalence you have is based on the notion of probabilities. In more concrete terms, probabilistic equivalence means that you know *perfectly* the odds of finding a difference between the two groups on any measure you observe. Notice that it doesn't mean that the two groups will be equal. It just means that you know the

program group In a comparative research design, like an experimental or quasi-experimental design, the program or treatment group receives the program of interest and is usually contrasted with a no-treatment comparison or control group or a group receiving another treatment.

treatment group In a comparative research design, like an experimental or quasi-experimental design, the program or treatment group receives the program of interest and is usually contrasted with a no-treatment comparison or control group or a group receiving another treatment.

comparison group In a comparative research design, like an experimental or quasi-experimental design, the control or comparison group is compared or contrasted with a group that receives the program or intervention of interest.

probabilistically equivalent The notion that two groups, if measured infinitely, would on average perform identically. Note that two groups that are probabilistically equivalent would seldom obtain the exact same average score.

Figure 9.4 Probabilistic equivalence does not mean that two randomly selected groups will obtain the exact same average score.

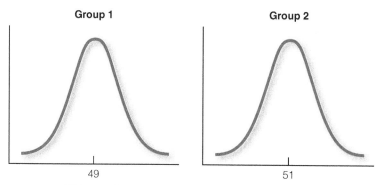

Group 1

Group 2

49

51

With α = .05, we expect that we will observe a pretest difference 5 times out of 100.

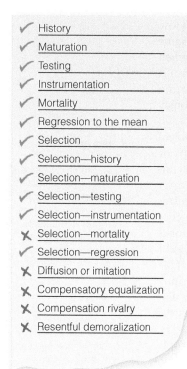

✔ History

✔ Maturation

✔ Testing

✔ Instrumentation

✔ Mortality

✔ Regression to the mean

✔ Selection

✔ Selection—history

✔ Selection—maturation

✔ Selection—testing

✔ Selection—instrumentation

✘ Selection—mortality

✔ Selection—regression

✘ Diffusion or imitation

✘ Compensatory equalization

✘ Compensation rivalry

✘ Resentful demoralization

Figure 9.5 Threats to internal validity for the posttest-only, randomized experimental design.

odds that they won't be equal. **Figure 9.4** shows two groups, one with a mean or average score of 49 on some measure, and the other with a mean of 51. Could these two groups be probabilistically equivalent even though their averages are different? Certainly!

When you randomly assign to groups, you assure probabilistic equivalence. In other words, when you assign subjects to groups randomly, groups can only differ due to chance. If, by chance, the groups differ on one variable, you have no reason to believe that they will automatically be different on any other. Even if you find that the groups differ on a pretest, you have no reason to suspect that they will differ on a posttest. Why? Because their pretest difference had to be due to chance! So, when you randomly assign, you are able to assume that the groups are equivalent. You don't expect them to be equal; but you can expect them to be probabilistically equal. Specifically, you can be confident that 95 out of 100 times, the pretest difference between groups will not be significant (the same probability idea is behind the 95% Confidence Interval and the 5% Type I or alpha error rate).

9.2b Experimental Design and Threats to Internal Validity

So far, we have established that experimental design offers two important advantages over other designs. First, they rely on having two groups, where one group serves the function of a control group. Second, random assignment ensures that the two groups are equivalent to each other (at least probabilistically) before a treatment is administered to any one group. Now, you might be thinking to yourself, how exactly do these two features lead to a high degree of internal validity? Put another way, can we think of the specific threats to internal validity that these features help address?

Figure 9.5 indicates (with a check mark) that the experimental design is strong against all the single-group threats to internal validity. Why? Because it's not a single group design! (This is where having a control group helps.) It's also strong against the multiple-group threats, except for selection mortality. For instance, it's strong against selection-history and selection-maturation threats because the two groups are created to be probabilistically equivalent to begin with. In addition, as will be described below, because experiments don't require use of a pretest, the method is also strong against selection-testing and selection-instrumentation threats when it doesn't use repeated (pre-post) measurement. However, the selection-mortality threat can be a problem if there are differential rates of dropouts in the two groups. This could result if the treatment or program is a noxious or negative one (such as a painful medical procedure like chemotherapy) or alternatively, if the control group condition is painful or difficult to tolerate. Note that an experimental design is susceptible to all of the social threats to internal validity. Because the design requires random assignment, in institutional settings such as schools it is more likely that research participants (students) would be aware of each other and of the conditions to which you have assigned them. This and other such limitations of experimental design will be discussed in detail at the end of this chapter. If you need a refresher on the distinctions between the various validity threats, it might help to review section 8-3b in Chapter 8.

R	X	O
R		O

Figure 9.6 Notation for the basic two-group posttest-only randomized experimental design.

9.2c Design Notation for a Two-Group Experimental Design

The simplest of all experimental designs is the **two-group posttest-only randomized experiment** (see **Figure 9.6**). Despite its simple structure, it is one of the best research designs for assessing cause-effect relationships. In terms of design notation, it has two lines—one for each group—with an R at the beginning of each line to indicate that the groups were randomly assigned.

One group gets the treatment or program (the X) and the other group is the comparison group and doesn't get the program. (Note that you could alternatively have the comparison (control) group receive the standard or typical treatment, in which case this study would be a relative comparison.)

A key point highlighted by Figure 9.6 is that a pretest is not required for this design. Usually, you include a pretest to determine whether groups are comparable prior to the program. However, because this design uses random assignment, you can assume that the two groups are probabilistically equivalent to begin with and the pretest is not required (although you'll see with covariance designs later in this chapter that a pretest may still be desirable in an experimental design).

In this design, you are most interested in determining whether the two groups are different after the program. Typically, you measure the groups on one or more measures (the "Os" in the notation) and you compare them by testing for the differences between the means using a *t*-test (if you have just two groups) or one-way Analysis of Variance (ANOVA) (if you have two or more groups). This kind of analysis enables you to make a decision about whether any difference you observed between the groups is likely to be due to chance or not. The details of conducting such an analysis are discussed in Chapter 12.

9.2d Difference between Random Selection and Assignment

While random selection and assignment may sound similar to each other, they are two very different concepts. Random selection is how you draw the sample of people for your study from a population (think sampling and external validity). Random assignment is how you assign the sample that you draw to different groups or treatments in your study (think design, especially experimental design, and internal validity).

two-group posttest-only randomized experiment A research design in which two randomly assigned groups participate. Only one group receives the program and both groups receive a posttest.

It is possible to have both random selection and assignment in a study. Let's say you drew a random sample of 100 clients from a population list of 1,000 current clients of your organization. That is random sampling. Now, let's say you randomly assign 50 of these clients to get some new program and the other 50 to be controls (e.g., get the "standard" approach). That's random assignment.

It is also possible to have only one of these (random selection *or* random assignment) but not the other in a study. For instance, if you do not randomly draw the 100 cases from your list of 1,000 but instead just take the first 100 on the list, you do not have random selection. You could, however, still randomly assign this nonrandom sample to treatment versus control. Or, you could randomly select 100 from your list of 1,000 and then nonrandomly assign them to treatment or control groups.

It's also possible to have neither random selection nor random assignment. In a typical nonequivalent-groups design (see Chapter 10, "Quasi-Experimental Design") you might choose two intact, preexisting groups to be in your study. This is nonrandom selection. Then, you could arbitrarily assign one group to get the new program and the other to be the comparison group. This is nonrandom (or nonequivalent) assignment.

9.3 Classifying Experimental Designs

Although many different types of experimental design variations exist, you can classify and organize them using a simple signal-to-noise ratio metaphor. In this metaphor, assume that what you observe or see in a research study can be divided into two components: the signal and the noise. Here is a literal example of what we mean. If you've ever tuned a radio, you know that some stations have a strong signal, some have a weaker signal, and that the process of tuning into the signal involves controlling the noise. (By the way, this is directly analogous to the discussion of signal and noise in the true-score theory of measurement discussed in Chapter 5, "Introduction to Measurement.") **Figure 9.7** shows a time series with a slightly downward slope (this is the signal that the time series is sending us). However, because there is so much **variability** or noise in the series, it is difficult even to detect the downward slope. When you divide the series into its two components, you can clearly see the negative slope.

Figure 9.7 An observed time series can be thought of as made up of two components, its signal and its noise.

variability The extent to which the values measured or observed for a variable differ.

What we observe can be divided into:

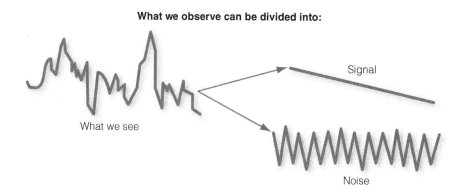

What we see

Signal

Noise

In most social science research, the signal is related to the key variable of interest—the construct you're trying to measure, or the effect of the program or treatment that's being implemented. The noise, on the other hand, consists of all of the random factors in the environment that make it harder to see the signal: for example, the lighting in the room, local distractions, how people felt that day, and so on. You can construct a ratio of these two by dividing the signal by the noise (see **Figure 9.8**). In order to see the true effect, you want the signal to be high relative to the noise. For instance, if you have a powerful treatment or program (meaning a strong signal) and good measurement (that is, low noise) you have a better chance of seeing the effect of the program than if you have either a strong program and weak measurement or a weak program and strong measurement.

$$\frac{\text{Signal}}{\text{Noise}}$$

Figure 9.8 The signal-to-noise ratio is simply a fraction where signal is divided by noise.

The signal-to-noise metaphor helps us classify experimental designs into two categories: signal enhancers or noise reducers. Doing either of these things—enhancing signal or reducing noise—improves the quality of the research and helps us see the effect (if any) more clearly. The *signal-enhancing experimental designs* are called the **factorial designs**. In these designs, the focus is almost entirely on the setup of the program or treatment, its components, and its major dimensions. In a typical factorial design, you would examine several different variations of a treatment. Factorial designs are discussed in the next section of this chapter.

The two major types of *noise-reducing experimental designs* are **covariance designs** and **blocking designs**. In these designs, you typically use information about the makeup of the sample or about pre-program variables to remove some of the noise in your study. Covariance and blocking designs are discussed following the section on factorial design.

factorial designs
Designs that focus on the program or treatment, its components, and its major dimensions, and enable you to determine whether the program has an effect, whether different subcomponents are effective, and whether there are interactions in the effects caused by subcomponents.

9.4 Signal Enhancing Designs: Factorial Designs

By directly manipulating your program or some features of your program, a **factorial design** focuses on the signal in your research. These designs are especially efficient because they enable us to examine which features or combinations of features of the program lead to a causal effect. Such manipulations of the program also serve to overcome the mono-operations bias to construct validity. This is an important way in which these designs function as signal enhancers. We'll start with the simplest factorial design, show you why it is such an efficient approach, explain how to interpret the results, and then move on to more advanced variations.

9.4a The Basic 2 × 2 Factorial Design

Probably the easiest way to begin understanding factorial designs is by looking at an example (see **Figure 9.9**). Imagine a design where you have an educational program in which you would like to look at a variety of program variations to see which works best in improving academic achievement in the subject that is being taught. Let the values in each cell in Figure 9.9 serve as group averages on this outcome variable. Assume that scores on this test range from 1 to 10, with higher values indicating greater achievement.

With regard to different kinds of programs, say you would like to vary the amount of time the children receive instruction, with one group getting one hour

covariance designs A type of randomized experimental design that helps minimize noise through the inclusion of one or more variables (called covariates) that account for some of the variability in the outcome measure or dependent variable.

blocking designs A type of randomized experimental design that helps minimize noise through the grouping of units (e.g., participants) into one or more classifications called blocks that account for some of the variability in the outcome measure or dependent variable.

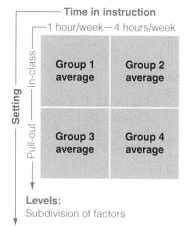

Levels:
Subdivision of factors

Factors:
Major independent variables

Figure 9.9 An example of a basic 2 × 2 factorial design.

R X_{11} O

R X_{12} O

R X_{21} O

R X_{22} O

Figure 9.10 Design notation for a 2 × 2 factorial design.

factor A major independent variable.

level A subdivision of a factor into components or features.

null case A situation in which the treatment has no effect.

of instruction per week and another getting four hours per week. Additionally, you'd like to vary the setting so that one group gets the instruction in class (probably pulled off into a corner of the classroom) and the other group is pulled out of the classroom for instruction in another room. You could think about having four separate studies to do this—where each cell combination reflects a different type of study design. But with factorial designs, you don't have to use four different studies to come up with an answer for which variation of your program works best. You can have it both ways if you cross each of your two-times-in-instruction conditions with each of your two settings, as in Figure 9.9.

It is very important that we are clear about the terms we are using to describe the elements of a factorial design. In these designs, a **factor** is the major independent variable. This example has two factors: time in instruction and program setting. A **level** is a subdivision of a factor. In this example, time in instruction has two levels (one vs. four hours) and program setting also has two levels (in vs. out of class).

Sometimes you depict a factorial design with a numbering notation. In this example, you can say that you have a 2 × 2 (spoken "two-by-two") factorial design. Here, the *number of numbers* tells you how many factors there are and the *values of the numbers* tell you how many levels. A 3 × 4 factorial design has 2 factors (one for each number) where one factor has 3 levels and the other has 4. The order of the numbers makes no difference and you could just as accurately call this a 4 × 3 factorial design. You can easily determine the number of different treatment groups that you have in any factorial design by multiplying through the number notation. For instance, the school study example has 2 × 2 = 4 groups. A 3 × 4 factorial design requires 3 × 4 = 12 groups.

You can also depict a factorial design in design notation. Because of the treatment-level combinations, it is useful to use subscripts on the treatment (X) symbol. Figure 9.10 shows that there are four groups, one for each combination of levels of factors. It also shows that the groups were randomly assigned and that this is a posttest-only design.

Now, let's look at a variety of different results you might get from this simple 2 × 2 factorial design. Each of the graphs in Figure 9.11 to Figure 9.16 describes a different possible outcome. Each outcome is shown in table form (the 2 × 2 table with the row and column averages of the outcome—academic achievement) and in graphic form (with each factor taking a turn on the horizontal axis and the outcome variable depicted on the vertical axis). Take the time to understand how and why the information in the tables agrees with the information in both of the graphs. Also, study the graphs and figures to verify that the pair of graphs in each figure show the exact same information graphed in two different ways. The lines in the graphs are technically not necessary; they are a visual aid that enables you to track where the averages for a single level go across levels of another factor.

The Null Outcome

The **null case** is a situation in which treatments have no effect. You might be thinking that this corresponds to what we call the "null hypothesis," and if so, you're exactly right. Figure 9.11 assumes that even if you didn't give the training, you would expect students to score a 5 on average on the outcome test. You can see in this hypothetical case that all four groups score an average of 5 and therefore the row and column averages must be 5. You can't see the lines for both levels in the graphs because one line falls right on top of the other.

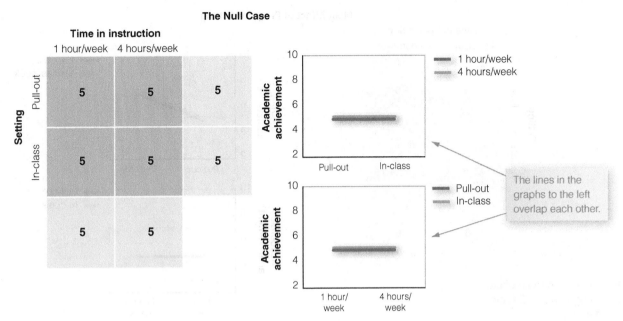

Figure 9.11 The null effects case in a 2 × 2 factorial design.

The Main Effects

A **main effect** is an outcome that is a consistent difference between levels of a factor. For instance, you would say there's a main effect for time if you find a statistical difference between the averages for the one-hour/week instruction and four-hours/week instruction at *both levels* of program setting (i.e., whether an in-class or a pull-out group). **Figure 9.12** depicts a main effect of time. For all settings, the four-hours/week condition worked better than the one-hour/week condition (in the former, the average academic score was 7, while in the latter it was 5).

main effect An outcome that shows consistent differences between all levels of a factor.

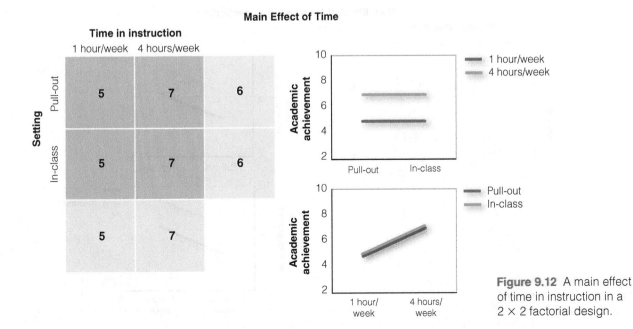

Figure 9.12 A main effect of time in instruction in a 2 × 2 factorial design.

Main Effect of Program Setting

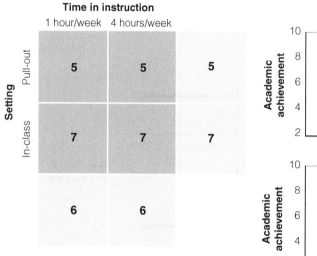

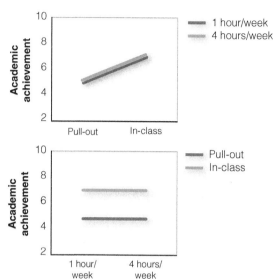

Figure 9.13 A main effect of setting in a 2 × 2 factorial design.

It is also possible to have a main effect for setting (and none for time). In the second main effect graph, shown in **Figure 9.13**, you see that in-class training was better (average academic score was 7) than pull-out training (average academic score was 5) for all levels of time.

Finally, it is possible to have a main effect on both variables simultaneously, as depicted in the third main effect (see **Figure 9.14**). In this instance, four hours/week always works better than one hour/week, and the in-class setting always works better than the pull-out setting.

Main Effect of Time and Program Setting

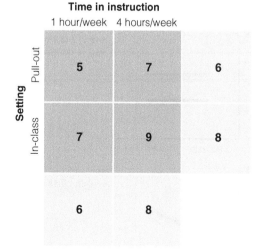

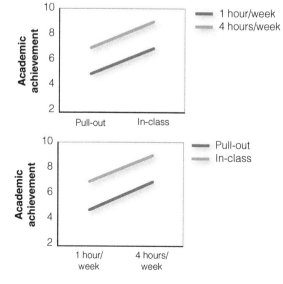

Figure 9.14 Main effects of both time and setting in a 2 × 2 factorial design.

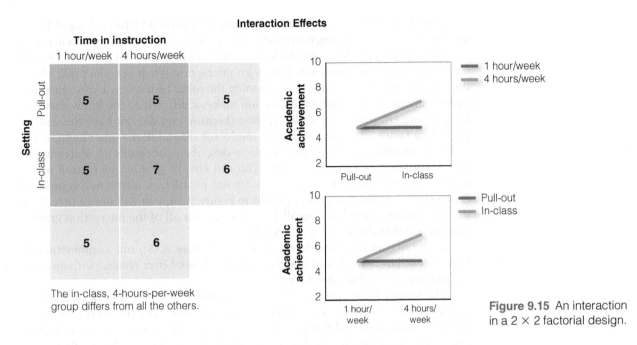

The in-class, 4-hours-per-week group differs from all the others.

Figure 9.15 An interaction in a 2 × 2 factorial design.

Interaction Effects

Because of the way you combine levels in factorial designs, these designs also enable you to examine the **interaction effects** that may exist between factors. An interaction effect exists when differences on one factor depend on which level you are on in another factor. It's important to recognize that an interaction is between factors, not levels. You wouldn't say there's an interaction between four-hours/week and in-class treatment. Instead, you would say that there's an interaction between time and setting, and then you would describe the specific levels involved.

interaction effect An effect that occurs when differences on one factor depend on which level you are on another factor.

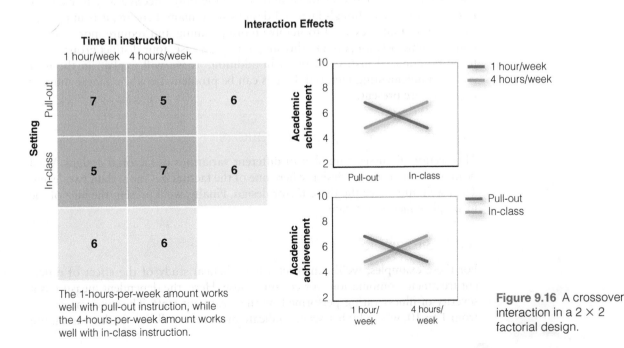

The 1-hours-per-week amount works well with pull-out instruction, while the 4-hours-per-week amount works well with in-class instruction.

Figure 9.16 A crossover interaction in a 2 × 2 factorial design.

How do you know whether there is an interaction in a factorial design? There are three ways you can determine whether an interaction exists. First, when you run the statistical analysis, the statistical table will report on all main effects and interactions. Second, you know there's an interaction when you can't talk about the effect on one factor without mentioning the other factor. If you can say at the end of your study that time in instruction makes a difference, you know that you have a main effect and not an interaction (because you did not have to mention the setting factor when describing the results for time). On the other hand, when you have an interaction, it is impossible to describe your results accurately without mentioning both factors. Finally, you can always spot an interaction in the graphs of group means; whenever lines are not parallel, an interaction is present! If you check out the main effect graphs in Figure 9.14, you will notice that all of the lines within a graph are parallel. In contrast, for all of the interaction graphs, you will see that the lines are not parallel.

In the first interaction effects graph (see **Figure 9.15**), one combination of levels—four-hours/week and in-class setting—shows better results (with an average outcome of 7) than the other three.

The second interaction (see **Figure 9.16**) shows a more complex interaction, called the crossover interaction. Here, a one-hour/week amount works well with pull-out instruction and the four-hours/week amount works well with in-class instruction. Furthermore, both of these combinations of levels do equally well.

9.4b Benefits and Limitations of Factorial Designs

A basic factorial design has several important features. First, it gives you great flexibility for exploring or enhancing the signal (treatment) in your studies. This extends the overall validity of your conclusions. Whenever you are interested in examining treatment variations, factorial designs should be strong candidates as the designs of choice. Second, factorial designs are efficient. Instead of conducting a series of independent studies, you are effectively able to combine these studies into one. Finally, factorial designs are the only effective way to examine interaction effects. Though factorial designs have many benefits, it is important to recognize that they tend to involve more planning and require more participants (mainly because, even in this simplest 2 × 2 case, four distinct groups need to be randomized instead of two). In addition, as we will explain in the next section, implementing factorial designs can be problematic when a large number of groups are present.

9.4c Factorial Design Variations

This section discusses a number of different variations of factorial designs. We'll begin with a two-factor design where one of the factors has more than two levels. Then we'll introduce the three-factor design. Finally, we'll present the idea of the incomplete factorial design.

A 2 × 3 Example

For these examples, we'll construct a hypothetical study of the effect of different treatment combinations for cocaine abuse. Here, the dependent measure is a severity-of-illness rating performed by the treatment staff. The outcome ranges from 1 to 10, where higher scores indicate more severe illness: in this case, more

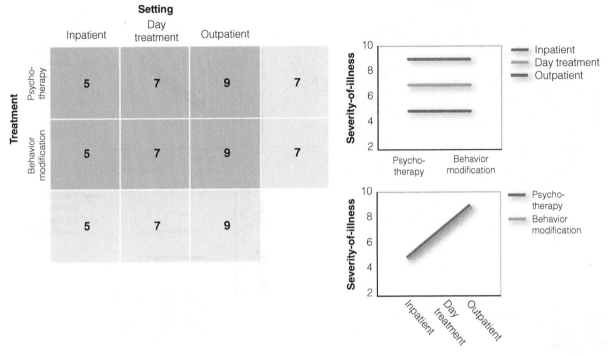

Figure 9.17 Main effect of setting in a 2 × 3 factorial design.

severe cocaine addiction. Furthermore, assume that the levels of treatment are as follows:

- Factor 1: Treatment
  - Psychotherapy
  - Behavior modification
- Factor 2: Setting
  - Inpatient
  - Day treatment
  - Outpatient

Note that the setting factor in this example has three levels.

Figure 9.17 shows what an effect for using a particular program setting might look like. You have to be careful when interpreting these results because higher scores mean the patient is doing *worse*. It's clear that inpatient treatment works best (average severity of illness is rated as 5), day treatment is next-best, and outpatient treatment is worst of the three. It's also clear that there is no difference between the two treatment levels (psychotherapy and behavior modification—both lead to an average severity-of-illness rating of 7). Even though both graphs in the figure depict the exact same data, it's easier to see the main effect for setting in the graph on the upper right, where setting is depicted with different lines on the graph rather than at different points along the horizontal axis (as in the graph on the lower right).

Figure 9.18 shows a main effect for treatment with psychotherapy performing better (remember the direction of the outcome variable) in all settings than behavior modification. The effect is clearer in the graph on the lower right, where treatment levels are used for the lines. Note that in both this and

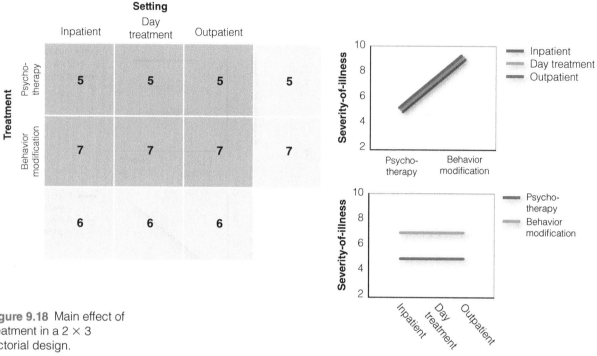

Figure 9.18 Main effect of treatment in a 2 × 3 factorial design.

Figure 9.17, the lines in all graphs are parallel, indicating that there are no interaction effects.

Figure 9.19 shows one possible interaction effect; psychotherapy works best with inpatient care, and behavior modification works best with outpatient

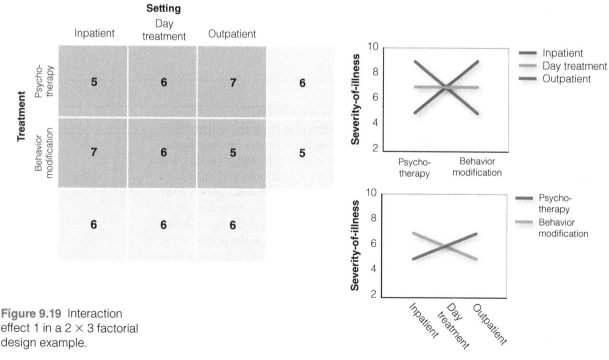

Figure 9.19 Interaction effect 1 in a 2 × 3 factorial design example.

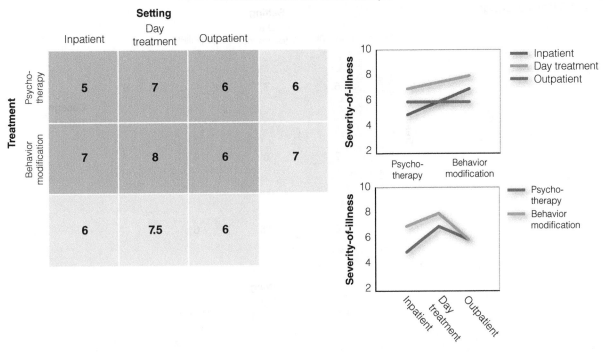

Figure 9.20 Interaction effect 2 in a 2 × 3 factorial design example.

care. The other interaction effect shown in **Figure 9.20** is a bit more compli-cated. Although there may be some main effects mixed in with the interac-tion, what's important here is that there is a unique combination of levels of factors that stands out as superior: psychotherapy done in the inpatient setting. After you identify a best combination like this, the main effects are virtually irrelevant.

A Three-Factor Example

Now let's examine what a three-factor study might look like. In addition to treat-ment and setting in the previous example, we'll include a new factor, "dosage." Further, assume that this factor has two levels. The factor structure in this 2 × 2 × 3 factorial experiment is as follows:

- Factor 1: Dosage
  - 100 mg.
  - 300 mg.
- Factor 2: Treatment
  - Psychotherapy
  - Behavior modification
- Factor 3: Setting
  - Inpatient
  - Day treatment
  - Outpatient

Notice that in this design you have 2 × 2 × 3 = 12 groups (see **Figure 9.21**). Although it's tempting in factorial studies to add more factors, the number of

Figure 9.21 Example of a
2 × 2 × 3 factorial design.

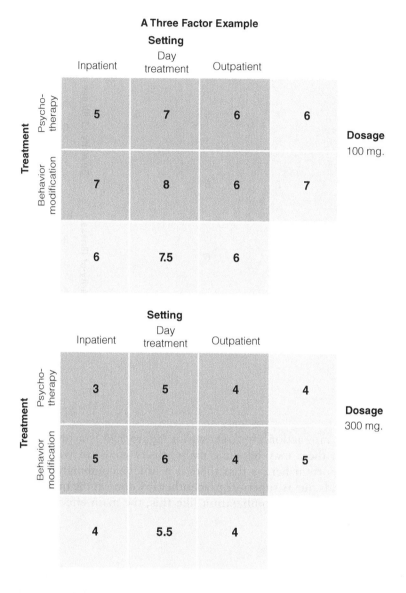

groups always increases multiplicatively and can quickly become unmanageable. Notice also that to show the tables of means, you have to have two tables now: each table shows a two-factor relationship. It's also difficult to graph the results in a study like this because there will be many different possible graphs. In the statistical analysis, you can look at the main effects for each of your three factors, the three two-way interactions (for example, treatment vs. dosage, treatment vs. setting, and setting vs. dosage) and at the one three-way interaction. Whatever else may be happening, it is clear that one combination of three levels works best: 300 mg. and psychotherapy in an inpatient setting. Thus, this study has a three-way interaction. If you were an administrator having to make a choice among the different treatment combinations, you would be best advised to select that one (assuming your patients and setting are comparable to the ones in this study).

Incomplete Factorial Design

It's clear that factorial designs can become cumbersome and have too many groups, even with only a few factors. In much research, you won't be interested in a **fully crossed factorial design**, like the ones shown previously that pair every combination of factor levels. Some of the combinations may not make sense from a policy or administrative perspective, or you simply may not have the funds to implement all combinations (**Figure 9.22**). In this case, you may decide to implement an **incomplete factorial design**. In this variation, some of the cells are intentionally left empty; you don't assign people to get those combinations of factors.

One of the most common uses of incomplete factorial design is to allow for a control or placebo group that receives no treatment. In this case, it is actually impossible to implement a study that simultaneously has several levels of treatment factors and includes a group that receives no treatment at all. So, you consider the control group to be its own cell in an incomplete factorial table, which allows you to conduct both relative and absolute treatment comparisons within a single study and to get a fairly precise look at different treatment combinations (see **Figure 9.23**).

"I've forgotten what this experiment is all about."

Farris, Joseph/CartoonStock

Figure 9.22 Sometimes complex factorial designs can become unwieldy.

Figure 9.23 An incomplete factorial design.

Incomplete Factorial Design

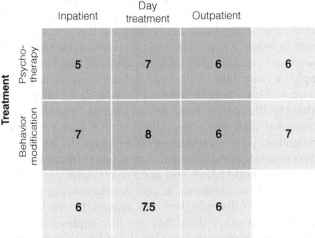

		Setting		
		Inpatient	Day treatment	Outpatient
Treatment	Psychotherapy	5	7	6	6
	Behavior modification	7	8	6	7
		6	7.5	6	
	No-treatment control group		9		

9.5 Noise-Reducing Designs: Randomized Block Designs

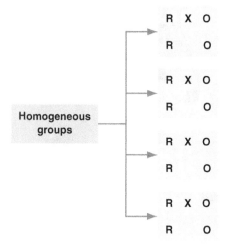

Figure 9.24 The basic randomized block design with four groups or blocks.

The **randomized block design** is research design's equivalent to stratified random sampling (see Chapter 4, "Sampling"). Like stratified sampling, randomized block designs are constructed to reduce noise or variance in the data. How do they do it? They require you to divide the sample into relatively homogeneous subgroups or blocks (analogous to "strata" in stratified sampling). Then, the experimental design you want to apply is implemented within each block or homogeneous subgroup. The key idea is that the variability within each block is less than the variability of the entire sample. As a result, each estimate of the treatment effect within a block is more efficient than estimates across the entire sample. When you pool these more efficient estimates across blocks, you should get a more efficient estimate overall than you would without blocking.

Figure 9.24 demonstrates a simple example. Let's assume that you originally intended to conduct a simple posttest-only randomized experimental design; but you recognized that your sample has several different homogeneous subgroups. For instance, in a study of college students, you might expect that students are relatively homogeneous with respect to class or year. So, you decide to block the sample into four groups: freshman, sophomore, junior, and senior. If your hunch is correct—that the variability within class is less than the variability for the entire sample—you will probably get more powerful estimates of the treatment effect within each block. Within each of your four blocks, you would implement the simple post-only randomized experiment.

Notice a couple of things about this strategy. First, to an external observer, it may not be apparent that you are blocking. You implement the same design in each block, and there is no reason that the people in different blocks need to be segregated or separated physically from each other when you carry out the experiment. In other words, blocking doesn't necessarily affect anything that you do with the research participants. Instead, blocking is a strategy for grouping people in your data analysis to reduce noise; it is an *analysis* strategy. Second, you will only benefit from a blocking design if you are correct in your hunch that the blocks are more homogeneous than the entire sample is. If you are wrong—if different college-level classes aren't relatively homogeneous with respect to your measures—you will actually be hurt by blocking. (You'll get a less powerful estimate of the treatment effect.) How do you know whether blocking is a good idea? You need to consider carefully whether the groups are relatively homogeneous. If you are measuring political attitudes, for instance, think carefully whether it is reasonable to believe that freshmen are more like each other than they are like sophomores or juniors. Do you have evidence from previous empirical studies to support your hunch? Would they be more homogeneous with respect to measures related to drug abuse? Ultimately, the decision to block involves judgment on the part of the researcher.

randomized block design Experimental designs in which the sample is grouped into relatively homogeneous subgroups or blocks within which your experiment is replicated. This procedure reduces noise or variance in the data.

9.6 Noise-Reducing Designs: Covariance Designs

The basic **Analysis of Covariance Design** (**ANCOVA**) is a pretest-posttest randomized experimental design. In contrast with the other designs analyzed so far, this design necessarily includes a pre-program measure. The notation shown in **Figure 9.25** suggests that the pre-program measure is the same one as the post-program measure (otherwise you would use subscripts to distinguish the two), and so it's appropriate to call this covariate a pretest. Note however that the pre-program covariate measure doesn't have to be a pretest; it can be any variable measured prior to the program intervention. It is also possible for a study to have more than one covariate.

The pre-program measure or pretest in this design is referred to as a covariate because of the way it's used in the data analysis; you co-vary it with the outcome variable or posttest to remove variability or noise. Thus, the ANCOVA design falls in the class of a noise-reduction experimental design.

For example, Spoth et al. (2000) employed ANCOVA to analyze the effects of a family intervention program on aggressive adolescent behavior. For this study, twenty-two schools in Iowa were randomized into treatment and control groups. The treatment included a seven-session intervention for parents and their sixth-grade children. In total, 846 families of sixth graders were recruited for treatment and control conditions.

The outcome measures included independent observer ratings of aggressive adolescent behavior, family member reports of adolescent aggressive and hostile behaviors, and adolescent self-reports of destructive behavior. In terms of a pretest, data on these outcome measures were collected from both groups before the seven-session intervention began in the treatment group. Thereafter, postintervention outcome measures were collected in sixth, seventh, eighth, and tenth grades.

ANCOVA was conducted to see if there was a difference in aggressive behavior of adolescents who did, versus those who did not, experience the family intervention program. In the ANCOVA analyses, pretest scores were included as covariates in the model along with adolescent gender. Essentially, controlling for pretest scores helps reduce the "noise" that might arise in the form of a previous history of aggressive behavior influencing current or future aggressive behavior.

Controlling for these factors, the authors did find evidence that a brief family intervention can reduce aggressive and hostile adolescent interactions with parents as well as adolescent aggressive behaviors outside of the home setting.

In social research, you frequently hear about statistical adjustments that attempt to "control" for important factors in your study. For instance, you might read that an analysis examined posttest performance after *adjusting for* the income and educational level of the participants. In this case, *income* and *education level* are covariates. A **covariate** is a variable you *adjust for* in your study. Sometimes the language that will be used is that of *removing the effects* of one variable from another. For instance, you might read that an analysis examined posttest performance after *removing the effect of* or *controlling for the effect of* income and educational level of the participants.

How does a covariate reduce noise? One of the most important ideas in social research is how you make a statistical adjustment—adjusting one variable based on its covariation with another variable. The adjustment for a covariate

```
R  O  X  O
R  O     O
```

Figure 9.25 Notation for the basic analysis of covariance design.

analysis of covariance (ANCOVA) An analysis that estimates the difference between groups on the posttest after adjusting for differences on the pretest.

covariate A measured variable that is correlated with the dependent variable. It can be included in an analysis to provide a more precise estimate of the relationship between an independent and dependent variable.

in the **ANCOVA** design is accomplished with statistical analysis. See the section "Analysis of Covariance" in Chapter 12 for details. The **ANCOVA** design is a noise-reducing experimental design. It *adjusts* posttest scores *for* variability on the covariate (pretest); this is what it means to *adjust for* the effects of one variable on another.

You can use *any* continuous variable as a covariate, but a pretest is usually best. Why? Because the pretest is usually the variable that is most highly correlated with the posttest. (A variable should correlate highly with itself at a different point in time, shouldn't it?) Because it's so highly correlated, when you "subtract it out" or remove its influence statistically, you're removing extraneous variability from the posttest. For example, you could remove the effect of baseline income in analyzing the impact of a particular program on outcome income. The rule in selecting covariates is to select the measure(s) that correlate most highly with the outcome and, if you are using multiple covariates, have little intercorrelation with each other. (Otherwise, you're simply adding redundant covariates—and you actually lose precision by doing that.) For example, you probably wouldn't want to use both gross and net income as two covariates in the same analysis because they are highly related and therefore redundant as adjustment variables.

9.7 Hybrid Designs: Switching-Replications Experimental Designs

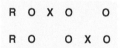

R O X O   O

R O   O X O

Figure 9.26 Notation for the switching-replications randomized experimental design.

switching-replications design A two-group design in two phases defined by three waves of measurement. The implementation of the treatment is repeated in both phases. In the repetition of the treatment, the two groups *switch* roles: The original control group in phase 1 becomes the treatment group in phase 2, whereas the original treatment group acts as the control. By the end of the study, all participants have received the treatment.

The **switching-replications design** is one of the strongest experimental designs. When the circumstances are right for this design, it addresses a major ethical concern in experimental designs: the denial of the program to control group participants due to random assignment. The design notation (see **Figure 9.26**) indicates that this is a two-group design with three waves of measurement. You might think of this as two pre-post, treatment-control designs grafted together. That is, the implementation of the treatment is repeated or *replicated*. In the repetition of the treatment, the two groups *switch* roles: the original control group becomes the treatment group in phase 2 and the original treatment acts as the control. By the end of the study, all participants would have received the treatment.

Dollahite et al. (2014) employed a switching-replications design to study the impact of a nutrition education program for low-income parents (the "Expanded Food and Nutrition Education Program" or EFNEP) on nutrition behaviors. A cohort of low-income parents of Head Start students was randomly divided into two groups. While the first group (let's call it the treatment group) received the EFNEP program over the first eight weeks of the study, the other group (i.e., the control group) was put on an eight-week waiting list. In the second eight-week period, this process was reversed—the original EFNEP parent group received no program while the original waiting list group received the EFNEP program. Thus, the experiment is "replicated" or "repeated" by switching the treatment and control group in a subsequent iteration of the experiment.

The switching-replications design is most feasible in organizational contexts where programs are repeated at regular intervals. For instance, it works

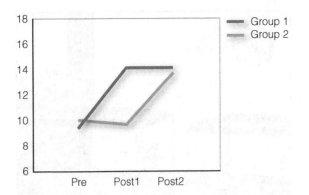

Figure 9.27 Switching-Replications design with a short-term persistent treatment effect.

especially well in schools that are on a semester system. All students are pre-tested at the beginning of the school year. During the first semester, Group 1 receives the treatment, and during the second semester, Group 2 gets it. The design also enhances organizational efficiency in resource allocation. Schools need to allocate only enough resources to give the program to half of the students at a time.

Let's look at two possible outcomes. In the first example, the program is given to the first group, and the recipients do better than the controls (see **Figure 9.27**). In the second phase, when the program is given to the original controls, they catch up to the original program group. Thus, you have a converge-diverge-reconverge outcome pattern. You might expect a result like this when the program covers specific content that the students master in the short term and where you don't expect them to continue improving as a result.

Now, look at the other example result (see **Figure 9.28**). During the first phase, you see the same result as before; the program group improves while the control group does not. As before, during the second phase, the original control group, in this case the program group, improved as much as the first program group did. This time, however, during phase two the original program group continued to increase even after it no longer received the program. Why would this happen? It could happen in circumstances where the program has continuing effects. For instance, if the program focused on learning skills, students might continue to improve their skills even after the formal program period, because they continue to apply the skills and improve in them.

We said earlier that the switching-replications design addressed specific threats to internal validity. But what threat does the switching-replications design address? Remember that in randomized experiments, especially when the groups are aware of each other, there is the potential for social threats to internal validity; compensatory rivalry, compensatory equalization, and resentful demoralization are all likely to be present in educational contexts where programs are given to some students and not to others. The switching-replications design helps lessen these threats because it ensures that everyone will eventually get the program. Additionally, it allocates who gets the program first in the fairest possible manner, through the lottery of random assignment (**Figure 9.29**).

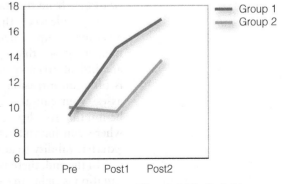

Figure 9.28 Switching-Replications design with a long-term continuing treatment effect. Group 1 received the treatment first, then Group 2.

Figure 9.29 What kind of validity threat do you see here?

"Well, I guess we're the control group."

9.8 Limitations of Experimental Design

Even though experimental designs are considered to be among the strongest of research designs, they are not without shortcomings. Many things can go wrong. First, you may find that your groups have differential dropout, thereby confounding the program with the outcome. This in effect diminishes the value of random assignment, because the initially probabilistically equivalent groups become less equivalent due to differential dropout. You might also be challenged on ethical grounds. After all, to use this approach you typically have to deny the program to some people who might be equally deserving of it as others. You could meet resistance from the staff members in your study who would like some of their favorite people to get the program. Alternatively, your design may call for assigning a certain group of people to a harmful program, which may not be plausible to implement on ethical grounds. The bottom line here is that experimental designs are difficult to carry out in many real-world contexts, and because an experiment is often an intrusion, you are setting up an artificial situation so that you can assess your causal relationship with high internal validity. As a result, you may be limiting the degree to which you can generalize your results to real contexts where you haven't set up an experiment. That is, you may have reduced your external validity to achieve greater internal validity.

In the end, there is no simple answer (no matter what anyone tells you). If the situation is right, an experiment is a strong design, but this may not necessarily be the case. Randomized experiments may only be appropriate in about 25–30 percent of the social research studies that attempt to assess causal relationships, depending on the nature of the intervention and of the context involved. Fortunately, there are many other research strategies, including the nearly limitless set of quasi-experimental designs that we'll review in the next chapter.

SUMMARY

This chapter introduced experimental designs. The basic idea of a randomized experiment was presented along with consideration of how it addresses internal validity, the key concepts of probabilistic equivalence, and the distinction between random selection and random assignment. Experimental designs can be classified broadly as signal enhancers or noise reducers. Factorial designs were presented as signal enhancers that emphasize studying different combinations of treatment (signal) features. Two types of noise-reducing strategies—randomized blocks and covariance designs—were presented along with descriptions of how each acts to reduce noise in the data. For each of the example designs, we reviewed some possible study results, portrayed in both tables and graphs. Among the possible results shown were main effects, when a factor has a significant effect by itself, and interactions, when factors combine to produce significant effects at different levels. We concluded the chapter by presenting some common limitations of experimental designs.

Key Terms

analysis of covariance
   (ANCOVA) p. 249
blocking designs p. 237
comparison group p. 233
covariance designs p. 237
covariate p. 249
factor p. 238
factorial designs p. 237
fully crossed factorial design p. 247

incomplete factorial
   design p. 247
interaction effect p. 241
level p. 238
main effect p. 239
null case p. 238
probabilistically
   equivalent p. 233
program group p. 233

random assignment p. 231
randomized block
   design p. 248
switching-replications
   design p. 250
treatment group p. 233
two-group posttest-only
   randomized experiment p. 235
variability p. 236

Suggested Websites

Penn State's Methodology Center Web Page on Factorial Experiments

http://methodology.psu.edu/ra/most/factorial

The Methodology Center has resources on many of the topics in this book, and many that are more advanced. The center is focused on development of new research methods, especially those that can be used to study and improve public health.

The NIH Center for Scientific Review

http://public.csr.nih.gov/Pages/default.aspx

This site was developed to help researchers understand how peer review of grant applications works. You can watch videos that may help you understand how the design considerations you are learning about are translated into proposals for research and then evaluated along with other aspects of the proposal.

Review Questions

1. The randomized experiment is considered the strongest design when you are interested in which validity type?

a. conclusion
b. internal
c. construct
d. external
(Reference: 9.1)

2. Subjects randomly assigned to a treatment and/or control group are assumed

a. to be similar.
b. to be probabilistically equivalent.
c. to share common backgrounds.
d. to have similar personalities.
(Reference: 9.2a)

3. The creation of an "artificial" laboratory experiment designed to increase _____ validity often decreases _____ validity.

a. construct, internal
b. construct, external
c. internal, external
d. external, internal
(Reference: 9.8)

4. Which of the following is *not* a way to determine whether an interaction exists in a 2 × 2 factorial design?

a. The levels of factors are exhaustive and mutually exclusive.
b. The statistical results that list all main effects and interactions.
c. If an interaction exists, you have to discuss both factors to explain the results.
d. When graphed, the group means form lines that are not parallel.
(Reference: 9.4a)

5. A researcher randomly assigns participants to two groups, administers a pretest, and then performs a t-test to assess their equivalence. What are the chances that the groups can have different means yet still be probabilistically equivalent, if this researcher sets his or her alpha level at .05?

a. They will be the same 5 times out of 100.
b. They will be the same 95 times out of 100.
c. They are different, nonequivalent groups if they have different means.
d. You cannot tell anything about the groups by administering a t-test on pretest measures.
(Reference: 9.2a)

6. Random selection is to _____ validity as random assignment is to _____ validity.

a. internal, external
b. external, internal
c. construct, internal
d. external, construct
(Reference: 9.2d)

7. Random selection involves _____, while random assignment is most related to _____.

a. sampling, research design
b. research design, research design
c. sampling, sampling
d. research design, sampling
(Reference: 9.2d)

8. Which two-group experimental design would increase the quality of research by "enhancing the signal" (making any treatment effects clearer)?

a. factorial design
b. covariance design
c. blocking design
d. both covariance and blocking design
(Reference: 9.4)

9. Which two-group experimental design would increase the quality of research results by "reducing the noise" that would surround the treatment effects?

a. factorial design
b. covariance design
c. blocking design
d. either covariance or blocking design
(Reference: 9.5)

10. How many factors are in a 2 × 3 factorial design?

a. one
b. two
c. three
d. six
(Reference: 9.4a)

11. The switching-replications design is a strong design both in terms of internal validity and ethics.

a. True
b. False
(Reference: 9.7)

12. Random selection and random assignment are equivalent terms.

a. True
b. False
(Reference: 9.2d)

13. An interaction effect is an outcome that is a consistent difference between levels of a factor.

a. True
b. False
(Reference: 9.4a)

14. The null case is a situation in which the treatments have no effect.

a. True
b. False
(Reference: 9.4a)

15. A 2 × 3 factorial design would have six groups of subjects or participants.

a. True
b. False
(Reference: 9.4a)

10

Quasi-Experimental Design

10.1 Foundations of Quasi-Experimental Design

Figure 10.1 Donald T. Campbell, one of the most important contributors to the development of the theory of validity and research design, particularly quasi-experimental design.

"Quasi" means "sort-of." So a "quasi-experiment" is a "sort of" experiment. Specifically, a quasi-experimental design is one that looks a bit like an experimental design (i.e., includes manipulation of an independent variable like a treatment or program) but lacks the key ingredient—random assignment. Donald T. Campbell, the distinguished methodologist who coined the term "quasi-experiment" and doctoral advisor of Professor Trochim, often referred to these designs as "queasy" experiments because they tend to give the experimental purists a queasy feeling (**Figure 10.1**). With respect to internal validity, quasi-experiments often appear to be inferior to randomized experiments. However, taken as a group, they are more frequently implemented than their randomized cousins. This is mainly because, in some circumstances, it is either impractical or unethical to randomly assign participants. For example, if we want to understand the impact of alcohol intake on fetus health, it would be highly unethical to assign pregnant women to consume different quantities of alcohol. In situations like these, a quasi-experiment that includes naturally existing groups may be designed to assess cause-and-effect hypotheses.

In this chapter, we are not going to cover all possible quasi-experimental designs comprehensively. Instead, we'll present two classic types of quasi-experimental designs in some detail. Probably the most commonly used quasi-experimental design (and it may be the most commonly used of all designs) is the nonequivalent-groups design (NEGD). In its simplest form, this design includes an existing group of participants who receive a treatment and another existing group that serves as a control or comparison group. Participants are not randomly assigned to either of the two conditions. A pretest and a posttest are collected for both groups before and after program implementation.

The second design we'll focus on is the regression-discontinuity design. We are not including it just because Professor Trochim did his dissertation on and wrote a book about it (although those were certainly factors weighing in its favor). We include it because it is an important (and often misunderstood) alternative to randomized experiments. Its distinguishing characteristic—assignment to treatment using a cutoff score on a pretreatment variable—allows you to assign to the program those who may "need" or "deserve" it most, without compromising internal validity. In addition, it is stronger in internal validity compared to many quasi-experimental designs—in fact, it can be thought of as a "randomized experiment at the cutoff point." Thus, given their ethical compatibility and strong degree of internal validity, regression-discontinuity designs are often favored (even compared to randomized experiments in some cases). However, it is important to remember that they cannot be applied in all situations—we will discuss this later in the chapter when we describe the basic regression-discontinuity design.

Finally, we'll briefly present an assortment of other common quasi-experiments that have specific applicability or noteworthy features. These designs include the proxy-pretest design, double-pretest design, nonequivalent dependent-variables design, pattern-matching design, and the regression point displacement design.

10.2 The Nonequivalent-Groups Design

The **nonequivalent-groups design (NEGD)** is probably the most frequently used design in social science research. Why? Because it is extremely intuitive in structure and can often be implemented relatively easily in practice. There exists a treatment group that gets a program and a control group that does not get the program. A pretest and posttest is conducted on both groups to measure a particular outcome before and after program implementation. In fact, the NEGD looks structurally exactly like a pretest-posttest randomized experiment. But it lacks the key feature of randomized designs—*random assignment*.

The lack of random assignment gives this design the term "nonequivalent" in its title. Nonequivalence, in this context, basically means that the researcher did not control the assignment to groups through the mechanism of random assignment. As a result, the treatment and control groups may be different prior to the study. That is, the NEGD is especially susceptible to the internal validity threat of selection (see Chapter 8, "Design"). Any prior differences between the groups may affect the outcome of the study. When a post-treatment difference between groups is observed, one cannot attribute that effect to the treatment with great confidence. It is possible that the groups' preexisting differences may have caused the observed post-treatment difference. Under the worst circumstances, the lack of random assignment can lead you to conclude that your program made a difference when in fact it didn't, or that it didn't make a difference when in fact it actually did.

The design notation for the basic NEGD is shown in **Figure 10.2**. The notation uses the letter N to indicate that the groups are nonequivalent.

You can try to mitigate the problem of "nonequivalence" by choosing treatment and control groups that are as similar to each other as possible. This could be done by choosing intact groups that appear to be similar—such as from all students in comparable classrooms, or all of the patients in similar hospitals. Selecting groups in this way allows you to compare the treated group with the comparison group more fairly, but you can never be fully sure that the groups are exactly comparable. Put another way, it's unlikely that the two groups would be as similar as they would if you assigned their members through a random lottery **Figure 10.3** shows an example of a quasi-experimental study that used intact groups to examine intervention effects in a nursing home setting. The study examined the effects of providing job resources and recovery opportunities on the health, well-being and performance of staff (Spoor, de Jonge, and Hamers, 2010).

10.2a Reaching Cause-and-Effect Conclusions with the NEGD

Let's begin our exploration of the NEGD by looking at an example. Suppose we want to study whether an after-school tutoring program positively impacts students' test scores. Imagine further that at the beginning of the school year, we announce the provision of the tutoring session and let students (say, all sixth-graders in a particular school) sign up for the session based on their interest. Those sixth-graders who sign up for the session form the treatment group, and

Figure 10.2 Notation for the Nonequivalent-Groups Design (NEGD).

nonequivalent-groups design (NEGD) A pre-post two-group quasi-experimental design structured like a pretest-posttest randomized experiment, but lacking random assignment to group.

Figure 10.3 The flowchart shows the quasi-experimental comparison of intact staff groups.

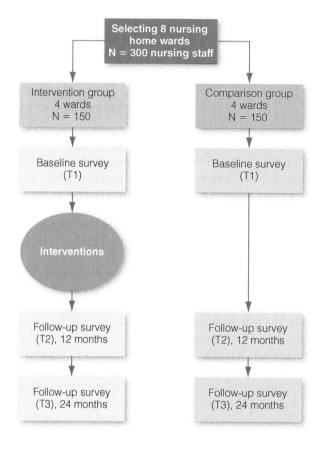

those who don't, automatically become the control or comparison group. Note, in this case, the groups are not randomly assigned to treatment or control—the researcher (you!) had no control over the assignment of treatment because the students selected themselves into the program. Before making the announcement regarding the availability of the tutoring session, we collect baseline test scores (or the pretest) from the entire class. Similarly, at the end of the school year, we administer the test again and collect test scores from both groups—this forms our posttest.

Figure 10.4 shows a bivariate distribution (which plots the relationship between two variables—in this case, the pretest and the posttest scores) from this simple example. The *treated cases* are indicated with +s and the *comparison cases* are indicated with Os. A couple of things should be obvious from the graph. To begin, you don't even need statistics to see that there is a whopping treatment effect. (However, statistics would help you estimate the size of that effect more precisely.) The program cases (+s) consistently score better on the posttest than the comparison cases (Os) do. Given the higher scores on the posttest for the treatment group, you may conclude that the program improved things.

However, as discussed previously, the fundamental threat to internal validity in the NEGD is selection—that the groups differed before the program. Does that appear to be the case here? In other words, did the program group participants

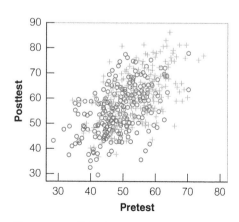

Figure 10.4 Bivariate distribution for a hypothetical example of a Nonequivalent-Groups Design.

start with an initial advantage—maybe their motivation to join the after-school tutoring session came from the fact that they were already better at the subject (something that can be checked from their pretest scores) and wanted to expand on their skills.

You can see the initial difference, the selection bias, when you look at the graph in **Figure 10.5**. It shows that the program group scored about five points higher than the comparison group on the pretest. The comparison group had a pretest average of about 50, whereas the program group averaged about 55. It also shows that the program group scored about fifteen points higher than the comparison group on the posttest. That is, the comparison group posttest score was again about 50, whereas this time the program group scored around 65. This suggests that program group participants did have an initial advantage on the pretest and that the positive results may be due in whole or in part to this initial difference.

These observations suggest that there is a potential selection threat, although the initial five-point difference doesn't explain why you observe a fifteen-point difference on the posttest. It may be that there is still a legitimate treatment effect here, even given the initial advantage of the program group.

Let's take a look at several different possible outcomes[*] from a NEGD to see how they might be interpreted. The important point here is that each of these outcomes would have a different story line. Some are more susceptible to threats to internal validity than others. In addition, different outcomes suggest different kinds of internal validity threats that might be operating. Before you read each of the descriptions, take a good look at the associated graph and try to figure out what the possible explanations for the results might be (other than the effect of the program). If you were a critic, what kinds of problems would you look for? Then, read the synopsis and see if it agrees with your perception.

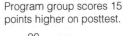

Program group scores 15 points higher on posttest.

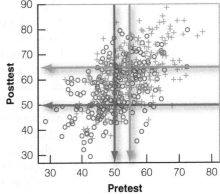

Program group has 5-point pretest advantage.

Figure 10.5 Nonequivalent-Groups Design with pretest and posttest averages marked for each group.

Possible Outcome 1

Sometimes it's useful to look at the means for the two groups. **Figure 10.6** plots the means for the distribution. The pre-post means of the program group scores are joined with a blue-dashed line and the pre-post means of the comparison group scores are joined with a red solid line. Here, you can see much more clearly both the original pretest difference of five points and the larger fifteen-point posttest difference.

How might you interpret these results? To begin, you need to recall that with the NEGD you are usually most concerned about selection threats. For this reason, one way to be more confident about the cause-and-effect conclusions from a NEGD is to provide evidence against various selection threats. In fact, this is the key to improving the internal validity of the NEGD, given its fundamental handicap of nonequivalent groups.

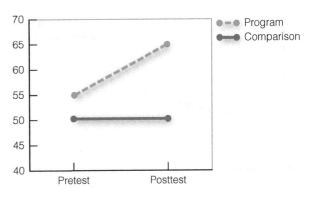

Figure 10.6 Plot of pretest and posttest means for possible outcome.

[*]The discussion of the five possible outcomes is based on the discussion in Cook, T. D., & Campbell, D. T. (1979). *Quasi-Experimentation: Design & Analysis Issues for Field Settings.* Boston: Houghton Mifflin, pp. 103–112.

Which selection threats might possibly be operating here? Notice, in this outcome, the comparison group did not change between the pretest and the posttest. The average test score for the comparison group at the start of the school year was 50, and it remained unchanged when the posttest was administered at the end of the school year. Therefore, it would be hard to argue that that the outcome is due to a *selection-maturation threat*. Why? Remember that a selection-maturation threat means that the groups are maturing at different rates and that this creates the illusion of a program effect when there is none. However, because the comparison group didn't mature (change) at all, it's hard to argue that differential maturation produced the outcome. What could have produced the outcome? A *selection-history threat* certainly seems plausible. Perhaps some event occurred (other than the program) that the program group reacted to and the comparison group didn't react to. Maybe a local event occurred for the program group (say, a study tour) but not for the comparison group. Notice how much more likely it is that outcome pattern 1 is caused by such a history threat than by a maturation difference. You could try to research if a selection-history threat is present by interviewing the treatment group students regarding other related activities they were engaged in during the academic year. You would then have to ensure that the control group was not engaged in the same activities over the given time course.

What about the possibility of *selection-regression*? This one actually works a lot like the selection-maturation threat. If the jump in the program group is due to *regression to the mean*, it would have to be because the program group was below the overall population pretest average and consequently regressed upwards on the posttest. However, if that's true, it should be even more the case for the comparison group, who started with an even lower pretest average. The fact that it doesn't appear to change or regress at all helps rule out the possibility that outcome 1 is the result of regression to the mean. Similarly, do you think *selection-testing* and *selection-mortality* could in any way cause the outcome you see?

Possible Outcome 2

The second hypothetical outcome (see **Figure 10.7**) presents a different picture. Here, both the program and comparison groups gain from pre to post, with the program group gaining at a slightly faster rate. This is almost the definition of a selection-maturation threat. The fact that the two groups differed to begin with suggests that they may already be maturing at different rates. The posttest scores don't do anything to help rule out that possibility. This outcome might also arise from a selection-history threat. If the two groups, because of their initial differences, react differently to some historical event, you might obtain the outcome pattern shown. Both *selection-testing* and *selection-instrumentation* are also possibilities, depending on the nature of the measures used. This pattern could indicate a *selection-mortality* problem if there are more low-scoring program cases that drop out between testings. What about selection-regression? It doesn't seem likely, for much the same reasoning as for outcome 1. If there were an upwards regression to the mean from pre to post, you would expect that regression would be greater for the comparison group because it has the lower pretest score.

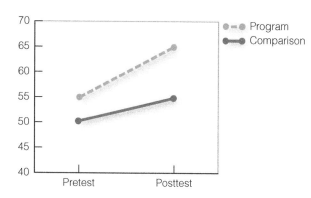

Figure 10.7 Plot of pretest and posttest means for possible outcome 2.

Possible Outcome 3

The third possible outcome (see **Figure 10.8**) cries out selection-regression! Or, at least it would if it *could* cry out. The regression scenario is that the program group was selected so that it was extremely high (relative to the population) on the pretest. The fact that the group scored lower, approaching the comparison group on the posttest, may simply be due to its regressing toward the population mean. You might observe an outcome like this when you study the effects of giving a scholarship or an award for academic performance. You give the award because students did well (in this case, on the pretest). When you observe the group's posttest performance, relative to an average group of students, it appears to perform worse. Pure regression! Notice how this outcome doesn't suggest a selection-maturation threat. What kind of maturation process would have to occur for the highly advantaged program group to decline while a comparison group evidences no change?

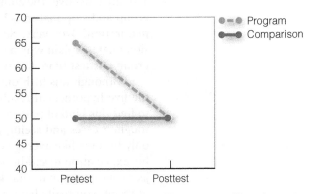

Figure 10.8 Plot of pretest and posttest means for possible outcome 3.

Possible Outcome 4

The fourth possible outcome also suggests a selection-regression threat (see **Figure 10.9**). Here, the program group is disadvantaged to begin with. The fact that it appears to pull closer to the comparison group on the posttest may be due to regression. This outcome pattern may be suspected in studies of compensatory programs—programs designed to help address some problem or deficiency. For instance, compensatory education programs are designed to help children who are doing poorly in some subject. They are likely to have lower pretest performance than more average comparison children. Consequently, they are likely to regress to the mean in a pattern similar to the one shown in outcome 4.

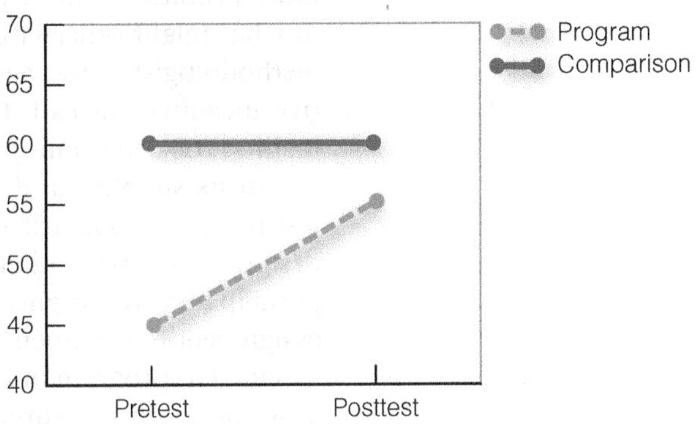

Figure 10.9 Plot of pretest and posttest means for possible outcome 4.

Possible Outcome 5

This last hypothetical outcome (see **Figure 10.10**) is sometimes referred to as a crossover pattern. Here, the comparison group doesn't appear to change from pre to post; but the program group does, starting out lower than the comparison group and ending up above it. This is the clearest pattern of evidence for the effectiveness of the program of all five of the hypothetical outcomes. It's hard to come up with a threat to internal validity that would be plausible here. Certainly, there is no evidence for selection-maturation here, unless you postulate that the two groups are involved in maturational processes that tend to start and stop, and just coincidentally you caught the program group maturing while the comparison group had gone dormant. However, if that were the case, why did the program group

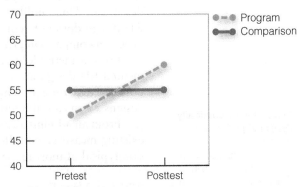

Figure 10.10 Plot of pretest and posttest means for possible outcome 5.

actually cross over the comparison group? Why didn't it approach the comparison group and stop maturing? How likely is this outcome as a description of normal maturation? Not very. Similarly, this isn't a selection-regression result. Regression might explain why a low-scoring program group approaches the comparison group posttest score (as in outcome 4), but it doesn't explain why it crosses over.

Although this fifth outcome is the strongest evidence for a program effect of the five hypothetical results, you can't very well construct your study expecting to find this kind of pattern. It would be a little bit like giving your program to the toughest cases and seeing whether you can improve them so much that they not only become like average cases, but actually outperform them. That's an awfully big expectation to saddle any program with. Typically, you wouldn't want to subject your program to that kind of expectation. If you do happen to find that kind of result, you really have a program effect that beats the odds.

10.3 The Regression-Discontinuity Design

What a terrible name for a research design! In everyday language, both parts of the term, regression and discontinuity, have primarily negative connotations. To most people, regression implies a reversion backwards or a return to some earlier, more primitive state, whereas discontinuity suggests an unnatural jump or shift in what might otherwise be a smoother, more continuous process. To a research methodologist, however, the term regression-discontinuity carries no such negative meaning. Instead, the **regression-discontinuity (RD) design** is seen as a useful method for determining whether a program or treatment is effective.

In its simplest and most traditional form, the RD design is a pretest-posttest program-comparison group strategy. The unique characteristic that sets RD designs apart from other pre-post group designs is the method by which research participants are assigned to conditions. As this is a quasi-experimental design, the assignment to treatment is obviously not random. In RD designs, participants are assigned to program or comparison groups solely on the basis of a cutoff score on a pre-program measure. Thus, the RD design is distinguished from randomized experiments (or randomized clinical trials) and from other quasi-experimental strategies by its unique method of assignment. This cutoff criterion also implies the major ethical and practical advantage of RD designs; they are appropriate when you want to target a program or treatment to those who most need or deserve it. Thus, unlike its randomized or quasi-experimental alternatives, the RD design does not require you to assign potentially needy individuals to a no-program comparison group to evaluate the effectiveness of a program.

From a methodological point of view, inferences drawn from a well-implemented RD design are comparable in internal validity to conclusions from randomized experiments. Thus, the RD design is a strong competitor to randomized designs when causal hypotheses are being investigated.

regression-discontinuity (RD) design A pretest-posttest program-comparison-group quasi-experimental design in which a cutoff criterion on the preprogram measure is the method of assignment to group.

From an administrative viewpoint, the RD design is often directly usable with existing measurement efforts, such as the regularly collected statistical information typical of most management-information systems.

10.3a The Basic RD Design

The basic RD design is a pretest-posttest two-group design. The term pretest-posttest implies that the same measure (or perhaps alternate forms of the same measure) is collected before and after some program or treatment. In fact, the RD design does not require that the pre and post measures be the same. In the RD design, we'll use the term pre-program measure instead of pretest. The term pre-program measure indicates that the before-and-after measures may be the same or different. The term "pretest" suggests that it is the same "test" that is in the "posttest" except that it is given before the program (i.e., pre). It is assumed that a cutoff value on the pretest or pre-program measure is the sole criterion being used to assign persons or other units to the program.

Two-group versions of the RD design might imply either that some treatment or program is being contrasted with a no-program condition or that two alternative programs are being compared. The description of the basic design as a two-group design implies that a single pretest-cutoff score is used to assign participants to either the program or comparison group. The term *"participants"* refers to the units assigned. In many cases, participants are individuals, but they could be any definable units such as hospital wards, hospitals, counties, and so on. The term *"program"* is used in this discussion of the RD design to refer to any program, treatment, intervention, or manipulation whose effects you want to examine. In notational form, the basic RD design might be depicted as shown **Figure 10.11**:

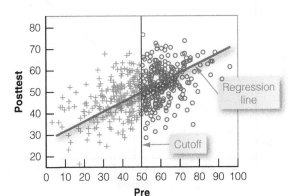

Figure 10.11 Notation for the Regression-Discontinuity (RD) design.

- *C* indicates that groups are assigned by means of a cutoff score on the pre-measure.
- An *O* stands for the administration of a measure to a group.
- An *X* depicts the implementation of a program.
- Each group is described on a single regression line (for example, program group on top and control group on the bottom).

To make this initial presentation more concrete, imagine a hypothetical study examining the effect of a new treatment protocol for inpatients with a particular diagnosis. For simplicity, assume that you want to try the new protocol on patients who are considered most ill, and that for each patient you have a continuous quantitative indicator of health that is a composite rating that takes values from 1 to 100, where high scores indicate greater health. Furthermore, assume that a pre-program measure (in this case, it is the pretest) cutoff score of 50 was (more or less arbitrarily) chosen as the assignment criterion, so that all those scoring lower than 50 on the pretest (those more ill) are to be given the new treatment protocol while those with scores greater than or equal to 50 are given the standard treatment.

It is useful to begin by considering what the data might look like if you did not administer the treatment protocol but instead only measured all participants at two points in time. **Figure 10.12** shows the hypothetical bivariate distribution for this situation. Each dot on the figure indicates a single person's pretest and posttest scores. The blue +s to the left of the cutoff show the program cases. They are more severely ill on both the pretest and the posttest. The red circles show the comparison group that is comparatively healthier on both measures. The vertical line at the pretest score of 50 indicates the cutoff point. (In Figure 10.12, the assumption is that no treatment has been given, so the cutoff point is for demonstration purposes only.) The solid red line through the bivariate distribution of +s and Os is the linear regression line, which we think

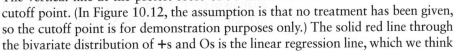

Figure 10.12 Pre-post distribution for an RD design with no treatment effect.

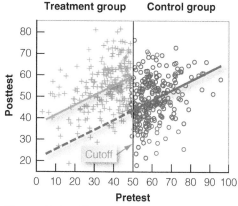

Figure 10.13 The RD design with ten-point treatment effect.

of as the "best fitting" line for the data. The distribution depicts a strong positive relationship between the pretest and posttest; this means that in general, the more healthy a person is at the pretest, the more healthy he or she is on the posttest, and conversely, the more severely ill a person is at the pretest, the more ill that person is on the posttest.

Consider what the outcome might look like if the new treatment protocol is administered (and the patients are selected into treatments based on the cutoff score) and it has a positive effect (see **Figure 10.13**). For simplicity, assume that the treatment had a constant effect that raised each treated person's health score by ten points.

Figure 10.13 is identical to Figure 10.12 except that all points to the left of the cutoff (that is, the treatment group) have been raised by ten points on the posttest. The dashed line in Figure 10.13 shows what you would expect the treated group's regression line to look like if the program had no effect (as was the case in Figure 10.12).

It is sometimes difficult to see the forest for the trees in these types of bivariate plots. So, let's remove the individual data points and look only at the regression lines. The plot of regression lines for the treatment effect case of Figure 10.13 is shown in **Figure 10.14**.

On the basis of Figure 10.14, you can now see how the RD design got its name; a program effect is suggested when you observe a jump or *discontinuity* in the *regression* lines at the cutoff point. This is illustrated more clearly in **Figure 10.15**.

10.3b The Role of the Comparison Group in RD Designs

In experimental or other quasi-experimental designs (such as the NEGD), you either assume or try to provide evidence that the program and comparison groups are equivalent prior to the program so that post-program differences can be attributed to the manipulation/treatment. The RD design involves no such assumption. Instead, with RD designs the key assumption is that in the absence of the program the pre-post relationship would be equivalent for the two groups (in other words, the treatment and comparison group distributions can be represented with the same regression line before the program is administered—see Figure 10.12). Thus, the strength of the RD design is dependent on two major factors. The first is the assumption that there is no artificial discontinuity in the pre-post relationship that happens to coincide with the cutoff point. The second factor concerns the degree to which you can know and correctly model the pre-post relationship. This constitutes the major problem in the statistical analysis of the RD design, which will be discussed in Chapter 12, "Inferential Analysis."

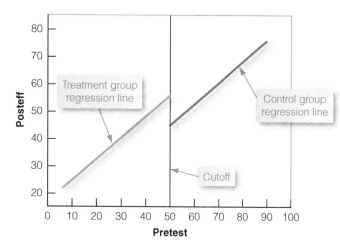

Figure 10.14 Regression lines for the data shown in Figure 10.13.

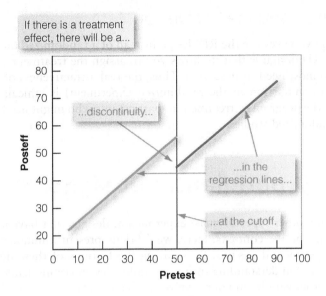

Figure 10.15 How the RD design got its name.

10.3c The Internal Validity of the RD Design

Although the RD design might initially seem susceptible to threats to internal validity, and especially to selection biases (after all, we are purposely selecting treatment and control groups that are different from each other—as in the example above, unhealthy patients and healthy patients), it is not. Only factors that would naturally induce a discontinuity in the pre-post relationship could be considered threats to the internal validity of inferences from the RD design. For instance, a selection-maturation threat would mean that the groups on either side of the cutoff point differed noncontinuously in maturation rates. It's likely that people differ in natural maturation, and even that this is related to pre-program levels. But, it's hard to imagine that maturation rates would differ abruptly between groups coincidentally at the cutoff. Basically, the key idea is that it would be hard to think of ways in which any of the threats to internal validity that we have already discussed could have created the observed discontinuity at exactly the cutoff point! In principle, then, the RD design is as strong in internal validity as its randomized experimental alternatives. In practice, however, the validity of the RD design depends directly on how well you can model the true pre-post relationship, certainly a serious statistical challenge.

10.3d Statistical Power and the RD Design

The previous discussion argues that the RD design is strong in internal validity. In fact, it is certainly stronger in internal validity than the NEGD, and perhaps as strong in internal validity as the randomized experimental design. However, RD designs are not as statistically powerful as randomized experiments (see Chapter 11 for a discussion of statistical power). That is, to achieve the same level of statistical accuracy, an RD design needs as many as 2.75 times the participants as a randomized experiment. For instance, if a randomized experiment needs 100 participants to achieve a certain level of power, the RD design might need as many as 275.

10.3e Ethics and the RD Design

So why would you ever use the RD design instead of a randomized one? The real allure of the RD design is that it allows you to assign the treatment or program to those who most need or deserve it. Thus, the real attractiveness of the design (especially in comparison to the randomized experiment) is ethical; you don't have to deny the program or treatment to participants who might need it most, as you do in randomized studies.

10.4 Other Quasi-Experimental Designs

There are many other types of quasi-experimental designs that have a variety of applications in specific contexts. Here, we'll briefly present a number of important quasi-experimental designs. By studying the features of these designs, you can gain a deeper understanding of how to tailor design components to address threats to internal validity in your own research contexts.

10.4a The Proxy Pretest Design

$$N \quad O_1 \quad X \quad O_2$$
$$N \quad O_1 \quad \quad O_2$$

Figure 10.16 The Proxy-Pretest design.

The **proxy-pretest design** (see **Figure 10.16**) looks like a standard pre-post design with an important difference. The pretest in this design is collected *after* the program is given! But how can you call it a pretest if it's collected after the program? Because you use a proxy variable to estimate where the groups would have been on the pretest. There are essentially two variations of this design. In the first, you ask the participants to estimate where their pretest level would have been. This can be called the recollection or retrospective design—participants are literally asked to "look back." For instance, you might ask participants to complete pretest measures by estimating how they would have answered the questions six months ago.

Hazan and Shaver's (1987) work on attachment theory provides an interesting example of the recollection method. In their research, the authors administer a "love quiz" and ask participants to choose one of the three possible descriptions that best captured the attachment they had with their parents as children. Thereafter, they ask participants to rate their most important romantic relationship on several dimensions. Comparing the subjects' retrospective feelings about parental attachment with later-life romantic attachments, the researchers found that people who reported secure relationships with parents when they were children tended to rate their romantic relationships as more trustworthy than participants who classified their relationship with their parents as insecure.

These designs are convenient (you capture both the pretest and the posttest information at the same time). However, a key disadvantage of the proxy pretest is that people may forget where they were at some prior time or they may distort the pretest estimates to make themselves look better now. However, at times, you might be interested not so much in where they were on the pretest but rather in where they think they were. The recollection proxy pretest would be a sensible way to assess participants' perceived gain or change.

The other proxy-pretest design uses archived records to stand in for the pretest. This design might be called the archived proxy-pretest design. For instance, imagine that you are studying the effects of an educational program on the math performance of eighth-graders. Unfortunately, you were brought in

proxy-pretest design A post-only design in which, after the fact, a pretest measure is constructed from preexisting data. This is usually done to make up for the fact that the research did not include a true pretest.

to do the study after the program had already been started (a too-frequent case, we're afraid). You are able to collect a posttest that shows math ability after training, but you have no pretest. Under these circumstances, your best bet might be to find a proxy variable from existing records that would estimate pretest performance. For instance, you might use the students' grade-point averages in math from the seventh grade as the proxy pretest.

The proxy-pretest design is not one you would typically ever select by choice; but, if you find yourself in a situation where you have to evaluate a program that has already begun, it may be the best you can do and would almost certainly be better than relying only on a posttest-only design.

10.4b The Separate Pre-Post Samples Design

The basic idea in the **separate pre-post samples design** (and its variations) is that the people you use for the pretest are not the same as the people you use for the posttest (see **Figure 10.17**). Take a close look at the design notation for the first variation of this design. There are four groups (indicated by the four lines), but two of the groups come from a single nonequivalent group and the other two also come from a single nonequivalent group (indicated by the subscripts next to N). Imagine that you have two agencies or organizations that you think are similar. You want to implement your study in one agency and use the other as a control. The program you are looking at is an agency-wide one and you expect the outcomes to be most noticeable at the agency level. For instance, let's say the program is designed to improve customer satisfaction. Because customers routinely cycle through your agency, you can't measure the same customers pre-post. Instead, you measure customer satisfaction in each agency at one point in time, implement your program, and then measure customer satisfaction in the agency at another point in time after the program. Notice that the customers will be different within each agency for the pretest and posttest. This design is not a particularly strong one because you cannot match individual participant responses from pre to post; you can only look at the change in average customer satisfaction (i.e., at the agency level). Here, you always run the risk that you have nonequivalence not only between the agencies but also between the pre and post groups as well. For instance, if you have different types of clients at different times of the year, this could bias the results. Another way of looking at this is as a proxy pretest on a different group of people.

The second example of the separate pre-post sample design is shown in design notation in **Figure 10.18**. Again, there are four groups in the study. This time, however, you are taking random samples from your agency or organization at each point in time. Therefore, this is the same design as the one in Figure 10.17 except for the random sampling. Probably the most sensible use of this design would be in situations where you routinely do sample surveys in an organization or community. For instance, assume that every year two similar communities do a community-wide survey of residents to ask about satisfaction with city services. Because of costs, you randomly sample each community each year. In one of the communities, you decide to institute a program of community policing, and you want to see whether residents feel safer and have changed in their attitudes toward police. You would use the results of last year's survey as the pretest in both communities and this year's results as the posttest. Again, this is not a particularly strong design. Even though you are taking random samples from each community each year, it may still be the case that the community changes fundamentally from

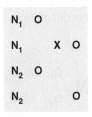

Figure 10.17 The Separate Pre-Post Samples design.

Figure 10.18 The Separate Pre-Post Samples design with random sampling of the pre-post groups from two nonequivalent agencies.

separate pre-post samples design A design in which the people who receive the pretest are not the same as the people who take the posttest.

Figure 10.19 The Double-Pretest design.

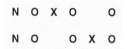

Figure 10.20 The Switching-Replications design.

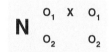

Figure 10.21 The NEDV design.

one year to the next and that the random samples within a community cannot be considered equivalent.

10.4c The Double-Pretest Design

The **double-pretest design** (see **Figure 10.19**) is a strong quasi-experimental design with respect to internal validity. Why? Recall that the pre-post NEGD is especially susceptible to selection threats to internal validity. In other words, the nonequivalent groups may be different in some way before the program is given, and you may incorrectly attribute posttest differences to the program. Although the pretest helps you assess the degree of pre-program similarity, it does not determine whether the groups are changing at similar rates prior to the program. Thus, the NEGD is especially susceptible to selection-maturation threats.

The double-pretest design includes two pretests prior to the program. Consequently, if the program and comparison group are maturing at different rates, you should detect this as a change from pretest 1 to pretest 2. Therefore, this design explicitly controls for selection-maturation threats. The design is also sometimes referred to as a dry-run, quasi-experimental design because the double pretests simulate what would happen when there is no treatment effect, which we refer to as the null case.

10.4d The Switching-Replications Design

The switching-replications quasi-experimental design is also strong with respect to internal validity, and because it allows for two independent implementations of the program, it may enhance external validity or generalizability (see **Figure 10.20**). The switching-replications design has two groups and three waves of measurement. In the first phase of the design, both groups are given pretests, one is given the program, and both are posttested. In the second phase of the design, the original comparison group is given the program while the original program group serves as the control. This design is identical in structure to its randomized experimental version (described in Chapter 9, "Experimental Design") but lacks the random assignment to group. It is certainly superior to the simple NEGD. In addition, because it ensures that all participants eventually get the program, it is probably one of the most ethically feasible quasi-experiments.

10.4e The Nonequivalent Dependent Variables (NEDV) Design

The **nonequivalent dependent variables (NEDV) design** is a deceptive one. In its simple form, it is an extremely weak design with respect to internal validity. However, in its pattern-matching variation (covered later in this chapter), it opens the door to an entirely different approach to causal assessment that is extremely powerful. The design notation shown in **Figure 10.21** is for the simple two-variable case. Notice that this design has only a single group of participants. The two lines in the notation indicate separate variables, not separate groups.

The idea in this design is that you have a program designed to change a specific outcome. For instance, assume you are training first-year high-school students in algebra. Your training program is designed to affect algebra scores; but it is not designed explicitly to affect geometry scores. You reasonably expect pre-post geometry performance to be affected by other internal validity factors such as history or maturation. In this case, the

double-pretest design
A design that includes two waves of measurement prior to the program.

nonequivalent dependent variables (NEDV) design
A single-group pre-post quasi-experimental design with two outcome measures where only one measure is theoretically predicted to be affected by the treatment and the other is not.

pre-post geometry performance acts like a control group; it models what would likely have happened to the algebra pre-post scores if the program hadn't been given. The key is that the control variable has to be similar enough to the target variable to be affected in the same way by history, maturation, and the other single-group internal validity threats, but not so similar that it is affected by the program.

Figure 10.22 shows the results you might get for the two-variable, algebra-geometry example. Note that this design works only if the geometry variable is a reasonable alternative for what would have happened on the algebra scores in the absence of the program. The real allure of this design is the possibility that you don't need a control group—the treatment group is effectively its own control group. In other words, you can give the program to your entire sample. The actual control, though, is the additional variable—the geometry score in the above example.

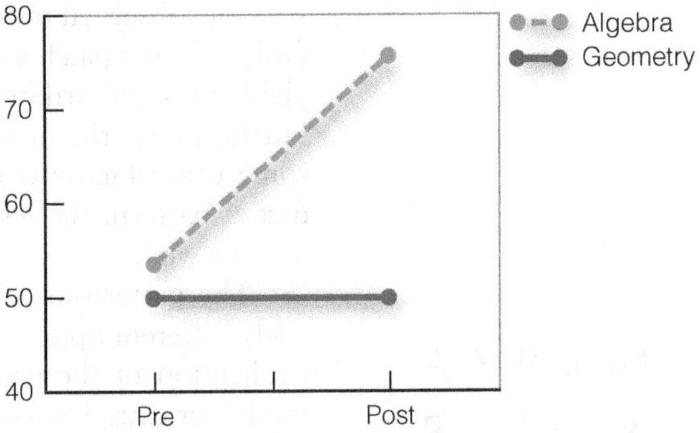

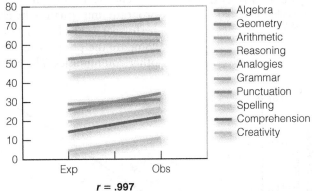

Figure 10.22 Example of a two-variable Nonequivalent Dependent Variables design.

The problem is that in its two-variable simple version, the assumption of the control variable (that it is similar enough to the target variable) is a difficult one to meet. (Note that a double-pretest version of this design would be considerably stronger.)

The Pattern-Matching NEDV Design

Although the two-variable NEDV design is quite weak, you can make it considerably stronger by adding multiple outcome variables in what might be called a pattern-matching NEDV design. In this variation, you need many outcome variables and a theory that tells *how affected* (from most to least) by the program each variable will be. Then you collect data that allow you to assess whether the observed pattern of data matches the theoretically predicted pattern. Let's reconsider the example from the algebra program in the previous discussion. Now, instead of having only an algebra and geometry score, imagine you have ten measures that you collect pre and post. You would expect the algebra measure to be most affected by the program (because that's what the program was most designed to affect). However, in this variation, you recognize that geometry might also be somewhat affected because training in algebra might be relevant, at least to some degree, to geometry skills. On the other hand, you might theorize that creativity would be much less affected, even indirectly, by training in algebra, and so you predict the creativity measure to be the least affected of the ten measures.

Figure 10.23 Example of a Pattern-Matching variation of the NEDV design.

Now, line up your theoretical expectations against your pre-post gains for each variable. You can see in **Figure 10.23** that the expected order of outcomes (on the left) is mirrored well in the actual outcomes (on the right).

Depending on the circumstances, the **pattern-matching NEDV design** can be quite strong with respect to internal validity. In general, the design is stronger if you have a larger set of variables and your expectation pattern matches well

pattern-matching NEDV design A single group pre-post quasi-experimental design with multiple outcome measures where there is a theoretically specified pattern of expected effects across the measures.

with the observed results. What are the threats to internal validity in this design? Only a factor (such as an historical event or a maturational pattern) that would yield the same ordered outcome pattern can act as an alternative explanation. Furthermore, the more complex the predicted pattern, the less likely it is that some other factor would yield it. The problem is, the more complex the predicted pattern, the less likely it is that you will find it matches your observed data as well.

The pattern-matching notion implicit in the NEDV design requires an entirely different approach to causal assessment, one that depends on detailed prior explication of the program and its expected effects. It suggests a much richer model for causal assessment than one that relies only on a simplistic dichotomous treatment-control model.

$$N_{(n = 1)} \quad O \quad X \quad O$$
$$N \qquad\quad O \qquad O$$

Figure 10.24 The Regression Point Displacement (RPD) design.

regression point displacement (RPD) design A pre-post quasi-experimental research design where the treatment is given to only one unit in the sample, with all remaining units acting as controls. This design is particularly useful in studying the effects of community-level interventions, where outcome data are routinely collected at the community level.

10.4f The Regression Point Displacement (RPD) Design

The **regression point displacement (RPD) design** is a simple quasi-experimental strategy that has important implications, especially for community-based research. The problem with community-level interventions is that it is difficult to do causal assessment to determine whether your program (as opposed to other potential factors) made a difference. Typically, in community-level interventions, program costs limit implementation of the program in more than one community. You look at pre-post indicators for the program community and see whether there is a change. If you're relatively enlightened, you seek out another similar community and use it as a comparison. However, because the intervention is at the community level, you have only a single unit of measurement for your program and comparison groups.

The RPD design (see **Figure 10.24**) attempts to enhance the single-program-unit situation by comparing the performance on that single unit with the performance of a large set of comparison units. In community research, you would compare the pre-post results for the intervention community with a large set of other communities. The advantage of doing this is that you don't rely on a single nonequivalent community; you attempt to use results from a heterogeneous set of nonequivalent communities to model the comparison condition and then compare your single site to this model. For typical community-based research, such an approach may greatly enhance your ability to make causal inferences.

We'll illustrate the RPD design with an example of a community-based AIDS education program to be implemented in one particular community in a state, perhaps a county. The state routinely publishes annual HIV positive rates by county for the entire state. So, the remaining counties in the state function as control counties. Instead of averaging all the control counties to obtain a single control score, you use them as separate units in the analysis. **Figure 10.25** shows the bivariate pre-post

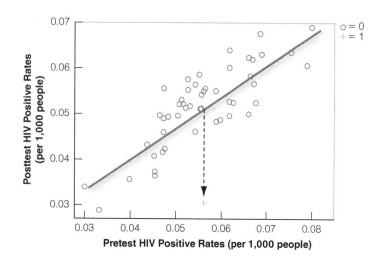

Figure 10.25 An example of the RPD design.

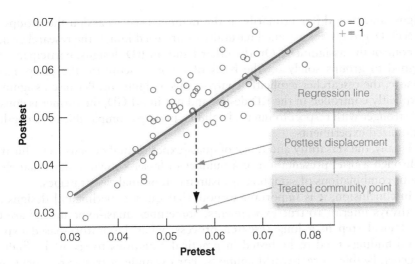

Figure 10.26 How the RPD design got its name.

distribution of HIV positive rates per 1,000 people for all the counties in the state. The program county—the one that gets the AIDS education program—is shown as a "+" and the remaining control counties are shown as Os. You compute a regression line for the control cases to model your predicted outcome for a county with any specific pretest rate. To estimate the effect of the program you test whether the displacement of the program county from the control county regression line is statistically significant.

Figure 10.26 shows why the RPD design was given its name. In this design, you know you have a treatment effect when there is a significant *displacement* of the program *point* from the control group *regression* line.

The RPD design is especially applicable in situations where a treatment or program is applied in a large coherent unit (such as a state, county, city, hospital, or company) instead of to an individual, where many other units are available as control cases, and where there is routine measurement (for example, monthly or annually) of relevant outcome variables. The analysis of the RPD design turns out to be a variation of the Analysis of Covariance (ANCOVA) model.

SUMMARY

We learned in previous chapters that randomized experiments are often considered the gold standard for causal inference (at least with respect to internal validity). However, many of the research questions that we would like to answer simply cannot be answered by relying on true experiments. This is usually because either we cannot randomly allocate participants to treatment conditions for practical reasons or it would be unethical to do so.

As an alternative to the experimental design, this chapter introduced the idea of quasi-experimental designs. These designs look a bit like their randomized or true experimental relatives (described in Chapter 9, "Experimental Design") but they lack their random assignment to groups. Two major types of quasi-experimental designs were explained in detail. Both are pre-post, two-group designs, and they differ primarily in the manner used to assign the groups. In the NEGD, groups are assigned naturally or are used intact; the researcher does not control the assignment. On the other hand, in RD designs, participants are assigned to groups solely on the basis of a cutoff score on the pre-program measure; the researcher explicitly controls this assignment. Because assignment is explicitly controlled in the RD design and not in NEGD, the former is considered stronger with respect to internal validity, perhaps comparable in strength to randomized experiments.

Finally, the versatility and range of quasi-experimental designs were illustrated through a brief presentation of a number of lesser-known designs that show various combinations of sampling, measurement, or analysis strategies.

In conclusion, it is important to note that quasi-experimental designs are not always inferior to true experiments. Sometimes quasi-experiments are the next logical step in a long research process, when laboratory-based experimental findings need to be tested in practical situations to see if the findings are generalizable to real-world contexts. For example, a true experiment may have demonstrated that a treatment (say, peer teaching) leads to a certain outcome (higher test score) under highly controlled artificial conditions. However, a larger external validity question remains unanswered—that is, whether peer teaching is generally a good thing for children in their schools. Because quasi-experimental studies are conducted in more natural settings, they can help minimize threats to external validity.

Key Terms

double-pretest design p. 270
nonequivalent-groups design (NEGD) p. 259
nonequivalent dependent variables (NEDV) design p. 270

pattern-matching NEDV design p. 271
proxy-pretest design p. 268
regression-discontinuity (RD) design p. 264

regression point displacement (RPD) design p. 272
separate pre-post samples design p. 269

Suggested Websites

American Evaluation Association

http://www.eval.org/

The American Evaluation Association maintains a very high-quality and informative website. Here you can get an overview of the field, including ways that applied social research has incorporated quasi-experimental designs in important evaluation studies.

TREND Statement

http://www.cdc.gov/trendstatement/

As more and more research synthesis happens (e.g., meta-analysis), the importance of complete and standardized reporting of research increases. The TREND (Transparent Reporting of Evaluations with Nonrandomized Designs) Statement provides a checklist and guidelines for clear and systematic reporting of quasi-experimental and other nonrandomized studies.

Review Questions

1. When a researcher uses two groups that already exist (such as two classrooms of students), gives a treatment to one and not the other, and compares the changes in group scores from pre to post, the researcher is using a

a. true experimental design.
b. nonequivalent-groups design.
c. randomized quasi-experimental design.
d. nonexperimental design.
(Reference: 10.2)

2. One of the greatest threats to internal validity in a quasi-experimental design is the likelihood that

a. one is comparing equivalent groups.
b. the nonequivalence of study groups may affect the results.
c. both groups will have similar pretest scores.
d. both groups will demonstrate regression toward the mean.
(Reference: 10.2)

3. The most important aspect of the term "nonequivalent" in a "nonequivalent-groups design" study is that

a. the groups are not equivalent.
b. research participants were not assessed to determine if they were equivalent.
c. research participants were not randomly assigned to groups.
d. the groups were haphazardly selected.
(Reference: 10.2)

4. Which of the following is the greatest threat to the internal validity of a quasi-experimental design?

a. selection bias (that the groups differed initially)
b. inadequate statistical power (not enough participants to see a difference)
c. inadequate preoperational explication of constructs (poor job of determining what is to be studied and why)
d. mortality (loss of participants during the course of the study)
(Reference: 10.2a)

5. If the pretest means of two groups in a quasi-experiment are analyzed and determined to be different, then there is likely a

a. maturation bias.
b. instrumentation bias.
c. selection bias.
d. testing bias.
(Reference: 10.2a)

6. Which quasi-experimental design is generally considered as strong with respect to internal validity as a randomized experiment?
a. bivariate distribution design
b. regression-discontinuity design
c. proxy-pretest design
d. switching-replications design
(Reference: 10.3c)

7. In which research design is a program given to the persons most in need?

a. bivariate distribution design
b. regression-discontinuity design
c. proxy-pretest design
d. switching-replications design
(Reference: 10.3)

8. In the regression-discontinuity design, a positive treatment effect can be inferred from

a. continuity between expected versus observed regression lines.
b. discontinuity between expected versus observed regression lines.
c. continuity between regression lines and additional knowledge, such as who received what treatment and how to interpret the outcome measures.
d. discontinuity between regression lines and additional knowledge of who received what treatment and how to interpret the outcome measures.
(Reference: 10.3a)

9. Which of the following quasi-experimental designs could be used if a researcher is brought into

a study after a program has been implemented and the researcher needs a pretest measure?

a. proxy-pretest design
b. double-pretest design
c. regression point displacement design
d. nonequivalent dependent variables design
(Reference: 10.4a)

10. Which of the following quasi-experimental designs is particularly good for addressing selection-maturation threats?

a. proxy-pretest design
b. double-pretest design
c. regression point displacement design
d. nonequivalent dependent variables design
(Reference: 10.4c)

11. The term *nonequivalent* in the nonequivalent-groups design refers to the fact that a pretest has determined that the groups are significantly different.

a. True
b. False
(Reference: 10.2)

12. Examining the bivariate distribution in a graph is useful because it allows you to see the relationship between two variables. When studying groups, as in a quasi-experimental design, the bivariate distribution in a graph allows you to readily see whether there may be a treatment effect even before you calculate any statistical results.

a. True
b. False
(Reference: 10.2a)

13. The regression-discontinuity design may be as strong as the randomized experiment in terms of internal validity, but the true experiment provides more statistical power (that is, you need fewer participants to determine with the same level of confidence whether there is an effect).

a. True
b. False
(Reference: 10.3d)

14. The nonequivalent dependent variables (NEDV) design can be a particularly powerful design with regard to causal analysis if viewed from the pattern-matching perspective when the dependent variables are reasonably well known in terms of how they should behave under various conditions (that is, they can be expected to respond in predictable ways to the independent variable).

a. True
b. False
(Reference: 10.4e)

15. If the groups in a nonequivalent-groups quasi-experiment seem to respond at different rates to a program, but in fact are maturing at different rates so that the apparent program effect is an illusion, we would say that the design suffers from a selection-mortality threat to validity.

a. True
b. False
(Reference: 10.2a)

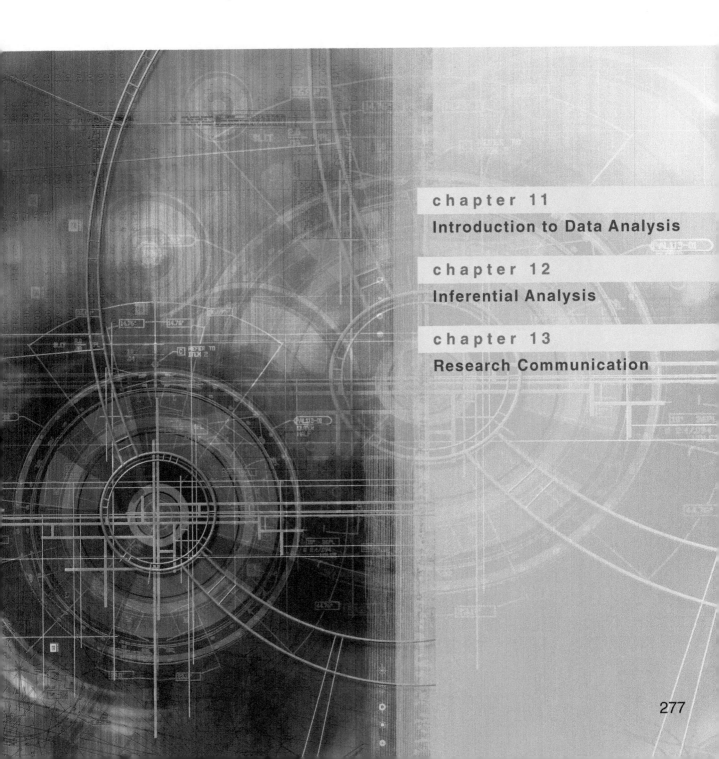

277

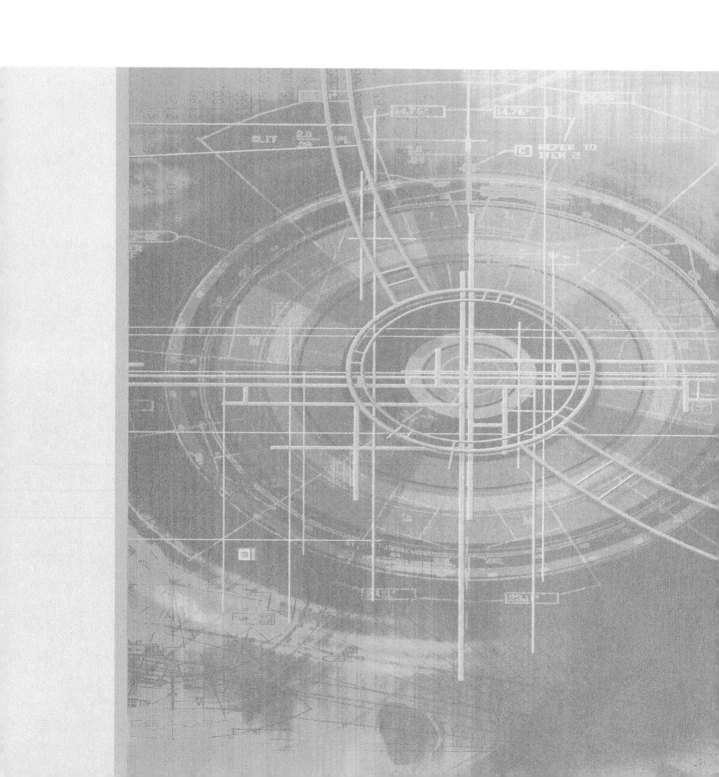

11

Introduction to Data Analysis

11.1 Foundations of Data Analysis

If there ever was something called a "research hurdle," then congratulations on having crossed over! By the time you get to the analysis of your data, most of the really difficult work has been done. It's much harder to come up with and define a research problem; obtain IRB approval; develop and implement a sampling plan; conceptualize, test, and operationalize your measures; develop a design; and, collect a clean and complete data set. If you have done this work well, then the analysis of the data should be relatively straightforward.

The main idea to keep in mind is that the process of data analysis has important implications for a key validity type—conclusion validity. Conclusion validity refers to the extent to which conclusions or inferences regarding relationships between the major variables in your research (e.g., the treatment and the outcome) are warranted.

The next thing to note is that in most social science research, data analysis involves three major steps, performed roughly in this order:

1. *Data preparation* involves logging the data in; making a codebook; entering the data into the computer; checking the data for accuracy; transforming the data; and, developing and documenting a database that integrates all of your measures.

2. *Descriptive statistics* describe the basic features of the data in a study. They provide meaningful summaries about the sample so that potential patterns might emerge from the data. Together with graphical analysis, they form the basis of virtually every form of quantitative analysis. With descriptive statistics, you are simply describing what the data show. Following the discussion of conclusion validity, you will learn about the basics of descriptive analysis in the rest of the chapter.

   Often, the descriptive statistics produced early on are voluminous because we first need to examine each variable individually. We need to know what the distributions of numbers look like, whether we have issues to deal with like extremely large or small values ("outliers"), and whether we have problems with missing data. The first step is usually a descriptive summary of characteristics of your sample. Then you move on to describing the basic characteristics of your study variables (the measures). You carefully select and organize these statistics into summary tables and graphs that show only the most relevant or important information. This is especially critical so that you don't "miss the forest for the trees." If you present too much detail, the reader may not be able to follow the central line of the results. More extensive analysis details are appropriately relegated to appendices—reserving only the most critical analysis summaries for the body of the report itself.

3. *Inferential statistical analysis* tests your specific research hypotheses. In descriptive and some relational studies, you may find that simple descriptive summaries like means, standard deviations, and correlations provide you with all of the information you need to answer your research question. In experimental and quasi-experimental designs, you will need to use more complex methods to determine whether the program or treatment has a statistically detectable effect. We save this type of statistical analysis for the next chapter—Chapter 12, "Inferential Analysis."

We'll remind you right now that this is not a statistics text. We'll cover lots of statistics, some elementary and some advanced, but we are not trying to teach you statistics here. Instead, we are trying to get you to think about data analysis and how it fits into the broader context of your research. Another goal is to help you achieve some measure of "statistical literacy," so that you understand the data analyses you need to think about, to set the stage for developing your own proposal.

We'll begin this chapter by discussing conclusion validity. After all, the point of any study is to reach a valid conclusion. This will give you an understanding of some of the key principles involved in data analysis. Then we'll cover the often-overlooked issue of data preparation. This includes all of the steps involved in cleaning and organizing the data for analysis. We then introduce the basic descriptive statistics and consider some general analysis issues. This sets the stage for consideration of the statistical analysis of the major research designs in Chapter 12.

11.2 Conclusion Validity

Conclusion validity is the degree to which conclusions you reach about relationships in your data are reasonable. The emphasis here is on the term *relationships*. The definition suggests that we are most concerned about conclusion validity when we are looking at how two or more variables are associated with each other.

It should be noted that conclusion validity is also relevant in qualitative research. For example, in an observational field study of homeless adolescents, a researcher might, based on field notes, see a pattern that suggests that teenagers on the street who use drugs are more likely to be involved in more complex social networks and to interact with a more varied group of people than non drug users. The relationship in this case would be between drug use and social network complexity. Even though this conclusion or inference may be based entirely on qualitative (observational or impressionistic) data, the conclusion validity of that relationship can still be assessed—that is, whether it is reasonable to conclude that there is a relationship between the two variables.

Similarly, in quantitative research, if you're doing a study that looks at the relationship between socioeconomic status (SES) and attitudes about capital punishment, you eventually want to reach some conclusion. Based on your data, you might conclude that there is a positive relationship—that persons with higher SES tend to be more in favor of capital punishment, whereas those with lower SES tend to be more opposed. Conclusion validity in this case is the degree to which that conclusion or inference is credible or believable.

Conclusion validity is relevant whenever you are looking at relationships between variables, including cause-and-effect relationships. Since causal relationships are the purview of internal validity (see Chapter 8), we have a potential confusion that we should clear up: How, then, do you distinguish conclusion validity from internal validity? Remember from the discussion of internal validity in Chapter 8 that in order to have a strong degree of internal validity in a causal study, three attributes must be present: covariation, temporal precedence of the presumed cause occurring prior to the presumed effect, and the absence of plausible alternative explanations. The first of these, covariation, is the degree to which the cause and effect are related. So when you are trying to establish internal validity, you first need to establish that the cause-effect relationship has conclusion validity. If our study is not concerned with cause-effect relationships, internal validity is irrelevant but conclusion validity still needs to be addressed. Therefore, conclusion validity is only concerned with whether or not a relationship exists; internal validity goes beyond that and assesses whether one variable in the relationship can be said to cause the other. So, in a sense, conclusion validity is needed before we can establish internal validity (but not the other way around).

For instance, in a program evaluation, you might conclude that there is a positive *relationship* between your educational program and achievement test scores because students in the program get higher scores and students who are not in the program get lower ones. Conclusion validity in this case is concerned with how reasonable your conclusion is. However, it is possible to conclude that, while a relationship exists between the program and outcome, the program itself didn't necessarily *cause* the outcome. Perhaps some other factor, and not the program, was responsible for the outcome in this study. The observed differences in the outcome could be due to the fact that the program group was smarter than the comparison group to begin with. Observed posttest differences between these groups could be due to this initial difference and not due to the result of your program. This issue—the possibility that some factor other than your program caused the outcome—is what internal validity is all about. It is possible that in a study you not only can conclude that your program and outcome are related (conclusion validity exists) but also that the outcome may have been caused by some factor other than the program (you don't have internal validity).

One underappreciated aspect of conclusion validity has to do with the context of the study. When we think about conclusion validity, we need to think about the context in which the research was carried out. Consider an example of two studies, one reporting a statistically significant and large effect and the other with a statistically significant but small effect. On the surface, the larger effect might seem to be more important. However, some relatively small effects can be very important, and some relatively big effects much less impressive in the context of real life. For example, a small effect on mortality (that is, life and death) would be more valuable than a large effect on something like a taste comparison of two kinds of frozen pizza (though we should not underestimate the importance of pizza quality, especially to its fanatics!). As you can see, conclusion validity is not purely a statistical issue; it is a matter of proper interpretation of the evidence a study produces.

Data analysis is much more than just drawing conclusions based simply on probability values associated with the null hypothesis (Cumming, 2012; Kline, 2013). We are now seeing a broader and more complete view of analysis in a context that includes power, measurement precision, effect sizes, confidence intervals, replication, and practical and clinical significance—all of which should enhance conclusion validity. We will have more to say about what "significance" means in Chapter 12.

11.2a Threats to Conclusion Validity

threats to conclusion validity Any factor that can lead you to reach an incorrect conclusion about a relationship in your observations.

Type I Error An erroneous interpretation of statistics in which you falsely conclude that there is a significant relationship between variables when in fact there is none. In other words, you reject the null hypothesis when it is true.

Threats to conclusion validity are any factors that can lead you to reach an incorrect conclusion about a relationship in your observations. You can essentially make two kinds of errors when talking about relationships:

- You conclude that there is a relationship when in fact there is not. (You're seeing things that aren't there!) This error is known as a **Type I Error**. This type of error is also referred to as a "false alarm" or a "false positive."

- **You conclude that there is no relationship when in fact there is.** (You missed the relationship or didn't see it.) This error is known as a **Type II Error**. This kind of error is also thought of as a "miss" or "false negative."

We'll classify the specific threats by the type of error with which they are associated.

Type I Error: Finding a Relationship When There Is Not One (or Seeing Things That Aren't There)

- *Level of Statistical Significance:* In statistical analysis, you attempt to determine the probability of whether your finding is either a real one or a chance event. You then compare this probability to a set criterion called the **alpha level** and decide whether to accept the statistical result as evidence that there is a relationship. In the social sciences, researchers conventionally use the rather arbitrary value, known as the **.05 level of significance,** as the criterion to decide whether their result is credible or could be considered a fluke. Essentially, the value .05 means that the result you got could be expected to occur by chance at least five times out of every 100 times you ran the statistical analysis, if the null hypothesis is in reality true.

  In other words, if you meet this criterion, it means that the probability that the conclusion about a significant result is false (Type I error) is at most 5 percent. As you can imagine, if you were to choose a bigger criterion as your alpha level, you would then increase your chances of committing a Type I error. For example, a .10 level of significance means that the result you got could be expected to occur by chance at least ten times out of every 100 times you ran the statistical analysis, when the null hypothesis is actually true. So, as compared to a 5 percent level of significance, a 10 percent level of significance increases the probability that the conclusion about a significant result is false. Therefore, because it is up to your discretion, it is important to choose the significance level carefully—as far as reducing the potential for making a Type I error goes, the lower the alpha level, the better it is.

- *Fishing and Error Rate Problem:* In anything but the most trivial research study, the researcher spends a considerable amount of time analyzing the data for relationships. Of course, it's important to conduct a thorough analysis, but most people are well aware of the fact that if you torture data long enough, you can usually turn up results that support or corroborate your hypotheses. In more everyday terms, you are "fishing" for a specific result by analyzing the data repeatedly under slightly differing conditions or assumptions.

  Let's think about why this is a problem. A major assumption that underlies most statistical analyses is that each analysis is independent of the other. However, that is usually not true when you conduct multiple analyses of the same data in the same study. For instance, let's say you conduct 20 statistical tests and for each one you use the .05 level criterion for deciding whether you are observing a relationship. For each test, the odds are 5 out of 100 that you will see a relationship even if there is not one there. Odds of 5 out of 100 are equal to the fraction 5/100 which is also equal to 1 out of 20. Now, in this study, you conduct 20 separate analyses. Let's say that you find that of the twenty results, only one is statistically significant at the .05 level. Does that mean you have found a real relationship? If you had only done the one analysis, you might conclude that you found a relationship in that result. However, if you did 20 analyses, you would expect to find one

Type II Error The failure to reject a null hypothesis when it is actually false.

alpha level The *p* value selected as the significance level. Specifically, alpha is the Type I error, or the probability of concluding that there is a treatment effect when, in reality, there is not.

.05 level of significance This is the probability of a Type I error. It is selected by the researcher before a statistical test is conducted. The .05 level has been considered the conventional level since it was identified by the statistician R. A. Fisher. See also *alpha level*.

fishing and the error rate problem A problem that occurs as a result of conducting multiple analyses and treating each one as independent.

effect size A standardized estimate of the strength of a relationship or effect. In planning a study, you need to have an idea of the minimum effect you want to detect in order to identify an appropriate sample size. In Meta-Analysis, effect sizes are used to obtain an overall estimate of outcomes across all relevant studies.

low reliability of measures A threat to conclusion validity that occurs because measures that have low reliability by definition have more noise than measures with higher reliability, and the greater noise makes it difficult to detect a relationship (i.e., to see the signal of the relationship relative to the noise).

poor reliability of treatment implementation A threat to conclusion validity that occurs in causal studies because treatments or programs with more inconsistent or unreliable implementation introduce more noise than ones that are consistently carried out, and the greater noise makes it difficult to detect a relationship (i.e., to see the signal of the relationship relative to the noise).

random irrelevancies in the setting A threat to conclusion validity that occurs when factors in the research setting that are irrelevant to the relationship being studied add noise to the environment, which makes it harder to detect a relationship between key variables, even if one is present.

of them significant by chance alone, even if no real relationship exists in the data. This threat to conclusion validity is called the **fishing and the error rate problem**.

The basic problem is that, like a fisherman, you repeatedly try to catch something—each analysis is like another cast into the water, and you keep going until to "catch" one. In fishing, it doesn't matter how many times you try to catch one (depending on your patience). In statistical analysis, it does matter because if you report just the one you caught, we would not know how many other analyses were run in order to catch that one. Actually, maybe this is also true for fishing. If you go out and catch one fish, the degree to which we think the fishing is "good" depends on whether you got that fish with the first cast or had to cast all day long to get it!

Instead, when you conduct multiple analyses, you should adjust the error rate (the significance level or alpha level) to reflect the number of analyses you are doing, and this should be planned from the time you develop your hypothesis. The bottom line is that you are more likely to see a relationship when there isn't one if you keep reanalyzing your data and don't take your fishing into account when drawing your conclusions.

Type II Error: Finding No Relationship When There Is One (or, Missing the Needle in the Haystack)

- *Small Effect Size:* When you're looking for the needle in the haystack, you essentially have two basic problems: the tiny needle and too much hay. You can think of this as a signal-to-noise ratio problem. What you are observing in research is composed of two major components: the signal, or the relationship you are trying to see; and the noise, or all of the factors that interfere with what you are looking at. This ratio of the signal to the noise (or needle to haystack) in your research is often called the **effect size.**

- *Sources of Noise:* There are several important sources of noise, each of which can be considered a threat to conclusion validity. One important threat is **low reliability of measures** (see the section "Reliability" in Chapter 5, "Introduction to Measurement"). This can be caused by many factors, including poor question wording, bad instrument design or layout, illegibility of field notes, and so on. In studies where you are evaluating a program, you can introduce noise through **poor reliability of treatment implementation.** If the program doesn't follow the prescribed procedures or is inconsistently carried out, it will be harder to see relationships between the program and other factors like the outcomes. Noise caused by **random irrelevancies in the setting** can also obscure your ability to see a relationship. For example, in a classroom context, the traffic outside the room, disturbances in the hallway, and countless other irrelevant events can distract the researcher or the participants. The types of people you have in your study can also make it harder to see relationships. The threat here is due to the **random heterogeneity of respondents.** If you have a diverse group of respondents, group members are likely to vary more widely on your measures or observations. Some of their variability may be related to the phenomenon you are looking at, but at least part of it is likely to constitute individual differences that are irrelevant to the relationship you observe. All of these threats add variability into the research context and contribute to the noise relative to the signal of the relationship you are looking for.

- *Source of Weak Signal:* Noise is only one part of the problem. You also have to consider the issue of the signal—the true strength of the relationship. A *low-strength intervention* could be a potential cause for a weak signal. For example, suppose you want to test the effectiveness of an after-school tutoring program on student test scores. One way to attain a strong signal, and thereby a larger effect size, is to design the intervention research with the strongest possible "dose" of the treatment. Thus, in this particular case, you would want to ensure that teachers delivering the after-school program are well trained and spend a significant amount of time delivering it. The main idea here is that if the intervention is effective, then a stronger dose will make it easier for us to detect the effect. That is, if the program is effective, then a higher dose of the intervention (two hours of tutoring) will create a bigger contrast in the treatment and control group outcomes compared to a weaker dose of the program (one hour of tutoring).

Problems That Can Lead to Either Conclusion Error

Every analysis is based on a variety of assumptions about the nature of the data, the procedures you use to conduct the analysis, and the match between these two. If you are not sensitive to the assumptions behind your analysis, you are likely to draw incorrect conclusions about relationships. In quantitative research, this threat is referred to as **violated assumptions of statistical tests**. For instance, many statistical analyses are based on the assumption that the data are distributed normally—that the population from which data are drawn would be distributed according to a normal or bell-shaped curve. If that assumption is not true for your data and you use that statistical test, you are likely to get an incorrect estimate of the true relationship. It's not always possible to predict what type of error you might make—seeing a relationship that isn't there or missing one that is. Similarly, if your analysis assumes random selection and/or random assignment, but if that assumption is not true, then you can end up increasing your likelihood of making Type I or Type II errors.

Similar problems can occur in qualitative research as well. There are assumptions, some of which you may not even realize, behind all qualitative methods. For instance, in interview situations you might assume that the respondents are free to say anything they wish. If that is not true—if the respondent is under covert pressure from supervisors to respond in a certain way—you may erroneously see relationships in the responses that aren't real and/or miss ones that are.

The threats discussed in this section illustrate some of the major difficulties and traps that are involved in one of the most basic areas of research—deciding whether there is a relationship in your data or observations. So, how do you attempt to deal with these threats? The following section details a number of strategies for improving conclusion validity by minimizing or eliminating these threats.

11.2b Improving Conclusion Validity

So let's say you have a potential problem ensuring that you reach credible conclusions about relationships in your data. What can you do about it? In general,

random heterogeneity of respondents A threat to statistical conclusion validity. If you have a very diverse group of respondents, they are likely to vary more widely on your measures or observations. Some of their variety may be related to the phenomenon you are looking at, but at least part of it is likely to just constitute individual differences that are irrelevant to the relationship being observed.

violated assumptions of statistical tests The threat to conclusion validity that arises when key assumptions required by a specific statistical analysis are not met in the data.

minimizing the various threats that increase the likelihood of making Type I and Type II errors (discussed in the previous section) can help improve the overall conclusion validity of your research.

One thing that can strengthen conclusion validity relates to improving **statistical power**. The concept of statistical power is central to conclusion validity and is related to both Type I and Type II errors—but more directly to Type II error. Statistical power is technically defined as the probability that you will conclude there is a relationship when in fact there is one. In other words, power is the odds of correctly finding the needle in the haystack. Like any probability, statistical power can be described as a number between 0 and 1. For instance, if we have statistical power of .8, then it means that the odds are 80 out of 100 that we will detect a relationship when it is really there. Or, it's the same as saying that the chances are 80 out of 100 that we will find the needle that's in the haystack. We want statistical power to be as high as possible. Power will be lower in our study if there is either more noise (a bigger haystack), a weaker signal (a smaller needle) or both. So, improving statistical power in your study usually involves important trade-offs and additional costs.

The rule of thumb in social research is that you want statistical power to be at least .8 in value. Several factors interact to affect power. Here are some general guidelines you can follow in designing your study that will help improve statistical power and thus the conclusion validity of your study.

- **Increase the sample size:** One thing you can usually do is collect more information—use a larger sample size. Of course, you have to weigh the gain in power against the time and expense of having more participants or gathering more data. There are now dedicated power analysis programs as well as an array of web-based calculators to enable you to determine a specific sample size for your analysis. These estimators ask you to identify your design, input your alpha level and the smallest effect that would be important to be able to see (sometimes referred to as the "minimally important difference"), and they provide you with a sample size estimate for a given level of power.

- **Increase the level of significance:** If you were to increase your risk of making a Type I error—increase the chance that you will find a relationship when it's not there—you would be improving statistical power of your study. In practical terms, you can do that statistically by raising the alpha level or level of significance. For instance, instead of using a .05 significance level, you might use .10 as your cutoff point. However, as you probably realize, this represents a trade-off. Because increasing the level of significance also makes it more likely for a Type I error to occur (which negatively affects conclusion validity), we recommend that you first try other steps to improve statistical power.

- **Increase the effect size:** Because the effect size is a ratio of the signal of the relationship to the noise in the context, there are two broad strategies here. To raise the signal, you can increase the salience of the relationship itself. This is especially true in experimental studies where you are looking at the effects of a program or treatment. If you increase the dosage of the program (for example, increase the hours spent in training or the number of training sessions), it should be easier to see an effect. The other option is to decrease the noise (or, put another way, increase reliability). In general, you can improve reliability by doing a better job of constructing measurement

statistical power The probability that you will conclude there is a relationship when in fact there is one. We typically want statistical power to be at least 0.80 in value.

instruments, by increasing the number of questions on a scale, or by reducing situational distractions in the measurement context. When you improve reliability, you reduce noise, which increases your statistical power and improves conclusion validity. Similarly, you can also reduce noise by ensuring good implementation. You accomplish this by training program operators and standardizing the protocols for administering the program and measuring the results.

11.3 Data Preparation

Now that you understand the basic concept of conclusion validity, it's time to discuss how we actually carry out data analysis. The first step in this process is data preparation. Data preparation involves acquiring or collecting the data; checking the data for accuracy; entering the data into the computer; transforming the data; and developing and documenting a database structure that integrates the various measures.

11.3a Logging the Data

In any research project, you might have data coming from several different sources at different times, such as:

- Survey returns
- Coded interview data
- Pretest or posttest data
- Observational data

In all but the simplest of studies, you need to set up a procedure for logging the information and keeping track of it until you are ready to do a comprehensive data analysis. Different researchers differ in how they keep track of incoming data. In most cases, you will want to set up a database that enables you to assess, at any time, which data are already entered and which still need to be entered. You could do this with any standard computerized spreadsheet (Microsoft Excel) or database (Microsoft Access, Filemaker) program. You can also accomplish this by using standard statistical programs (for example, SPSS, SAS, Minitab, Datadesk, etc.) by running simple descriptive analyses to get reports on data status.

It is also critical that the data analyst retains and archives the original data records—returned surveys, field notes, test protocols, and so on—for a reasonable period of time. Most professional researchers retain such records for at least five to seven years. For important or expensive studies, the original data might be stored in a formal data archive. The data analyst should always be able to trace a result from a data analysis back to the original forms on which the data were collected. Most IRBs now require researchers to keep information that could identify a participant separate from the data files. All data and consent forms should be kept in a secure location with password protection and encryption whenever possible. A database for logging incoming data is a critical component in good research recordkeeping.

11.3b Checking the Data for Accuracy

As soon as you receive the data, you should screen them for accuracy. In some circumstances, doing this right away allows you to go back to the sample to clarify any problems or errors. You should ask the following questions as part of this initial data screening:

- Are the responses legible/readable?
- Are all important questions answered?
- Are the responses complete?
- Is all relevant contextual information included (for example, date, time, place, and researcher)?

In most social research, the quality of data collection is a major issue. Ensuring that the data-collection process does not contribute inaccuracies helps ensure the overall quality of subsequent analyses.

11.3c Developing a Database Structure

The database structure is the system you use to store the data for the study so that it can be accessed in subsequent data analyses. You might use the same structure you used for logging in the data; or in large, complex studies, you might have one structure for logging data and another for storing it. As mentioned previously, there are generally two options for storing data on a computer: database programs and statistical programs. Usually database programs are the more complex of the two to learn and operate, but generally they allow you greater flexibility in manipulating the data.

In every research project, you should generate a printed **codebook** that describes each variable in the data and indicates where and how it can be accessed. Minimally the codebook should include the following items for each variable:

- **Variable name**
- **Variable description**
- **Variable format (number, data, text)**
- **Instrument/method of collection**
- **Date collected**
- **Respondent or group**
- **Variable location (in database)**
- **Notes**

The codebook is an indispensable tool for the analysis team. Together with the database, it should provide comprehensive documentation that enables other researchers who might subsequently want to analyze the data to do so without any additional information.

11.3d Entering the Data into the Computer

If you administer an electronic survey then you don't have to enter the data into a computer—the data are already in the computer! However, if you decide to use paper measures, you can enter data into a computer in a variety of ways. Probably the easiest is to just type in the data directly. You could enter the data in a word processor (like Microsoft Word) or in a spreadsheet program (like Microsoft Excel). You could also use a database or a statistical program for data entry. Note that most statistical programs allow you to import spreadsheets directly into data

codebook A written description of the data that describes each variable and indicates where and how it can be accessed.

files for the purposes of analysis, so sometimes it makes more sense to enter the data in a spreadsheet (the "database" it will be stored in) and then import it into the statistical program.

It is very important that your data file is arranged by a unique ID number for each case. Each case's ID number on the computer should be recorded on its paper form (so that you can trace the data back to the original if you need to). When you key-in the data, each case (or each unique ID number) would typically have its own separate row. Each column usually represents a different variable. If you input your data in a word processor, you usually need to use a delimiter value to separate one variable from the next on a line. In some cases, you might use the comma, in what is known as a comma-separated file. For instance, if two people entered 1-to-5 ratings on five different items, we might enter their data into a word processing program like this:

001, 4, 5, 3, 5, 4

002, 4, 4, 3, 4, 4

Here, the first number in each row is the ID number for the respondent. The remaining numbers are their five 1-to-5 ratings. You would save this file as a text file (not in the word processor's native file format) and then it could be imported directly into a statistics package or database. The only problem with this is that you may have times when one of your qualitative variables can include a comma within a variable. For instance, if you have a variable that includes city and state you might have a respondent who is from "New York, NY." If you enter this as is into the word processor and save it as a comma-delimited file, there will be a problem when you import it into a statistics program or spreadsheet because the comma will incorrectly split that field into two. So, in cases like this, you either need to not include comma values in the variable or you need to use a delimiter that will not be confusing. In this case, you might use a tab to delimit each variable within a line. Consider the above example with the addition of the city/state variable where we use a tab ($\rightarrow$) to delimit the fields:

001→New York, NY→4→5→3→5→4

002→Hartford, CT→4→4→3→4→4

When you transfer your text-based input into a spreadsheet or a statistical program for analysis purposes, the delimiter will tell the program that it has just finished reading the value for one variable and that now the value for the next variable is coming up.

To ensure a high level of data accuracy for quantitative data, you can use a procedure called **double entry**. In this procedure, you enter the data once. Then, you use a special program that allows you to enter the data a second time and then checks the second entries against the first. If there is a discrepancy, the program immediately notifies you and enables you to determine which is the correct entry. This double-entry procedure significantly reduces entry errors. However, these double-entry programs are not widely available and require some training.

An alternative is to enter the data once and set up a procedure for checking the data for accuracy. These procedures might include rules that limit the data that can be entered into the program, typically thought of as a "validation rule." For example, you might set up a rule indicating that values for the variable Gender

double entry An automated method for checking data-entry accuracy in which you enter data once and then enter them a second time, with the software automatically stopping each time a discrepancy is detected until the data enterer resolves the discrepancy. This procedure assures extremely high rates of data-entry accuracy, although it requires twice as long for data entry.

can only be 1 or 2. Then, if you accidentally try to enter a 3, the program will not accept it. Once data are entered, you might spot-check records on a random basis. In cases where you have two different data-entry workers entering the same data into computer files, you can use Microsoft Word's "Document Compare" functionality to compare the data files. The disadvantage of this approach is that it can only compare entries when they are completed (unlike traditional double-entry verifiers that stop the data enterer immediately when a discrepancy is detected).

If you do not have a program with built-in validation rules, you need to examine the data carefully so that you can check that all the data fall within acceptable limits and boundaries. For instance, simple summary reports would enable you to spot whether there are persons whose age is 601 or whether anyone entered a 7 where you expected a 1-to-5 response.

11.3e Data Transformations

After the data are entered, it is often necessary to transform the original data into variables that are more usable. There are a variety of transformations that you might perform. The following are some of the more common ones:

Missing values: Many analysis programs automatically treat blank values as missing. In others, you need to designate specific values to represent missing values. For instance, you might use a value that could not be valid for your variable (e.g., -99) to indicate that the item is missing. You need to check the specific analysis program you are using to determine how to handle missing values, and be sure that the program correctly identifies the missing values so they are not accidentally included in your analysis. Some measures come with scoring manuals that include procedures for prorating scales in which less than 100 percent of the data are available. In other cases, no such procedures exist and the researcher must decide how to handle missing data. Many articles and several books have been written on how to best estimate a missing value from other available data, as well as when estimation of missing values is not advisable. You may not need to delete entire participants from your study just because they have missing values, especially if values are not missing at random.

Item reversals: On scales and surveys, the use of reversal items (see Chapter 6, "Scales, Tests and Indexes") can help reduce the possibility of a response set. When you analyze the data, you want all scores for questions or scale items to be in the same direction, where high scores mean the same thing and low scores mean the same thing. In such cases, you may have to reverse the ratings for some of the scale items to get them in the same direction as the others. For instance, let's say you had a five-point response scale for a self-esteem measure where 1 meant strongly disagree and 5 meant strongly agree. One item is "I generally feel good about myself." If respondents strongly agree with this item, they will put a 5, and this value would be indicative of higher self-esteem. Alternatively, consider an item like "Sometimes I feel like I'm not worth much as a person." Here, if a respondent strongly agrees by rating this a 5, it would indicate low self-esteem. To compare these two items, you would reverse the scores. (Probably you'd reverse the latter item so that higher values always indicate higher self-esteem.) You want a transformation where, if the original value was 1, it's changed to 5; 2 is changed to 4; 3 remains the same; 4 is changed to 2; and 5 is changed to 1. Although you could program these changes as separate statements in most programs, it's easier to do this with a simple formula like the following:

$$\text{New Value} = (\text{High Value} + 1) - \text{Original Value}$$

In our example, the *high value* for the scale is 5; so to get the new (transformed) scale value, you simply subtract the *original value* on each reversal item from 6 (that is, 5 + 1).

Scale and subscale totals: After you transform any individual scale items, you will often want to add or average across individual items to get scores for any subscales and a total score for the scale.

Categories: You may want to collapse one or more variables into categories. For instance, you may want to collapse income estimates (in dollar amounts) into income ranges.

Variable transformations: In order to meet assumptions of certain statistical methods, we often need to transform particular variables. Depending on the data you have, you might transform them by expressing them in logarithm or square-root form. For example, if data on a particular variable are skewed in the positive direction, then taking its square root can make it look closer to a normal distribution—a key assumption for many statistical analyses. You should be careful to check if your transformation produced an erroneous value—for example, in the above case, if you proposed transforming a variable with some negative values (one cannot take a square root of a negative number!). Finally, you should be careful in interpreting the results of your statistical analysis when using transformed variables—remember, with transformed variables, you are no longer analyzing the relationship between the original variable and some other variable on the same scale that you started with.

11.4 Descriptive Statistics

Descriptive statistics describe the basic features of the data in a study. They provide simple summaries about the sample and the measures. Together with simple graphical analysis, they form the basis of virtually every quantitative analysis of data.

descriptive statistics
Statistics used to describe the basic features of the data in a study.

Descriptive statistics present quantitative descriptions in a manageable form. In a research study, you may have many measures, or you might measure a large number of people on any given measure. Descriptive statistics help you summarize large amounts of data in a sensible way. Each descriptive statistic reduces data into a simpler summary. For instance, consider a simple number used to summarize how well a batter is performing in baseball, the batting average. This single number is the number of hits divided by the number of times at bat (reported to three significant digits). A batter who is hitting .333 is getting a hit one time in every three at-bats. One batting .250 is hitting one time in four. The single number describes a large number of discrete events. Or, consider the scourge of many students: the grade-point average (GPA). This single number describes the general performance of a student across a potentially wide range of course experiences.

Every time you try to describe a large set of observations with a single indicator, you run the risk of distorting the original data or losing important detail (see **Figure 11.1**). The batting average doesn't tell you

Figure 11.1 Do you see any conclusion validity problems in the descriptive summary of Hillsville?

Dana Fradon/Condé Nast

whether batters hit home runs or singles. It doesn't tell whether they've been in a slump or on a streak. The GPAs don't tell you whether the students were in difficult courses or easy ones, or whether the courses were in their major field or in other disciplines. Even given these limitations, descriptive statistics provide a powerful summary that enables comparisons across people or other units.

A single variable has three major characteristics that are typically described:

- The distribution
- The central tendency
- The dispersion

In most situations, you would describe all three of these characteristics for each of the variables in your study.

11.4a The Distribution

The **distribution** is a summary of the frequency of individual values or ranges of values for a variable. The simplest distribution lists every value of a variable and the number of persons who had each value. For instance, a typical way to describe the distribution of college students is by year in college, listing the number or percent of students at each of the four years. Or, you describe gender by listing the number or percent of males and females. In these cases, the variable has few enough values that you can list each one and summarize how many sample cases had the value. But what do you do for a variable like income or GPA? These variables have a large number of possible values, with relatively few people having each one. In this case, you group the raw scores into categories according to ranges of values. For instance, you might look at GPA according to the letter-grade ranges, or you might group income into four or five ranges of income values.

One of the most common ways to describe a single variable is with a **frequency distribution**. Depending on the particular variable, all of the data values might be represented, or you might group the values into categories first. For example, with age, price, or temperature variables, it is usually not sensible to determine the frequencies for each value. Rather, the values are grouped into ranges and the frequencies determined. In many situations, we are able to determine that a distribution is approximately normal by examining a graph. But we can also use statistical estimates of skew (leaning toward one end or the other) and kurtosis (peaks and flatness of the distribution) to make judgments about deviations from a normal distribution.

Frequency distributions can be depicted in two ways, as a table or as a graph. **Figure 11.2** shows an age frequency distribution with five categories of age ranges defined. The same frequency distribution can be depicted in a graph as shown in **Figure 11.3**. This type of graph is often referred to as a histogram or bar chart.

Distributions can also be displayed using percentages. For example, you could use percentages to describe the following:

- Percentage of people in different income levels
- Percentage of people in different age ranges
- Percentage of people in different ranges of standardized test scores

A frequency distribution
in table form

Category	Percent
Under 35	9%
36–45	21%
46–55	**45%**
56–65	19%
66+	6%

Figure 11.2 A frequency distribution in table form.

distribution The manner in which a variable takes different values in your data.

frequency distribution A summary of the frequency of individual values or ranges of values for a variable.

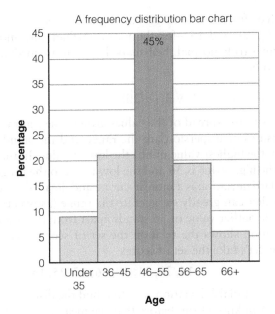

A frequency distribution bar chart

Figure 11.3 A frequency distribution bar chart.

11.4b Central Tendency

The **central tendency** of a distribution is an estimate of the center of a distribution of values. There are three major types of estimates of central tendency:

- Mean
- Median
- Mode

The **mean** or average is probably the most commonly used method of describing central tendency. To compute the mean, all you do is add up all the values and divide by the number of values. For example, the mean or average quiz score is determined by summing all the scores and dividing by the number of students taking the exam. Consider the test score values:

$$15, 20, 21, 20, 36, 15, 25, 15$$

The sum of these eight values is 167, so the mean is $167/8 = 20.875$.

The **median** is the score found at the exact middle of the set of values. One way to compute the median is to list all scores in numerical order and then locate the score in the center of the sample. For example, if there are 500 scores in the list, score number 250 would be the median. If you order the eight scores shown previously, you would get

$$15, 15, 15, 20, 20, 21, 25, 36$$

There are eight scores and score number 4 and number 5 represent the halfway point. Since both of these scores are 20, the median is 20. If the two middle scores had different values, you find the value midway between them to determine the median.

The **mode** is the most frequently occurring value in the set of scores. To determine the mode, you might again order the scores as shown previously and then count each one. The most frequently occurring value is the mode. In our example, the value 15 occurs three times and is the mode. In some distributions, there is more than one modal value. For instance, in a bimodal distribution, two values occur most frequently.

central tendency An estimate of the center of a distribution of values. The most usual measures of central tendency are the mean, median, and mode.

mean A description of the central tendency in which you add all the values and divide by the number of values.

median The score found at the exact middle or 50th percentile of the set of values. One way to compute the median is to list all scores in numerical order and then locate the score in the center of the sample.

mode The most frequently occurring value in the set of scores.

dispersion The spread of the values around the central tendency. The two common measures of dispersion are the range and the standard deviation.

range The highest value minus the lowest value.

variance A statistic that describes the variability in the data for a variable. The variance is the spread of the scores around the mean of a distribution. Specifically, the variance is the sum of the squared deviations from the mean divided by the number of observations minus 1. The standard deviation and variance both measure dispersion, but because the standard deviation is measured in the same units as the original measure and the variance is measured in squared units, the standard deviation is usually more directly interpretable and meaningful.

Notice that for the same set of eight scores, we got three different values—20.875, 20, and 15—for the mean, median, and mode, respectively. If the distribution is truly normal (bell-shaped), the mean, median, and mode are all equal to each other.

11.4c Dispersion or Variability

Dispersion refers to the spread of the values around the central tendency. The two common measures of dispersion are the range and the standard deviation. The **range** is simply the highest value minus the lowest value. In the previous example distribution, the high value is 36 and the low is 15, so the range is $36 - 15 = 21$.

The standard deviation is a more accurate and detailed estimate of dispersion, because an outlier can greatly exaggerate the range (as was true in this example where the single outlier value of 36 stands apart from the rest of the values). The standard deviation shows the relation the set of scores has to the mean of the variable. Again let's take the set of scores:

$$15, 20, 21, 20, 36, 15, 25, 15$$

To compute the standard deviation, you first find the distance between each value and the mean. You know from before that the mean for the data in this example is 20.875. So, the differences from the mean are:

$$15 - 20.875 = -5.875$$
$$20 - 20.875 = -0.875$$
$$21 - 20.875 = +0.125$$
$$20 - 20.875 = -0.875$$
$$36 - 20.875 = 15.125$$
$$15 - 20.875 = -5.875$$
$$25 - 20.875 = +4.125$$
$$15 - 20.875 = -5.875$$

Notice that values that are below the mean have negative discrepancies and values above it have positive ones. Next, you square each discrepancy:

$$-5.875 \times -5.875 = 34.515625$$
$$-0.875 \times -0.875 = 0.765625$$
$$+0.125 \times +0.125 = 0.015625$$
$$-0.875 \times -0.875 = 0.765625$$
$$15.125 \times 15.125 = 228.765625$$
$$-5.875 \times -5.875 = 34.515625$$
$$+4.125 \times +4.125 = 17.015625$$
$$-5.875 \times -5.875 = 34.515625$$

$$\sqrt{\frac{\Sigma (X - \bar{X})^2}{(n-1)}}$$

where:
X = Each score
$\bar{X}$ = The mean or average
n = The number of values
Σ = Sum of the values

Now, you take these squares and sum them to get the Sum of Squares (SS) value. Here, the sum is 350.875. Next, you divide this sum by the number of scores minus 1. Here, the result is $350.875/7 = 50.125$. This value is known as the **variance**. To get the standard deviation, you take the square root of the variance (remember that you squared the deviations earlier). This would be SQRT(50.125) = 7.079901129253.

Although this computation may seem convoluted, it's actually quite simple and is automatically computed in statistical programs. To see this, consider the formula for the standard deviation shown in **Figure 11.4**. In the top part of the ratio, the numerator, notice that each score has the mean subtracted from it, the

Figure 11.4 Formula for the standard deviation.

Table 11.1 Table of descriptive statistics.

N	8
Mean	20.8750
Median	20.0000
Mode	15.00
Std. deviation	7.0799
Variance	50.1250
Range	21.00

difference is squared, and the squares are summed. In the bottom part, you take the number of scores minus 1. The ratio is the variance and the square root is the standard deviation. In English, the standard deviation is described as follows:

The square root of the sum of the squared deviations from the mean divided by the number of scores minus one.

Although you can calculate these univariate statistics by hand, it becomes quite tedious when you have more than a few values and variables. For instance, we put the eight scores into a commonly used statistics program (SPSS) and got the results shown in Table 11.1. This table confirms the calculations we did by hand previously.

The standard deviation allows you to reach some conclusions about specific scores in your distribution. Assuming that the distribution of scores is normal or bell-shaped (or close to it), you can reach conclusions like the following:

- **Approximately 68 percent of the scores in the sample fall within one standard deviation of the mean.**
- **Approximately 95 percent of the scores in the sample fall within two standard deviations of the mean.**
- **Approximately 99 percent of the scores in the sample fall within three standard deviations of the mean.**

For instance, since the mean in our example is 20.875 and the standard deviation is 7.0799, you can use the statement listed previously to estimate that approximately 95 percent of the scores will fall in the range of 20.875—(2 × 7.0799) to 20.875 + (2 × 7.0799) or between 6.7152 and 35.0348. This kind of information is critical in enabling you to compare the performance of individuals on one variable with their performance on another, even when the variables are measured on entirely different scales.

11.4d Correlation

Correlation is one of the most common and useful measures in statistics. A correlation is a single number that describes the degree of relationship between two variables. Let's work through an example to show how this statistic is computed.

Correlation Example

Let's assume that you want to look at the relationship between two variables, height and self-esteem. Perhaps you have a hypothesis that how tall you are is related to your self-esteem. (Incidentally, we don't think you have to worry about

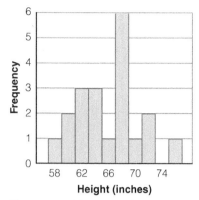

Figure 11.5 Histogram for the height variable in the example correlation calculation.

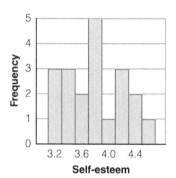

Figure 11.6 Histogram for the self-esteem variable in the example correlation calculation.

Table 11.2 Hypothetical data to demonstrate the correlation between height and self-esteem.

Person	Height	Self-Esteem
1	68	4.1
2	71	4.6
3	62	3.8
4	75	4.4
5	58	3.2
6	60	3.1
7	67	3.8
8	68	4.1
9	71	4.3
10	69	3.7
11	68	3.5
12	67	3.2
13	63	3.7
14	62	3.3
15	60	3.4
16	63	4.0
17	65	4.1
18	67	3.8
19	63	3.4
20	61	3.6

the direction of causality here; it's not likely that self-esteem causes your height.) Let's say you collect some information on twenty individuals—all male. (The average height differs for males and females; so, to keep this example simple, we'll just use males.) Height is measured in inches. Self-esteem is measured based on the average of 10, 1-to-5 rating items (where higher scores mean higher self-esteem). See Table 11.2 for the data for the 20 cases. (Don't take this too seriously; we made this data up to illustrate what correlation is.)

Now, let's take a quick look at the bar chart (histogram) for each variable (see **Figure 11.5** and **Figure 11.6**).

Table 11.3 shows the descriptive statistics.

Finally, look at the simple bivariate (two-variable) plot (see **Figure 11.7**). You should immediately see in the bivariate plot that the relationship between the variables is a positive one, because as you move from lower to higher on one variable, the values on the other variable tend to move from lower to higher as well. If you were to fit a single straight line through the dots, it would have a positive slope or move up from left to right. (If you can't see the

Table 11.3 Descriptive statistics for correlation calculation example.

Variable	Mean	St. Dev.	Variance	Sum	Min.	Max.	Range
Height	65.4	4.40574	19.4105	1308	58	75	17
Self-Esteem	3.755	0.426090	0.181553	75.1	3.1	4.6	1.5

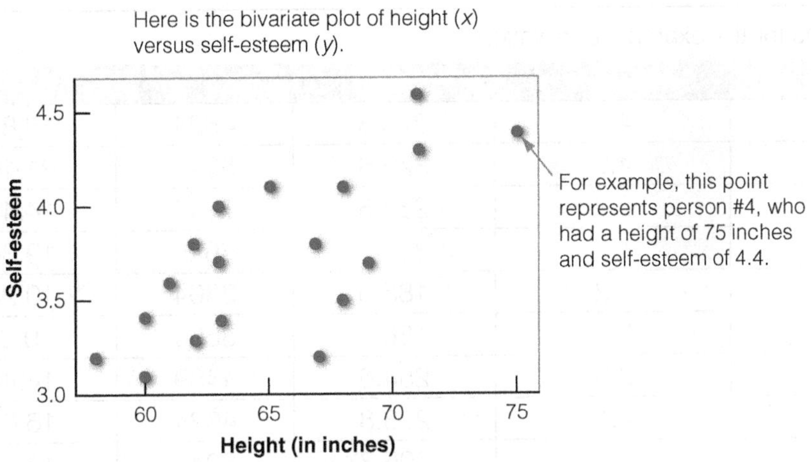

Here is the bivariate plot of height (x) versus self-esteem (y).

For example, this point represents person #4, who had a height of 75 inches and self-esteem of 4.4.

Figure 11.7 Bivariate plot for the example correlation calculation.

positive relationship, review the section "Types of Relationships" in Chapter 1.) Since the correlation is nothing more than a quantitative estimate of the relationship, you would expect a positive correlation.

What does a positive relationship mean in this context? It means that, in general, higher scores on one variable tend to be paired with higher scores on the other, and that lower scores on one variable tend to be paired with lower scores on the other. You should confirm visually that this is generally true in the plot in Figure 11.7.

Calculating the Correlation

Now you're ready to compute the correlation value. The formula for the correlation is shown in **Figure 11.8**.

The symbol r stands for the correlation. Through the magic of mathematics, it turns out that r will always be between -1.0 and $+1.0$. If the correlation is negative, you have a negative relationship; if it's positive, the relationship is positive. (Pretty clever, huh?) You don't need to know how we came up with this formula unless you want to be a statistician. But you probably will need to know how the formula relates to real data—how you can use the formula to compute the correlation. Let's look at the data you need for the formula. Table 11.4 shows the original data with the other necessary columns.

The first three columns are the same as those in Table 11.2. The next three columns are simple computations based on the height and self-esteem data in the first three columns. The bottom row consists of the sum of each column. This is all the information you need to compute the correlation. **Figure 11.9** shows the values from the bottom row of the table (where N is 20 people) as they are related to the symbols in the formula:

$$r = \frac{N\Sigma xy - (\Sigma x)(\Sigma y)}{\sqrt{[N\Sigma x^2 - (\Sigma x)^2][N\Sigma y^2 - (\Sigma y)^2]}}$$

where:

N = Number of pairs of scores

Σxy = Sum of the products of paired scores

Σx = Sum of the x scores

Σy = Sum of the y scores

Σx^2 = Sum of squared x scores

Σy^2 = Sum of squared y scores

Figure 11.8 The formula for correlation.

N = 20

Σxy = 4937.6

Σx = 1308

Σy = 75.1

Σx^2 = 85912

Σy^2 = 285.45

Figure 11.9 The parts of the correlation formula with the numerical values from the example.

Table 11.4 Computations for the example correlation.

Person	Height (x)	Self-Esteem (y)	xy	x^2	y^2
1	68	4.1	278.8	4624	16.81
2	71	4.6	326.6	5041	21.16
3	62	3.8	235.6	3844	14.44
4	75	4.4	330	5625	19.36
5	58	3.2	185.6	3364	10.24
6	60	3.1	186	3600	9.61
7	67	3.8	254.6	4489	14.44
8	68	4.1	278.8	4624	16.81
9	71	4.3	305.3	5041	18.49
10	69	3.7	255.3	4761	13.69
11	68	3.5	238	4624	12.25
12	67	3.2	214.4	4489	10.24
13	63	3.7	233.1	3969	13.69
14	62	3.3	204.6	3844	10.89
15	60	3.4	204	3600	11.56
16	63	4	252	3969	16
17	65	4.1	266.5	4225	16.81
18	67	3.8	254.6	4489	14.44
19	63	3.4	214.2	3969	11.56
20	61	3.6	219.6	3721	12.96
Sum =	**1308**	**75.1**	**4937.6**	**85912**	**285.45**

Now, when you plug these values into the formula in Figure 11.8, you get the following. (We show it here tediously, one step at a time in **Figure 11.10**.)

So, the correlation for the 20 cases is .73, which is a fairly strong positive relationship. It seems like there is a relationship between height and self-esteem, at least in this made-up data!

Figure 11.10 Example of the computation of the correlation.

The Correlation Formula

$$r = \frac{N\Sigma xy - (\Sigma x)(\Sigma y)}{\sqrt{[N\Sigma x^2 - (\Sigma x)^2][N\Sigma y^2 - (\Sigma y)^2]}}$$

$$r = \frac{20(4937.6) - (1308)(75.1)}{\sqrt{[20(85912) - (1308 * 1308)][20(285.45) - (75.1 * 75.1)]}}$$

$$r = \frac{98752 - 98230.8}{\sqrt{[1718240 - 1710864][5709 - 5640.01]}}$$

$$r = \frac{521.2}{\sqrt{[7376][68.99]}} = \frac{521.2}{\sqrt{508870.2}} = \frac{521.2}{713.3514} = .73$$

Testing the Significance of a Correlation

After you've computed a correlation, you can determine the probability that the observed correlation occurred by chance. That is, you can conduct a significance test. Most often, you are interested in determining the probability that the correlation is a real one and not a chance occurrence. When you are interested in that, you are testing the mutually exclusive hypotheses:

$$H_0: r = 0$$
$$H_1: r \neq 0$$

In effect, you are testing whether the real correlation is zero or not. If you are doing your analysis by hand, the easiest way to test this hypothesis is to look up a table of critical values of r online or in a statistics text. As in all hypothesis testing, you need to determine first the significance level you will use for the test. Here, we'll use the common significance level of $\alpha = .05$. This means that we are conducting a test where the odds that the correlation occurred by chance are no more than 5 out of 100. Before we look up the critical value in a table, we also have to compute the **degrees of freedom or *df***. The *df* for a correlation is simply equal to $N - 2$ or, in this example, is $20 - 2 = 18$. Finally, we have to decide whether we are doing a one-tailed or two-tailed test (see the discussion in Chapter 1, "Foundations"). In this example, since we have no strong prior theory to suggest whether the relationship between height and self-esteem would be positive or negative, we'll opt for the two-tailed test. With these three pieces of information—the significance level (alpha = .05), degrees of freedom (*df* = 18), and type of test (two-tailed)—we can now test the significance of the correlation we found. When we look up this value in the handy little table, we find that the critical value is .4438. This means that if our correlation is greater than .4438 or less than −.4438 (remember, this is a two-tailed test) we can conclude that the odds are less than 5 out of 100 that this is a chance occurrence. Since our correlation of .73 is actually quite a bit higher, we conclude that it is not a chance finding and that the correlation is statistically significant (given the parameters of the test) and different from no correlation ($r = 0$). We can reject the null hypothesis and accept the alternative—we have a statistically significant correlation. Not only would we conclude that this estimate of the correlation is statistically significant, we would also conclude that the relationship between height and self-esteem is a strong one. Conventionally, a correlation over .50 would be considered large or strong.

The Correlation Matrix

All we've shown you so far is how to compute a correlation between two variables. In most studies, you usually have more than two variables. Let's say you have a study with ten interval-level variables and you want to estimate the relationships among all of them (between all possible pairs of variables). In this instance, you have forty-five unique correlations to estimate (more later about how we knew that). You could do the computations just completed forty-five times to obtain the correlations, or you could use just about any statistics program to automatically compute all forty-five with a simple click of the mouse.

We used a simple statistics program to generate random data for ten variables with twenty cases (persons) for each variable. Then, we told the program to compute the correlations among these variables. The results are shown in Table 11.5.

This type of table is called a **correlation matrix**. It lists the variable names (in this case, C1 through C10) down the first column and across the first row. The diagonal of a correlation matrix (the numbers that go from the upper-left corner to the lower right) always consists of ones because these are the correlations

degrees of freedom (df) A statistical term that is a function of the sample size. In the t-test formula, for instance, the *df* is the number of persons in both groups minus 2.

correlation matrix A table of all inter-correlations for a set of variables.

Table 11.5 Hypothetical correlation matrix for ten variables.

	C1	C2	C3	C4	C5	C6	C7	C8	C9	C10
C1	1.000									
C2	.274	1.000								
C3	−.134	−.269	1.000							
C4	.201	−.153	.075	1.000						
C5	−.129	−.166	.278	−.011	1.000					
C6	−.095	.280	−.348	−.378	−.009	1.000				
C7	.171	−.122	.288	.086	.193	.002	1.000			
C8	.219	.242	−.380	−.227	−.551	.324	−.082	1.000		
C9	.518	.238	.002	.082	−.015	.304	.347	−.013	1.000	
C10	.299	.568	.165	−.122	−.106	−.169	.243	.014	.352	1.000

$$\frac{N*(N-1)}{2}$$

Figure 11.11 Formula for determining the number of unique correlations given the number of variables.

between each variable and itself (and a variable is always perfectly correlated with itself). The statistical program we used shows only the lower triangle of the correlation matrix. In every correlation matrix, there are two triangles: the values below and to the left of the diagonal (lower triangle) and above and to the right of the diagonal (upper triangle). There is no reason to print both triangles because the two triangles of a correlation matrix are always mirror images of each other. (The correlation of variable x with variable y is always equal to the correlation of variable y with variable x.) When a matrix has this mirror-image quality above and below the diagonal, it is referred to as a symmetric matrix. A correlation matrix is always a symmetric matrix.

To locate the correlation for any pair of variables, find the value in the table for the row and column intersection for those two variables. For instance, to find the correlation between variables C5 and C2, look for where row C2 and column C5 is (in this case, it's blank because it falls in the upper triangle area) and where row C5 and column C2 is and, in the second case, the correlation is −.166.

Okay, so how did we know that there are forty-five unique correlations when there are ten variables? There's a simple little formula that tells how many pairs (correlations) there are for any number of variables (see **Figure 11.11**). N is the number of variables. In the example, we had 10 variables, so we know we have (10 * 9)/2 = 90/2 = 45 pairs.

Other Correlations

The specific type of correlation we've illustrated here is known as the **Pearson product moment correlation**, named for its inventor, Karl Pearson. It is appropriate when both variables are measured at an interval level (see the discussion of level of measurement in Chapter 5, "Introduction to Measurement"). However there are other types of correlations for other circumstances. For instance, if you have two ordinal variables, you could use the Spearman Rank Order Correlation (rho) or the Kendall Rank Order Correlation (tau). When one measure is a continuous, interval level one, and the other is dichotomous (two-category), you can use the Point-Biserial Correlation. Statistical programs will allow you to select which type of correlation you want to use. The formulas for these various correlations differ because of the type of data you're feeding into the formulas, but the idea is the same; they estimate the relationship between two variables as a number between −1 and +1.

Pearson Product Moment Correlation A particular type of correlation used when both variables can be assumed to be measured at an interval level of measurement.

SUMMARY

This chapter introduced the basics involved in data analysis. Conclusion validity is the degree to which inferences about relationships in data are reasonable. You can make two types of errors when reaching conclusions about relationships. A Type I error occurs when you conclude there is a relationship when in fact there is not (seeing something that's not there). A Type II error occurs when you conclude there is no effect when in fact there is (missing the needle in the haystack). There are a variety of strategies for reducing the possibilities for making these errors and, thus, for improving conclusion validity, including ways to reduce the size of the haystack and increase the size of the needle. Data preparation involves logging the data in; checking the data for accuracy; entering the data into the computer; transforming the data; and developing and documenting a database structure that provides the key characteristics of the various measures. Descriptive statistics describe the basic features of the data in a study. The basic descriptive statistics include summaries of the data distributions, measures of central tendency and dispersion or variability, and the different forms of correlation, all of which can and should be examined in numeric and graphic form.

Key Terms

.05 level of significance p. 283
alpha level p. 283
central tendency p. 293
codebook p. 288
correlation matrix p. 299
degrees of freedom or
    df p. 299
descriptive statistics p. 291
dispersion p. 294
distribution p. 292
double entry p. 289
effect size p. 284

fishing and the error rate
    problem p. 284
frequency distribution p. 292
low reliability of measures
    p. 284
mean p. 293
median p. 293
mode p. 293
Pearson product moment
    correlation p. 300
poor reliability of treatment
    implementation p. 284

random heterogeneity of
    respondents p. 284
random irrelevancies in the
    setting p. 284
range p. 294
statistical power p. 286
Type I Error p. 282
Type II Error p. 283
threats to conclusion validity p. 282
variance p. 294
violated assumptions of
    statistical tests p. 285

Suggested Websites

The American Statistical Association:

http://www.amstat.org/

The ASA is the world's largest organization for statisticians. The website includes a huge array of professional resources as well as information on statistics education at all levels.

David Lane's online statistics text:

http://onlinestatbook.com/2/index.html

This is a complete online statistics text with many interesting applets and short videos to demonstrate concepts and procedures.

Review Questions

1. Which of the following is *not* one of the three steps involved in data analysis?

a. data preparation
b. data generation
c. descriptive statistics
d. inferential statistical analysis of the research questions
(Reference: 11.1)

2. Conclusion validity involves the degree to which researchers
a. draw accurate conclusions regarding relationships in the data
b. can accurately infer a causal relationship between treatment and outcome variables
c. can generalize individual research results to the population of interest
d. accurately conclude that a measure reflects what it is intended to measure
(Reference: 11.2)

3. Conclusion validity is

a. only a statistical inference issue
b. only a judgmental, qualitative issue
c. relevant in both quantitative and qualitative research
d. unnecessary except in the most complex research designs
(Reference: 11.2)

4. Conclusion validity is separate from internal validity in that conclusion validity

a. is less critical to evaluation of causal relationships
b. is concerned only with generalizing
c. is only concerned with whether a program or treatment caused an outcome
d. is only relevant in observational studies
(Reference: 11.2)

5. A researcher can improve conclusion validity by using

a. survey measures
b. reliable measures
c. quantitative measures
d. qualitative measures
(Reference: 11.2b)

6. Which of the following would *not* improve statistical power?

a. increasing sample size
b. decreasing the intensity of the treatment
c. decreasing the effect size
d. decreasing the significance level
(Reference: 11.2b)

7. The probability of concluding that there is a relationship when, in fact, there is not, is called

a. Type I error
b. Type II error
c. the confidence interval
d. power
(Reference: 11.2a)

8. What type of condition(s) may create a threat to conclusion validity?

a. poor reliability of measures
b. poor implementation of program procedures
c. irrelevant events that distract participants
d. anything that adds variability to the research context (that is, all of the above)
(Reference: 11.2a)

9. What threat to conclusion validity occurs when a researcher "fishes" for a significant result, using a series of statistical techniques on the same data, while assuming each analysis is independent?

a. finding no relationship when there is one
b. finding a relationship when there is none
c. this does not present a threat to conclusion validity
d. low statistical power
(Reference: 11.2a)

10. What kind of error is committed when a researcher reports that no relationship exists when, in fact, there is one?

a. Type I error
b. Type II error
c. Type III error
d. the third variable error
(Reference: 11.2a)

11. Which of the following is a good conceptual definition of the standard deviation?

a. the most infrequent score in a distribution of scores
b. the average amount of variation from the mean
c. the sum of the median, mode and mean
d. the total amount of variation in a distribution of scores
(Reference: 11.4c)

12. The value 0.05 means that the result you got could be expected to occur by chance at least 5 times out of every 100 times you run the statistical analysis when the null hypothesis is actually true.

a. True
b. False
(Reference: 11.2a)

13. The alpha level and level of significance refer to the same thing.

a. True
b. False
(Reference: 11.2a)

14. In research, as in life, it is best to be patient when "fishing" for a good result.

a. True
b. False
(Reference: 11.2a)

15. Power will be lower in a study if there is more noise (a bigger haystack), a smaller needle (a weaker signal), or both.

a. True
b. False
(Reference: 11.2b)

16. The rule of thumb in social research is that you want statistical power to be at least 0.8 in value.

a. True
b. False
(Reference: 11.2b)

12

Inferential Analysis

305

12.1 Foundations of Analysis for Research Design

The heart of the data analysis—the part where you answer the major research questions—is inextricably linked to research design. The design frames the entire research endeavor because it specifies how measures and participants are brought together. So, it shouldn't surprise you that the research design also frames data analysis (especially in causal research) by determining the types of analysis that you can and cannot do.

This chapter describes the relationship between design and quantitative analysis. We will define inferential statistics, which differ from descriptive statistics in that they are explicitly constructed to address a research question or hypothesis. We then present the General Linear Model (GLM)—the GLM underlies all of the analyses presented here, and is the major framework for statistical modeling in social research, so if you get a good understanding of what that's all about, the rest should be a little easier to handle. For this reason, even though each specific design has its own unique quirks and idiosyncrasies, things aren't as confusing or complicated as they may seem at first. Also, we have emphasized the concepts and basic formulas rather than specific statistical program output, because this way you will have some understanding of what is going on "under the hood" of the statistical programs that all use some variation of this model.

We then move on to specifics—first by considering the basic randomized experimental designs, starting with the simplest—the two-group posttest-only experiment—and then moving on to more complex designs. Finally, we briefly discuss the world of quasi-experimental analysis where the quasi nature of the design leads to all types of analytic problems (some of which may even make you queasy). By the time you're through with all of this, we hope that you'll have a pretty firm grasp on at least two things: 1) how analysis is linked to your research design, and 2) the perils of applying a seemingly obvious analysis to the wrong design structure.

12.2 Inferential Statistics

Inferential statistics is the process of trying to reach conclusions that extend beyond the immediate data. Such approaches are undertaken when you are trying to use the data as the basis for drawing broader inferences (thus, the name) or generalizations. That is, you use inferential statistics to try to infer from the sample data to the larger population. Or, you use inferential statistics to make judgments to the probability that an observed difference between groups is a dependable one. At this stage, you are also able to estimate the size of the effect or relationship in your data. Thus, you use descriptive statistics simply to describe what's going on in the data from your sample; you use inferential statistics to draw conclusions from your data to the more general case.

In this chapter, we concentrate on inferential statistics which are useful in experimental and quasi-experimental research design as well as program outcome evaluation. To understand how inferential statistics are used to analyze data for various research designs, there are several foundational issues we need to take up.

First, you need to understand what is meant by the term **general linear model (GLM)**. Virtually all of the major inferential statistics come from the family of statistical models known as the GLM. Given the importance of the GLM, it's a good idea for any serious social researcher to become familiar with the basic idea of it. The discussion of the GLM here is elementary and considers only the

inferential statistics
Statistical analyses used to reach conclusions that extend beyond the immediate data alone.

general linear model (GLM) A system of equations that is used as the mathematical framework for many of the statistical analyses used in applied social research.

simplest of these models; but it will familiarize you with the idea of the linear model and how it is adapted to meet the analysis needs of different research designs.

Second, one of the keys to understanding how the GLM can be adapted for the analysis of specific research designs is to learn what a dummy variable is and how it is used. Perhaps these variables would be better named as stand-in variables. Essentially a **dummy variable** is one that uses discrete numbers, usually 0 and 1, to represent different groups in your study in the equations of the GLM. The concept of dummy variables is a simple one that enables some complicated things to happen. For instance, by including a simple dummy variable in a model, you can model two separate groups (e.g., a treatment and comparison group) within a single equation.

Third, we need to return to the idea of conclusion validity in the context of statistical analysis. Despite what you may have heard, making a valid inference involves more than just a statistical test and its associated probability (p value). This is because a statistical test by itself addresses only one aspect of the study outcome, and a very limited one at that. The p value is the probability of a result given that the null hypothesis is true. And that is all it is. To really answer a research question fully, we also need to know about the size of the effect (the effect size, ES), and the relative precision of our estimate of the effect (with a confidence interval, CI).

We have discussed effect sizes as a signal-to-noise ratio, in which the signal is the effect (difference in means, strength of correlation, etc.) and the noise is error variance in our measurements. Effect Sizes provide a common metric for comparing and integrating results across studies. Effect Sizes are the basis of meta-analysis and they enable us to have a cumulative kind of science, as discussed in Chapter 1. There are many books, statistical programs, and online calculators to help researchers translate results into appropriate Effect Sizes.

dummy variable A variable that uses discrete numbers, usually 0 and 1, to represent different groups in your study. Dummy variables are often used in equations in the General Linear Model (GLM).

We have also discussed confidence intervals earlier. Way back in Chapter 4 we discussed sampling error, the standard error of an estimate, and the "68, 95, 99 rule" that can be applied when looking at the standard error in normally distributed data. Not surprisingly, researchers have adopted the 95 percent confidence interval (two standard errors) as an acceptable boundary for an estimate. So if we report an effect size with a 95 percent confidence interval, we are providing readers with not only a standardized estimate of our study outcome, but also an idea about the range of likely outcomes in 95 percent of possible replications of our study.

Lastly, there is the matter of how the statistical result translates to real life. Even if the result is statistically significant, is it of any practical significance in real life? Just because we can see that a program or intervention leads to statistically significant improvement doesn't mean the amount of improvement is meaningful or all that valuable (**Figure 12.1**). For instance, if we can conduct a study that shows that a

Bernard Schoenbaum/The New Yorker Collection/Cartoon Bank

"Oh, if only it were so simple."

Figure 12.1 To really understand complex study results, we have to know something about practical as well as statistical significance.

Table 12.1 Possible outcomes of a study with regard to statistical and practical significance.

		Implications for Significance	
		Not Practically Significant	Practically Significant
Possible Outcomes	Not statistically significant	**Back to the drawing board** (reconsider everything: theory, sample, measures, etc.)	**A definite maybe** (study may be underpowered; consider more participants, better measures?)
	Statistically significant	**Juice not worth the squeeze** (increase the "dosage," better implementation?)	**What we want!**

post-incarceration program can reduce recidivism into jail from 26 percent to 24 percent and we determine that this is a statistically significant amount, it doesn't mean that 2 percent is all that practically valuable to us or that it is worth the cost of the program. Practical importance means the degree to which a statistical result translates to a meaningful difference in daily life. One famous example of this comes from the classic meta-analysis by Smith and Glass (1980) on the benefits of psychotherapy. They combined results of 375 studies and determined that the overall effect size was .68 (more than two-thirds of standard deviation improvement). In terms of practical importance, this meant that the median treated client (at the 50th percentile at the beginning of the study) ended up better off than 75 percent of those untreated. Table 12.1 provides a summary of the possible combinations of statistical and practical significance.

The GLM that you'll read about in the balance of this chapter does the main work of providing a model to test your hypothesis. As you can see, putting statistically significant results from the model in the context of Effect Sizes bounded by confidence intervals and translated into practical importance provides a much more complete understanding of your data than the statistical test alone.

12.3 General Linear Model

The GLM is the foundation for the *t*-test, Analysis of Variance (ANOVA), Analysis of Covariance (ANCOVA), regression analysis, and many of the multivariate methods including factor analysis, cluster analysis, multidimensional scaling, discriminant function analysis, canonical correlation, and others. That is, all of the analyses we just mentioned can be described using the same family of equations known as the GLM. Who cares? Well, have you ever wondered why the best quantitative researchers seem to know so much about different types of analyses? Sure, they spent years slogging away in advanced statistics classes in graduate school. But the real reason they can address so many different situations with some authority statistically is that they are really applying one overall approach, the GLM, which has lots of variations. Learn the general approach and you will understand, at least intuitively, the range of analyses (and more) described above.

Before getting into applications of the GLM, we should note that all forms of it include some general assumptions about the nature of the data: 1) the relationships between variables are linear 2) samples are random and independently drawn from the population 3) variables have equal (homogeneous) variances, and 4) variables have normally distributed error. In any statistical analysis, it is important to know and check the assumptions of your data. The first and most direct way to do this is to look at simple plots of your data like histograms and scatterplots to alert you to anything that might put your analysis on shaky ground. There are also statistical diagnostics that you can look at to further evaluate assumptions. In the examples in this chapter we will assume that the statistical test assumptions are true. As you go further in your study of data analysis, you will find that there are hundreds of books and websites that are entirely devoted to the GLM and to specific ways of testing assumptions and employing it to answer various kinds of questions.

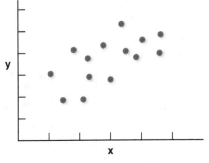

Figure 12.2 A bivariate plot.

12.3a The Two-Variable Linear Model

Let's begin by trying to understand the simplest variation of the GLM, the two-variable case. **Figure 12.2** shows a bivariate plot of two variables. These may be any two continuous variables, but in the discussion that follows, think of them as a pretest (on the x-axis) and a posttest (on the y-axis). Each dot on the plot represents the pretest and posttest score for an individual. The pattern clearly shows a positive relationship; in general, people with higher pretest scores also have higher posttests, and vice versa.

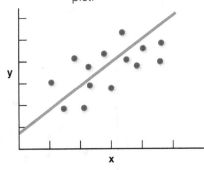

Figure 12.3 A straight-line summary of the data.

The goal in data analysis is to summarize or describe accurately what is happening in the data. The bivariate plot shows the data. How might you best summarize these data? **Figure 12.3** shows that a straight line through the cloud of data points would effectively describe the pattern in the bivariate plot. Although the line does not perfectly describe any specific point (because no point falls precisely on the line), it does accurately describe the pattern in the data. When you fit a line to data, you are using a **linear model**. The term "linear" refers to the fact that you are fitting a line to the data. The term "model" refers to the equation summarizing the line that you fit. A line like the one shown in Figure 12.3 is often referred to as a **regression line** (a description of the relationship between two variables) and the analysis that produces this regression line is often called **regression analysis**.

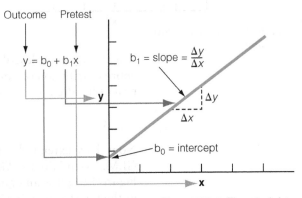

Figure 12.4 The straight-line model.

linear model Any statistical model that uses equations to estimate lines.

regression line A line that describes the relationship between two or more variables.

Figure 12.4 shows the equation for a straight line. You may recognize this equation from your high school algebra classes where it is often stated in the form $y = mx + b$. This is exactly the same as the equation in Figure 12.4. It just uses b_0 in place of b and b_1 instead of m. The equation in Figure 12.4 has the following components:

y = the y-axis variable, the dependent variable, outcome or posttest
x = the x-axis variable, independent variable, predictor or pretest
b_0 = the intercept (value of y when $x = 0$)
b_1 = the slope of the line

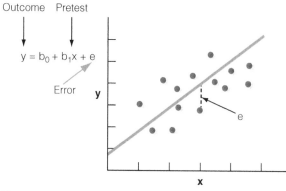

Figure 12.5 The two-variable linear model.

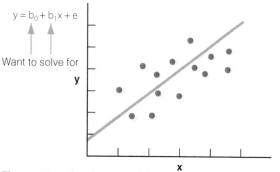

Figure 12.6 What the model estimates.

The slope of the line can be described as the ratio of change in posttest scores to change in pretest scores. As mentioned previously, this equation does not perfectly fit the cloud of points in Figure 12.2. If it did, every point would fall on the line. You need one more component to describe the way this line fits to the bivariate plot.

Figure 12.5 shows this equation for the two-variable or bivariate linear model. The piece added to the equation in Figure 12.5 is an **error term** that describes the vertical distance from the straight line to each point. This component is called "error" because it is the degree to which the line is in error in describing each point. When you fit the two-variable linear model to your data, you have an x and y score for each person in your study. You input these value pairs into a computer program. The program estimates the b_0 and b_1 values, as indicated in **Figure 12.6**. You actually get two numbers back in the computer results for those two values.

You can think of the two-variable regression line like any other descriptive statistic; it simply describes the relationship between two variables, much as a mean describes the central tendency of a single variable. Just as the mean does not accurately represent every value in a distribution, the regression line does not accurately represent every value in the bivariate distribution. You use regression lines as summaries because they show the general patterns in your data and allow you to describe these patterns in more concise ways than showing the entire distribution of points. In one simple equation, you can summarize the pattern of hundreds or thousands of points in a scatterplot.

12.3b The "General" in the General Linear Model

Now, let's extend the two-variable case to the multiple variable one—the GLM. Essentially the GLM looks similar to the two-variable model shown in Figure 12.6; it is an equation. The big difference is that each of the four terms in the GLM stands for a whole set of variables instead of representing only a single variable. So, the general linear model can be written as follows:

$$y = b_0 + bx + e$$

where

y = a set of outcome variables
x = a set of predictor variables or covariates
b_0 = the set of intercepts (value of each y when each $x = 0$)
b = a set of coefficients, one each for each x
e = error term

This model allows you to summarize an enormous amount of information. In an experimental or quasi-experimental study, you would represent the independent variable (e.g., program or treatment) with one or more dummy-coded variables (see next section), each represented in the equation as an additional x-value. (However, the convention is to use the symbol z to indicate that the variable is a

regression analysis A general statistical analysis that enables us to model relationships in data and test for treatment effects. In regression analysis, we model relationships that can be depicted in graphic form with lines that are called *regression lines*.

error term A term in a regression equation that captures the degree to which the regression line is in error (that is, the residual) in describing each point.

dummy-coded x.) If your study has multiple outcome variables, you can include them as a set of y-values. If you have multiple predictors, you can include them as a set of x-values. For each x-value (and each z-value), you estimate a b-value that represents an x, y relationship. The estimates of these b-values and the statistical testing of these estimates is what enables you to test specific research hypotheses about *relationships* between variables or differences between groups. This kind of analysis is what lies at the heart of inferential statistics.

The GLM allows you to summarize a wide variety of research outcomes. The major problem for the researcher who uses the GLM is **model specification**—how to identify the equation that best summarizes the data for a study. If the model is misspecified, the estimates of the coefficients (the b-values) that you get from the analysis are likely to be biased (wrong). In complex situations, this model specification problem can be a serious and difficult one.

The GLM is one of the most important tools in the statistical analysis of data. It represents a major achievement in the advancement of social research in the twentieth century. All of the analyses described in this chapter will be expressed using specific forms of the GLM.

model specification
The process of stating the equation that you believe best summarizes the data for a study.

12.3c Dummy Variables

A dummy variable is a numerical variable used in the GLM (especially in regression analysis) to represent subgroups of the sample in your study. It is not a variable used by dummies. In fact, you have to be pretty smart to figure out how to use dummy variables. In research design, a dummy variable is typically used to distinguish different treatment or program groups. In the simplest case, you would use a 0,1 dummy variable, where a person is given a value of 1 if in the treated group or a 0 if placed in the control group.

Dummy variables are useful because they enable you to use a single regression equation to represent multiple groups. This means that you don't need to write out separate equation models for each subgroup. The dummy variables act like *switches* that turn various values on and off in an equation. Another advantage of a 0,1 dummy-coded variable is that even though it is a nominal-level variable, you can treat it statistically like an interval-level variable. (If this made no sense to you, you probably should refresh your memory on levels of measurement covered in Chapter 5, "Introduction to Measurement") For instance, if you take an interval-level analysis like computing an average of a 0,1 variable, you will get a meaningful result—the proportion of 1s in the distribution.

$$y_i = \beta_0 + \beta_1 z_i + e_i$$

where:
y_i = Outcome score for the ith unit
β_0 = Coefficient for the intercept
β_1 = Coefficient for the slope
z_i = 1 if i^{th} unit is in the treatment group; 0 if i^{th} unit is in the control group
e_i = Residual for the ith unit

Figure 12.7 Use of a dummy variable in a regression equation.

To illustrate how dummy variables work, consider a simple regression model that could be used for analysis of a posttest-only two-group randomized experiment shown in **Figure 12.7**. Because this is a posttest-only design, there is no pretest variable in the model. Instead, there is a dummy variable, z, that represents the two groups, and there is the posttest variable, y. The key term in the model is β_1, the estimate of the difference between the groups. To see why this is the case, and how dummy variables work, we'll use this simple model to show you how dummy variables can be used to pull out the separate subequations for each subgroup. Then we'll show how to estimate the difference between the subgroups by subtracting their respective equations. You'll see that you can pack an enormous amount of information into a single equation using dummy variables. In the end, you should be able to understand why β_1 is actually the difference between the treatment and control groups in this model.

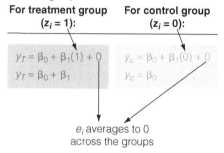

First, determine effect for each group.

For treatment group $(z_i = 1)$:	For control group $(z_i = 0)$:
$y_T = \beta_0 + \beta_1(1) + 0$	$y_c = \beta_0 + \beta_1(0) + 0$
$y_T = \beta_0 + \beta_1$	$y_c = \beta_0$

e_i averages to 0 across the groups

Figure 12.8 Using a dummy variable to create separate equations for each dummy variable value.

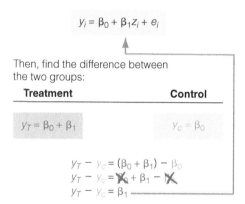

$$y_i = \beta_0 + \beta_1 z_i + e_i$$

Then, find the difference between the two groups:

Treatment	Control
$y_T = \beta_0 + \beta_1$	$y_c = \beta_0$

$y_T - y_c = (\beta_0 + \beta_1) - \beta_0$
$y_T - y_c = \beta_0 + \beta_1 - \beta_0$
$y_T - y_c = \beta_1$

Figure 12.9 Determine the difference between two groups by subtracting the equations generated through their dummy variables.

To begin understanding how dummy variables work, let's develop an equation specifically for the control group, using the model in Figure 12.7. For the control group, the β_1 term drops out of the equation because β_1 times 0 (which is what Z is for the control group) equals 0. Furthermore, we always assume that the error term e_i averages to 0 (that there will be as many errors above the line as below it), so the e_i term also drops out. Thus, when we take the equation in Figure 12.6 and use it to determine a specific equation for the control group, we see that the predicted value for the control group is β_0 as shown in **Figure 12.8**.

Now, to figure out the treatment-group line, you substitute the value of 1 for Z, again recognizing that, by assumption, the error term averages to 0. The equation for the treatment group indicates that the treatment group value is the sum of the two beta values.

Now you're ready to move on to the second step—computing the difference between the groups. How do you determine that? Well, the difference must be the difference between the equations for the two groups that you worked out previously. In other words, to find the difference between the groups, you find the difference between the equations for the two groups! It should be obvious in **Figure 12.9** that the difference is β_1. Think about what this means: The difference between the groups is β_1. Okay, one more time just for the sheer heck of it: the difference between the groups in this model is β_1!

Whenever you have a regression model with dummy variables, you can always see how the variables are being used to represent multiple subgroup equations by following these steps:

- Create separate equations for each subgroup by substituting the dummy values (as in Figure 12.8).
- Find the difference between groups by finding the difference between their equations (as in Figure 12.9).

12.3d The *t*-Test

Okay, here's where we'll try to pull together the previous sections on the general linear model and the use of dummy variables. We'll show you how the general linear model uses dummy variables to create one of the simplest tests in research analysis, the *t*-test. If you can understand how this works, you're well on your way to grasping how all research designs discussed here can be analyzed. The *t*-test assesses whether the means of two groups (for example, the treatment and control groups) are *statistically* different from each other. Why is it called the *t*-test? Well, the statistician who invented the test was William Gosset who worked at the Guinness Brewery in Dublin, Ireland, and published under the pseudonym "Student" because of company policies regarding research publication (even though he did the work at home on his own time). Mr. Gosset actually used the letter z in his original paper, but in later years z came to be very widely known as a symbol for a standardized score, so writers of textbooks like this one began using t to differentiate the two (Salsburg, 2001). Whatever the reason, don't lose any sleep over it. The *t*-test is just a name and, and, as the bard says, what's in a name?

Before you can proceed to the analysis itself, it is useful to understand what the term "difference" means in the question, "Is there a *difference* between the groups?" Each group can be represented by a bell-shaped curve that describes the group's distribution on a single variable. You can think of the bell curve as a smoothed histogram or bar graph describing the frequency of each possible measurement response.

Figure 12.10 shows the distributions for the treated and control groups in a study. Actually, the figure shows the idealized or smoothed distribution—the actual distribution would usually be depicted with a histogram or bar graph. The figure indicates where the control and treatment group means are located. A *t*-test addresses whether the two means are statistically different.

What does it mean to say that the averages for two groups are statistically different? Consider the three situations shown in **Figure 12.11**. The first thing to notice about the three situations is that *the difference between the means is the same in all three*. But, you should also notice that the three situations don't look the same; they tell different stories. The top example shows a case with moderate variability of scores within each group (variability is reflected in the spread, width, or range of the bell-shaped distribution). The second situation shows the high-variability case. The third shows the case with low variability. Clearly, you would conclude that the two groups appear most different or distinct in the bottom or low-variability case. Why? Because there is relatively little overlap between the two bell-shaped curves. In the high-variability case, the group difference appears least striking (even though it is identical) because the two bell-shaped distributions overlap so much.

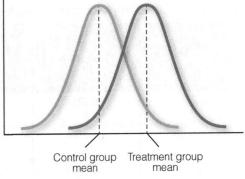

Is there a *difference*?

Figure 12.10 Idealized distributions for treated and control group post-test values.

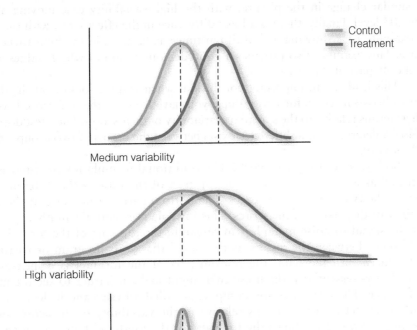

Medium variability

High variability

Low variability

Control
Treatment

Figure 12.11 Three scenarios for differences between means.

Table 12.2 Example data matching the Low, Medium and High variability graphs shown in Figure 12.11.

Variability	Experimental M (SD)	Control M (SD)	Raw Score Difference [95% CI]	t (p)	Standardized Difference (Cohen's d) [95% CI]
Low	60 (2)	50 (2)	10 [8.97, 11.03]	19.365 (<.0001)	5 [3.97, 6.03]
Medium	60 (5)	50 (5)	10 [7.42, 12.58]	7.746 (<.0001)	2 [1.38, 2.62]
High	60 (20)	50 (20)	10 [.34, 20.34]	1.936 (.0576)	.5 [.01, 1.01]

In order to concretize this idea, we have constructed a table that corresponds to the three distributions in Figure 12.11. In Table 12.2, you see the low-, medium-, and high-variability scenarios represented with data that correspond to each case. Notice that the means and mean differences are the same for all three rows. But while the mean difference is 10 points for each, the standard deviations increase from 2 to 5 to 20. Now look at the effects of this change in the confidence intervals for the difference, the test statistics, and the effect sizes (all of which reflect the signal to noise ratio). First of all, we see that the 95 percent CIs for the difference grow immensely from the low- to high-variability situations. Next, when we look at the t statistics and associated p values, we see a similar change in the picture, with the high-variability case moving above the .05 level. Finally, there is a huge difference in the effect sizes, with the low-variability case showing a d of 5, ten times as large as in the high-variability case. This example also provides a good demonstration of why p values alone tell only part of this story.

This leads to an important conclusion: when you are looking at the differences between scores for two groups, you have to judge the difference between their means relative to the spread or variability of their scores. The t-test does just this—it determines if a difference exists between the means of two groups (think 't' for two).

So how does the t-test work? The traditional formula for the t-test is expressed as a ratio or fraction. The top part of the ratio is the difference between the two means or averages. The bottom part is a measure of the variability or dispersion of the scores. This formula is essentially another example of the signal-to-noise metaphor in research; the top part of the formula, the difference between the means, is the signal that your program or treatment introduced into the data; the bottom part of the formula is a measure of variability—essentially, the noise that might make it harder to see the group difference. The ratio that you compute is called a **t-value** and it describes the difference between the groups relative to the variability of the scores in the groups. **Figure 12.12** shows the formula for the t-test and how the numerator and denominator are related to the distributions.

The top part of the formula is easy to compute—just find the difference between the means. The bottom part is called the **standard error of the difference**. To compute it, take the variance (see Chapter 11, "Introduction to Data Analysis") for each group and divide it by the number of people in that group. You add these two values and then take their square root. The specific formula is given in

t-value The estimate of the difference between the groups relative to the variability of the scores in the groups.

standard error of the difference A statistical estimate of the standard deviation one would obtain from the distribution of an infinite number of estimates of the difference between the means of two groups.

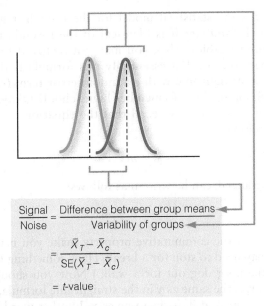

$$SE(\bar{X}_T - \bar{X}_c) = \sqrt{\frac{var_T}{n_T} + \frac{var_c}{n_c}}$$

Figure 12.13 Formula for the standard error of the difference between the means.

Signal / Noise = Difference between group means / Variability of groups

$$= \frac{\bar{X}_T - \bar{X}_c}{SE(\bar{X}_T - \bar{X}_c)}$$

$$= t\text{-value}$$

Figure 12.12 Formula for the *t*-test.

$$t = \frac{\bar{X}_T - \bar{X}_c}{\sqrt{\frac{var_T}{n_T} + \frac{var_c}{n_c}}}$$

Figure 12.14 Final formula for the *t*-test.

Figure 12.13. Remember, that the variance is simply the square of the standard deviation. The final formula for the *t*-test is shown in **Figure 12.14**. Before you leave the formula, notice how important the sample size (*n*) is in this model. As the sample size increases, the standard error decreases. This means that, relatively speaking the degree of difference in means in the numerator is going to appear larger with bigger sample sizes. In fact, this is one of the reasons that statistical significance is not the whole story; a significant *p* value is always possible with a large enough sample size.

The *t*-value will be positive if the first mean is larger than the second value and negative if it is smaller. To test the significance, you need to set a risk level (as previously mentioned, called the alpha level). In most social research, the rule of thumb is to set the alpha level at .05. This means that five times out of a hundred, you would find a statistically significant difference between the means even if there were none (meaning that you'd get a misleading result that often in repeated studies if the null hypothesis was actually true). You also need to determine the degrees of freedom (df) for the test. In the *t*-test, the *df* is the sum of the persons in both groups minus two. Given the alpha level, the *df*, and the *t*-value, you can determine whether the *t*-value is large enough to be significant. If it is, you can conclude that the difference between the means for the two groups is different (given the variability). Statistical computer programs routinely provide the significance test results, but if you ever wanted to conduct a test like this by hand, you could check a table online or in the back of a statistics book. These are just like the tables that have been in existence since the time of Gosset, Pearson, Fisher, and the other pioneers of statistical science who developed them.

Now that you understand the basic idea of the *t*-test, we want to show you how it relates to the general linear model, regression analysis, and dummy variables. If you can understand the *t*-test, you will be able to understand how you can change the model to address more complex research designs. One tool, the general linear model, can open the door to understanding how almost all research designs are analyzed. At least that's the theory. Let's see if we can achieve this level of understanding.

$$y_i = \beta_0 + \beta_1 z_i + e_i$$

Figure 12.15 The regression formula for the *t*-test (and also the two-group one-way posttest-only Analysis of Variance or ANOVA model).

Okay, so here's the statistical model for the *t*-test in regression form (see Figure 12.15). Look familiar? It is identical to the formula in Figure 12.7 to introduce dummy variables. Also, you may not realize it (although we hope against hope that you do), but essentially this formula is the equation from high school for a straight line with a random error term (e_i) thrown in. Remember high school algebra? Remember high school? Okay, for those of you with faulty memories, you may recall that the equation for a straight line is often given as follows:

$$y = mx + b$$

which, when rearranged, can be written as follows:

$$y = b + mx$$

(The complexities of the commutative property make you nervous? If this gets too tricky, you may need to stop for a break. Have something to eat, make some coffee, or take the poor dog out for a walk.) Now you should see that in the statistical model, y_i is the same as y in the straight-line formula, β_0 is the same as b, β_1 is the same as m, and Z_i is the same as x. In other words, in the statistical formula, β_0 is the intercept and β_1 is the slope (see **Figure 12.16**).

Wait a minute. β_1 is the same thing as the slope? Earlier we said that β_1 was the difference between the two groups. How can it be the slope of a line? What line? How can a slope also be a difference between means? To see this, you have to look at the graph of what's going on in Figure 12.16. The graph shows the measure on the vertical axis. This is exactly the same as the two bell-shaped curves shown in Figure 12.10 and Figure 12.11, except that here they're turned on their sides and are graphed on the vertical dimension. On the horizontal axis, the Z variable, our dummy variable, is plotted. This variable has only two possible values: a 0 if the person is in the control group or a 1 if the person is in the program group. This kind of variable is a dummy variable because with its two values it is a stand-in variable that represents the two groups that received the different program or treatment conditions (see the discussion of dummy variables earlier in this chapter). The two points in the graph indicate the average (posttest) value for the control ($Z = 0$) and treated ($Z = 1$) cases. The line that

Figure 12.16 The elements of the equation in Figure 12.15 in graphic form.

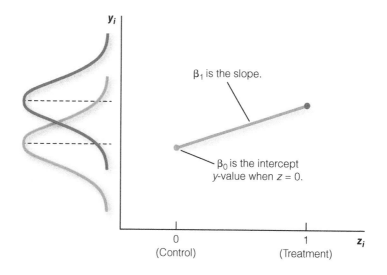

connects the two dots is only included for visual enhancement purposes; because there are no Z values between 0 and 1, there can be no values plotted where the line is. Nevertheless, you can meaningfully speak about the slope of this line—the line that would connect the posttest means for the two values of Z. Do you remember the definition of slope? (Here we go again, back to high school!) The **slope** is the change in y over the change in x (or, in this case, Z). Stated differently, it is the change in y for a one-unit change in x (which in this example is z). But because it is a dummy variable, the change in Z between the groups is always equal to 1. Therefore, the slope of the line must be equal to the difference between the average y-values for the two groups. That's what we set out to show (reread the first three sentences of this paragraph). β_1 is the same value that you would get if you subtracted the two means from each other. (In this case, because the treatment group equals 1, you are subtracting the control group out of the treatment-group value. A positive value implies that the treatment-group mean is higher than the control-group mean; a negative value means it's lower.)

But remember at the beginning of this discussion, we pointed out that just knowing the difference between the means was not good enough for estimating the treatment effect because it doesn't take into account the variability or spread of the scores. So how do you do that here? Every regression-analysis program will give, in addition to the β values, a report on whether each β value is statistically significant. They test whether the β value differs from zero. It turns out that the t-value for the β_1 coefficient is the exact same number that you would get if you did a t-test for independent groups.

Now let's think of a slightly different case. Suppose, instead of two groups, we wish to compare the difference in means across three or more groups. One way to do this would be to run a series of t-tests between all the different group pairs. Say, we want to compare annual salary packages for managers across three companies, so we could run a t-test for each pair of companies. However, as you have seen from previous examples presented in this chapter, statisticians don't like to do things in steps—they are always finding ways to analyze all the data in one group! The model in this case is the same as before, but it is now called a one-way analysis of variance (ANOVA), and the test statistic is the F ratio (instead of the t-value). So the t-test is just a special case of ANOVA: if you analyze the means of two groups by ANOVA, you get the same results as doing it with a t-test. In other words, ANOVA generalizes the t-test to more than two groups.

You might ask, if ANOVA is used to compare *means* of different groups, why then does it have analysis of *variance* in its title? Despite its name, remember that ANOVA is actually concerned with differences between *means* of groups, not differences *between* variances. The name analysis of *variance* comes from the way the procedure uses variances to decide whether the means are different. This is how it works: In ANOVA, the statistical program looks at the absolute difference in means between samples, but it also looks to see what variation exists *within* each of the groups. The intuition is that if the difference *between* treatments is a lot bigger than the difference *within* treatments, then a real effect is present and, therefore, you conclude you have achieved statistical significance. Because this method compares variation between groups to variation within groups, it is called analysis of *variance*.

Here are a few conclusions from all this:

- The t-test, one-way ANOVA, and regression analysis are mathematically equivalent and yield the *same* results in a two-group case.
- The regression-analysis method of the t-test utilizes a dummy variable (Z) for treatment.

slope The change in y for a change in x of one unit.

- Regression analysis is the most useful model of the three because of its generality. Remember that regression analysis is a part of the general linear model. If you understand how it works for the *t*-test you should be able to grasp that, by adding more terms to the *t*-test version of the regression formula, you can create more complex analyses that address virtually all of the major research designs.

Below, we will discuss how GLM is used in the analysis of specific research designs—both experimental and quasi-experimental.

12.4 Experimental Analysis

Now we're ready to turn to the discussion of the major experimental designs and how they are analyzed. The simplest experimental design, a two-group posttest-only randomized experiment, is usually analyzed with the simple *t*-test (which is also the simplest variation of the one-way ANOVA) that we described above. The factorial experimental designs are usually analyzed with the ANOVA model. Randomized block designs (RD) use a special form of the ANOVA-blocking model that uses dummy-coded variables to represent the blocks. The Analysis of Covariance experimental design uses, not surprisingly, the Analysis of Covariance statistical model. Each of these analyses can be expressed with a single equation that is a variation of the general linear model.

12.4a The Two-Group Posttest-Only Randomized Experiment

To analyze the two-group posttest-only randomized experimental design you need an analysis that meets the following requirements:

- Has two groups
- Uses a post-only measure
- Has a distribution for each group on the response measure, each with an average and variation
- Assesses treatment effect as the statistical (non-chance) difference between the groups

You can estimate the treatment effect for the posttest-only randomized experiment in three ways. All three yield mathematically equivalent results, a fancy way of saying that they give you the exact same answer. So why are there three different ones? In large part, these three approaches evolved independently and only afterwards was it clear that they are essentially three ways of doing the same thing. So, what are the three ways? First, you can compute an independent *t*-test as described earlier. Second, you could compute a one-way ANOVA between two independent groups. Finally, you can use regression analysis to regress the posttest values onto a dummy-coded treatment variable. Of these three, the regression-analysis approach is the most general. In fact, we describe the statistical models for all the experimental and quasi-experimental designs in regression-model terms. You just need to be aware that the results from all three methods are identical. So, the analysis model for the simple two-group posttest only randomized experimental design is the *t*-test, which is expressed as a regression formula in Figure 12.15 above.

12.4b Factorial Design Analysis

Now that you have some understanding of the GLM and dummy variables, we can present the models for other more complex experimental designs rather easily. **Figure 12.17** shows the regression model for a simple 2 × 2 factorial design.

If you remember from our discussions in Chapter 9, factorial designs allow you to manipulate certain features of your program and thus belong to the signal-enhancing category of experimental designs. These designs are especially efficient because they enable us to examine which features or combinations of features of the program lead to a causal effect.

In the simplest factorial designs, you have two factors, each of which has two levels. The model uses a dummy variable (represented by a Z) for each factor. In two-way factorial designs like this, you have two main effects and one interaction effect. In this model, the main effects are the statistics associated with the beta values that are adjacent to the Z variables. The interaction effect is the statistic associated with β_3 (that is, the t-value for this coefficient) because it is adjacent in the formula to the multiplication of (interaction of) the dummy-coded Z variables for the two factors. Because there are two dummy-coded variables, and each has two values, you can write out 2 × 2 = 4 separate equations from this one general model. (Go ahead, we dare you. If you need to refresh your memory, check back to the discussion of dummy variables presented earlier in this chapter.) You might want to see if you can write out the equations for the four cells. Then, look at some of the differences between the groups. You can also write two equations for each Z variable. These equations represent the main effect equations. To see the difference between levels of a factor, subtract the equations from each other.

$$y_i = \beta_0 + \beta_1 z_{1i} + \beta_2 z_{2i} + \beta_3 z_{1i} z_{2i} + e_i$$

where:

y_i = Outcome score for the ith unit
β_0 = Coefficient for the intercept
β_1 = Mean difference on factor 1
β_2 = Mean difference for factor 2
β_3 = Interaction of factor 1 and factor 2
z_{1i} = Dummy variable for factor 1
    (0 = 1 hour/week, 1 = 4 hours/week)
z_{2i} = Dummy variable for factor 2
    (0 = in-class, 1 = pull-out)
e_i = Residual for the ith unit

Figure 12.17 Regression model for a 2 × 2 factorial design.

$$y_i = \beta_0 + \beta_1 z_{1i} + \beta_2 z_{2i} + \beta_3 z_{3i} + \beta_4 z_{4i} + e_i$$

where:

y_i = Outcome score for the ith unit
β_0 = Coefficient for the intercept
β_1 = Mean difference for treatment
β_2 = Blocking coefficient for block 2
β_3 = Blocking coefficient for block 3
β_4 = Blocking coefficient for block 4
z_{1i} = Dummy variable for treatment 1
    (0 = control, 1 = treatment)
z_{2i} = 1 if block 2, 0 otherwise
z_{3i} = 1 if block 3, 0 otherwise
z_{4i} = 1 if block 4, 0 otherwise
e_i = Residual for the ith unit

Figure 12.18 Regression model for a randomized block design.

12.4c Randomized Block Analysis

To jog your memory, randomized block designs are a particular type of noise-reducing research design (see Chapter 9, "Experimental Design"). If you remember, we said that randomized block designs were research design's equivalent to stratified random sampling—they require you to divide the sample into relatively homogeneous subgroups or blocks (analogous to "strata" in stratified sampling) and then the experimental design you want to apply is implemented within each block or homogeneous subgroup.

The statistical model for the randomized block design can also be presented in regression-analysis notation. **Figure 12.18** shows the model for a case where there are four blocks or homogeneous subgroups.

Notice that a number of dummy variables are used to specify this model. The dummy variable Z_1 represents the treatment group. The dummy variables Z_2, Z_3, and Z_4 indicate Blocks 2, 3, and 4, respectively. Analogously, the beta values (βs) reflect the treatment and Blocks 2, 3, and 4. What happened to Block 1 in this model? To see what the equation for the Block 1 comparison group is, fill in your dummy variables and multiply through. In this case, all four Zs are equal to 0, and you should see that the intercept (β_0) is the estimate for the Block 1 control group. For the Block 1 treatment group, $Z_1 = 1$ and the estimate is equal to $\beta_0 + \beta_1$. By substituting the appropriate dummy variable switches,

$$y_i = \beta_0 + \beta_1 x_i + \beta_2 z_i + e_i$$

where:

y_i = Outcome score for the ith unit

β_0 = Coefficient for the intercept

β_1 = Pretest coefficient

x_i = Covariate

β_2 = Mean difference for treatment

z_i = Dummy variable for treatment
    (0 = control, 1 = treatment)

e_i = Residual for the ith unit

Figure 12.19 Regression model for the ANCOVA.

you should be able to figure out the equation for any block or treatment group.

The data that is entered into this analysis would consist of five columns and as many rows as you have participants: the posttest data and one column of 0s or 1s for each of the four dummy variables.

12.4d Analysis of Covariance

The statistical model for the Analysis of Covariance or ANCOVA, which estimates the difference between the groups on the posttest after adjusting for differences on the pretest, can also be given in regression-analysis notation. The model shown in **Figure 12.19** is for a case where there is a single covariate, a treated group, and a control group.

The dummy variable Z_i represents the treatment group. The beta values (βs) are the parameters being estimated. The value β_0 represents the intercept. In this model, it is the predicted posttest value for the control group for a given X value (and, when X = 0, it is the intercept for the control-group regression line). Why? Because a control group case has a Z = 0, and since the Z variable is multiplied with β_2, that whole term would drop out.

The data that is entered into this analysis would consist of three columns—the posttest data y_i, one column, Z_i of 0s or 1s to indicate which treatment group the participant is in, and the **covariate** score, X_i—and as many rows as you have participants.

This model assumes that the data in the two groups are well described by straight lines that have the same slope. If this does not appear to be the case, you have to modify the model appropriately. How do you do that? Well, we'll tell you the short answer, but for the complete one you need to take an advanced statistics course. The short answer is that you add the term $\beta_i X_i Z_i$ to the model in Figure 12.19. If you've been following along, you should be able to create the separate equations for the different values of the dummy variable Z_i and convince yourself that the addition of this term will allow for the two groups to have different slopes.

12.5 Quasi-Experimental Analysis

The quasi-experimental designs differ from the experimental ones in that they don't use random assignment to assign units (people) to groups. The lack of random assignment in these designs tends to complicate their analysis considerably, but the principles are the same and the general linear model is still used. For example, to analyze the Nonequivalent Groups design (NEGD), you have to adjust the pretest scores for measurement error in what is often called a Reliability-Corrected Analysis of Covariance model. In the simplest case of the RD design, you use a variation of the ANCOVA model. The Regression Point Displacement design (RPD) has only a single treated unit. Nevertheless, the analysis of the RPD design is based directly on the traditional ANCOVA model.

The experimental designs discussed previously have comparatively straightforward analysis models. You'll see here that you pay a price for not using random assignment like they do; the analyses generally become more complex.

12.5a Nonequivalent Groups Analysis

The Nonequivalent Groups Design (NEGD) has two groups—a program and comparison group—each measured pre and post (see **Figure 12.20**). The statistical model that you might intuitively expect to be used in this situation would include pretest variable, posttest variable, and a *dummy variable* that describes which group the person is in. These three variables would be the input for the statistical analysis.

N	O	X	O
N	O		O

Figure 12.20 Design notation for the NEGD.

In this example, assume you are interested in estimating the difference between the groups on the posttest after adjusting for differences on the pretest. This is essentially the ANCOVA model as described in connection with randomized experiments. (Review the discussion "Covariance Designs" and how to adjust for pretest differences in Chapter 9, "Experimental Design.") There's only one major problem with this model when used with the NEGD; it doesn't work! If you fit the traditional ANCOVA model you are likely to get the wrong answer because the model is biased when nonrandom assignment is use. Why? Well, the detailed story is too complex to go into here, but we can skip to the conclusion. It turns out that the bias is due to two major factors: measurement error in the pretest coupled with the initial nonequivalence between the groups on the pretest. It does not occur in randomized experiments because there is no pretest nonequivalence (at least not probabilistically). The bias will be greater with greater nonequivalence between groups; the less similar the groups, the bigger the problem.

$$x_{adj} = \bar{x} + r(x - \bar{x})$$

where:

x_{adj} = Adjusted pretest value
$\bar{x}$ = Group mean
x = Original pretest value
r = Reliability

Figure 12.21 Formula for adjusting pretest values for unreliability in the reliability-corrected ANCOVA.

How do we fix the problem? A simple way is to adjust the pretest for the amount of measurement error. Since reliability is related to measurement error—the higher the reliability, the lower the error and vice versa—we use an estimate of reliability to do the correction. The formula for the adjustment is simple (see **Figure 12.21**).

The idea in this formula is that you are going to construct new pretest scores for each person. These new scores will be adjusted for pretest unreliability by an amount proportional to the reliability. Once you've made this adjustment, all you need to do is substitute this adjusted pretest score for the original pretest score in the ANCOVA model (see **Figure 12.22**).

$$y_i = \beta_0 + \beta_1 x_{adj} + \beta_2 z_i + e_i$$

where:

y_i = Outcome score for the ith unit
β_0 = Coefficient for the intercept
β_1 = Pretest coefficient
β_2 = Mean difference for treatment
x_{adj} = Transformed pretest
z_i = Dummy variable for treatment (0 = control, 1 = treatment)
e_i = Residual for the ith unit

Figure 12.22 The regression model for the reliability-corrected ANCOVA.

Notice that the only difference is that the X in the original ANCOVA is changed to the term X_{adj}. Of course, things are never as simple as they seem. There are some complications even for this simple reliability-corrected ANCOVA model. For example, in the reliability-adjustment formula we did not tell you which of the many methods for estimating reliability should be used (see Chapter 5). Since they are likely to differ and we never know which is the "correct" estimate, we often run the analysis multiple ways, using different reliability estimates each time to try to bracket the "correct" answer. But don't let all of these details get in your way. The important point here is that this design, like all of the others discussed in this chapter, uses the same basic approach to analysis, a variation of the general linear model.

12.5b Regression-Discontinuity Analysis

The basic RD design is also a two-group pretest-posttest model. As in other versions of this design structure (see the sections the Analysis of Covariance Randomized Experiment and the Nonequivalent Groups Design in this chapter), you will need a statistical model that includes a term for the pretest, one for the posttest, and a dummy-coded variable to represent the program.

$$\bar{x}_i = x_i - x_c$$

where:

$\bar{x}_i$ = Pretest-cutoff value

x_i = Pretest

x_c = Cutoff value

Figure 12.23 Adjusting the pretest by subtracting the cutoff in the Regression-Discontinuity (RD) analysis model.

$$y_i = \beta_0 + \beta_1 \bar{x}_i + \beta_2 z_i + e_i$$

where:

y_i = Outcome score for the ith unit

β_0 = Coefficient for the intercept

β_1 = Pretest coefficient

β_2 = Mean difference for treatment

$\bar{x}_i$ = Pretest – cutoff value

z_i = Dummy variable for treatment
(0 = control, 1 = treatment)

e_i = Residual for the ith unit

Figure 12.24 The regression model for the basic regression-discontinuity design.

$$y_i = \beta_0 + \beta_1 x_i + \beta_2 z_i + e_i$$

where:

y_i = Outcome score for the ith unit

β_0 = Coefficient for the intercept

β_1 = Pretest coefficient

β_2 = Mean difference for treatment

x_i = Covariate

z_i = Dummy variable for treatment
(0 = control, 1 = treatment)

e_i = Residual for the ith unit

Figure 12.25 The regression model for the RPD design assuming a linear pre-post relationship.

For the simplest case where we can assume that a straight line with the same slope fits well the pre-post data for both groups, the only analysis complication has to do with the cutoff point which is unique to the RD design. And the fix is fairly straightforward. As with the NEGD, all you need to do is adjust the pretest scores, in this case by subtracting the cutoff value from each pretest, as shown in the formula in **Figure 12.23**. This transformation has the effect of "centering" the regression model on the pretest cutoff, because by subtracting the cutoff we are effectively setting the intercept of the model to the cutoff value. (Don't get that? Don't worry! That's for a more advanced discussion than we're doing here.)

Once this adjustment is done, the adjusted pretest score is substituted for the original pretest in the same old ANCOVA model, as shown in **Figure 12.24**.

Okay, so that's the good news. The analysis of the simplest RD design is essentially a very basic variation of the same old ANCOVA regression model. The bad news? As with all of the quasi-experiments, things are seldom as simple in reality as they are in this simple model. For instance, in many situations, it will not be reasonable to assume that a straight line with the same slope is appropriate for both groups. In fact, if the data require curved lines, the modeling can be considerably more complex. But, however complex, it will almost certainly involve some variation of the general linear model. That's the key message that recurs throughout this chapter.

12.5c Regression Point Displacement Analysis

We come now to the last of the quasi-experimental designs we want to discuss in reference to analysis—the Regression Point Displacement (RDP) design. At this point in the chapter, you should be able to anticipate the kind of analysis we are going to suggest. You'll see that the principles are the same here as for all of the other analyses, especially in that this analysis also relies on the GLM and regression analysis.

The RPD design, like many of the others, involves a pretest, posttest, and a dummy variable to represent the treatment group (where 0 = comparison and 1 = program). These requirements are identical to the requirements for the ANCOVA model and should look very familiar by now. The only difference is that the RPD design has only a single treated group score. That is, only one pretest-posttest score pair reflects the treated group.

The analysis model for the RPD design is the now-familiar ANCOVA model stated in regression model form (see **Figure 12.25**).

The only unusual feature is that for the dummy variable there will only be one treated pre-post unit. Of course, the usual warnings about reality apply here as well. If the pre-post distribution is not well described by a straight line, then we will need to adjust the model to allow for the appropriate curvilinearity. But these nuances are for more advanced texts. The important lesson is, once again, that a simple variation of the general linear model can be used for the RPD design, just as it is used for the all the others.

SUMMARY

Phew! We're hoping for your sake that you weren't assigned to read this chapter in a single night. And, we're hoping you didn't put this off until the night before the exam. However, in case you did, let us summarize the salient points.

Table 12.3. Summary of the statistical models for the experimental and quasi-experimental research designs.

Design	Analysis	Notes
Experimental Designs		
The two-group posttest-only randomized experiment	$y_i = \beta_0 + \beta_1 z_i + e_i$	No pretest term x_i needed. This is equivalent to the t-test.
Factorial design (2 ← 2)	$y_i = \beta_0 + \beta_1 z_{1i} + \beta_2 z_{2i} + \beta_3 z_{1i} z_{2i} + e_i$	No pretest term x_i needed; the z terms are dummy variables for each of the two factors.
Factorial blocks	$y_i = \beta_0 + \beta_1 z_{1i} + \beta_2 z_{2i} + \beta_3 z_{3i} + \beta_4 z_{4i} + e_i$	No pretest term x_i needed; z_1 is dummy for treatment; z_{2-4} are dummies for blocks 2−4 of 4.
Analysis of covariance (ANCOVA)	$y_i = \beta_0 + \beta_1 x_i + \beta_2 z_i + e_i$	The analysis of covariance (ANCOVA) model
Quasi-Experimental Designs		
Nonequivalent groups design (NEGD)	$y_i = \beta_0 + \beta_1 x_{adj} + \beta_2 z_i + e_i$	$x_{adj} = \bar{x} + r(x - \bar{x})$ where $\bar{x}$ is the pretest average and r is the reliability estimate.
Regression discontinuity design (RD)	$y_i = \beta_0 + \beta_1 \tilde{x} + \beta_2 z_i + e_i$	$\tilde{x} = $ pretest $(x_i) - $ cutoff value
Regression point displacement design (RPD)	$y_i = \beta_0 + \beta_1 x_i + \beta_2 z_i + e_i$	Identical to ANCOVA except that for dummy treatment variable, there is only one pre-post score pair where $z_i = 1$ (treated case).

This chapter described data analysis for the basic research designs. The key to understanding such analyses, a branch of inferential statistics, is the General Linear Model (GLM). The GLM is a general formula and approach to modeling data, and every one of the designs discussed in this chapter can be analyzed using a basic variation of the GLM. So, if you can learn about the GLM, you'll have a general tool that can be applied in almost any inferential analysis of a research design. There are several key concepts needed to understand the GLM. Especially critical are the ideas of a regression line and its formula, and the notion of a dummy variable and how it can be manipulated in a formula. If you can get the general sense of these ideas, you should be able to appreciate how the GLM can be applied and adapted to all of the designs.

To help make this point, and to summarize all of the models, we'll throw every one of the analysis models into a single table, Table 12.3.

Now, don't panic when you look at this just because there are a bunch of formulas there. Take a deep breath and look at the formulas. Scan down the column of formulas and look for similarities. Notice how each has the same symbol, y_i, to the left of the equal sign. That's because every one of these designs has a posttest score and y_i is the symbol we use to represent the posttest in the formula. Notice that each has a β_0 immediately to the right of the equals sign. In every model, that symbol is the intercept. If we drew graphs of the lines we were fitting to the data—and we could do so, after analyzing the results draw such lines for each one—the β_0 symbol would always represent the intercept of those lines (i.e., where the line is when $X = 0$). And notice that every formula has at least one Z term. The Z

term represents different groups in the design. It is typically (but not always) the dummy variable, where 1 represents the treated group and 0 represents the comparison or control. Notice also that many of the formulas have an X in them. The X represents a pretest value, sometimes called a covariate. And, finally, notice that each formula ends with an e_i value that represents the presumably random component that remains in the data when we fit the rest of the model. That's it. In one sense, it's really not that hard. Even if you only have the beginning intuitive sense of how these models work, the table illustrates succinctly that all of the formulas are variations on a similar theme. They're all variations of the GLM and, for these designs, they're all variations of a specific subclass of the GLM known as regression analysis. So, when you take that statistics course down the road, you may want to keep this in mind when you cover regression analysis: That one statistical tool can provide you with a general way to do a lot of very useful specific things.

Key Terms

dummy variable p. 307
error term p. 310
general linear model
   (GLM) p. 306
inferential statistics p. 306

interaction effect p. 319
linear model p. 309
model specification p. 311
regression analysis p. 309
regression line p. 309

slope p. 317
standard error of the difference
   p. 314
t-value p. 314

Suggested Websites

Web Interface for Statistics Education (WISE)

http://wise.cgu.edu/

Professor Dale Berger's website at Claremont Graduate University provides a wide array of resources for statistics education, including papers, data sets, and a very nice compilation of interactive tutorials and applets, all thoroughly documented.

Statistics Hell

http://www.statisticshell.com/html/cocytus.html

Professor Andy Field at the University of Sussex has written a number of very popular, entertaining, and helpful books on doing statistical analysis. Now he has added another great resource for students and researchers alike, complete with his characteristic humor.

Review Questions

1. One of the key assumptions of the general linear model that helps us look at differences between groups is

a. use of an experimental analysis.
b. the existence of a linear relationship in variables.
c. the existence and use of dummy variables.
d. limited overlap in bell-shaped curves.
(Reference: 12.3)

2. A dummy variable is one used in regression analysis to represent

a. participants.
b. subgroups of participants, such as treatment and control groups.
c. levels of measurement.
d. sublevels of measurement, such as dependent variables.
(Reference: 12.3c)

3. Which of the following does *not* describe the function of a *t*-test?

a. determination of whether the difference between two groups is important in practical terms
b. measurement of the ratio of a program's signal to its noise
c. determination of whether there is a statistically significant difference between the means of a treatment group and a control group
d. analysis of the difference between group means divided by the standard error of the difference
(Reference: 12.3d)

4. In addition to an alpha level and a *t*-value, what other information must a researcher have in order to look up the probability associated with a particular *t*-value in a standard table of significance?

a. standard deviation
b. standard error
c. degrees of freedom
d. dummy variables
(Reference: 12.3d)

5. Which distribution of scores between a control and a treatment group will most likely result in a statistically significant difference being discovered?

a. similar means
b. different means
c. different means and significant overlap of scores
d. different means and little overlap of scores
(Reference: 12.3d)

6. A straight line fitted to data in a statistical analysis is called

a. the spline.
b. the analysis of variance.
c. a regression line.
d. a linear model.
(Reference: 12.3a)

7. In general, the "error" in a general linear model is

a. the squared sum of the distance of each score from the regression line.
b. the vertical distance of each score from the regression line.
c. the horizontal distance of each score from the regression line.
d. the correlation value applied to the regression line.
(Reference: 12.3a)

8. The ratio used to find the *t*-value representing the difference between two groups is composed of

a. the correlation of the groups over the standard deviation.
b. the variance of the groups over the standard deviation.
c. the difference between groups over the standard error of the difference.
d. the sum of the groups over the sum of the differences between groups.
(Reference: 12.3d)

9. Which of the three different ways to estimate a treatment effect for a researcher using the posttest-only two-group randomized experiment will produce equivalent results?

a. independent *t*-test, one-way analysis of variance (ANOVA), and regression analysis
b. independent *t*-test, two-way analysis of variance (ANOVA), and factorial design
c. two-way analysis of variance (ANOVA), factorial design, and regression analysis
d. two-way analysis of variance (ANOVA), randomized block analysis, and regression analysis
(Reference: 12.3d)

10. In a regression equation, the dummy variable

a. represents the slope of the line.
b. specifies the difference in group means.
c. is treated statistically like an interval-level variable.
d. is an error term.
(Reference: 12.3c)

11. The main difference between inferential and descriptive statistics is that inferential statistics are used to test a specific hypothesis.

a. True
b. False
(Reference: 12.1)

12. The *t*-test, analysis of variance and covariance, and regression analysis are all very different kinds of analysis based on different underlying statistical models.

a. True
b. False
(Reference: 12.3)

13. One reason that it is good to examine a plot of your data is to determine if the general picture is one of a relatively straight line, suggesting that the relationships observed are in fact linear and appropriate for one of the general linear model (GLM) analyses.

a. True
b. False
(Reference: 12.3)

14. The slope of a regression line can be described as the change in the posttest given in pretest units.

a. True
b. False
(Reference: 12.3a)

15. It is possible to use regression analysis to study the results of an experiment in which you have a treatment and control group.

a. True
b. False
(Reference: 12.4a)

13

Research Communication

13.1 Research Communication

"Research is complete only when the results are shared with the scientific community" (APA, 2010, p. 9). This means that even though you have already put countless hours into your study, the work is still far from done. In fact, this final stage—telling the story of your study—may be one of the most difficult steps on the Research Road. Developing a concise and effective report is an art form in itself. In many research projects, you may need to compose multiple presentations and reports that show the results at different levels of detail for various types of audiences. We will first address research communication in the broad framework of the research enterprise and then provide guidance on composing a report of an individual study.

13.1a Research Communication and the Research-Practice Continuum

If you look back at the Research-Practice Continuum model presented in Chapter 1, you can see the key role of research communication in translational research and evidence-based practice. Research communication is essential throughout the research-practice continuum—when it works, that is. As we have done throughout the book, we will consider some ways that things can potentially go wrong. It is beyond our scope to consider all of the ways that translation of research can go wrong when reported in mainstream media, or the ways that the public might misunderstand research reports. Our focus is on research methods, and in validity terms, we are concerned about threats to translational research validity. There are two primary ways that bias creeps into the system. One way is underrepresentation of negative or null findings, and the other is overrepresentation of so-called positive outcomes. When these sources of bias are compounded by the influence of market forces (i.e., money), truly disastrous results may ensue, as we see over and over again in the lawsuits related to pharmaceutical products that turn out to be more harmful than beneficial.

In terms of underrepresentation, unless the study is shared, there can be no influence on research synthesis or guideline development. There is a term for the missing studies in an area of research called the **file drawer problem** (Rosenthal, 1979). The term is dated now, but it refers to the unknown studies that were never submitted or published, left instead to the private archives and file drawers of those who did the studies (this sentence may have made your professor a little bit ill if, like most researchers, she or he has such an archive). The problem of overrepresentation of positive results is likely to be an even bigger problem for two reasons: 1) As noted in our discussion of research ethics and data analysis, there is a tendency for researchers to equate a report of statistical significance with the success of their study, potentially leading to misleading reporting and 2) There is also a well-known bias toward acceptance and publication of studies that report statistically significant findings.

These issues are receiving increasing attention by researchers, journals, funding agencies, and other entities in the broad research enterprise (e.g., Ioannidis, 2005). At the system level, there are a number of possible improvements, including registration of clinical trials prior to launching them (so that things cannot be changed later to make the study "turn out better" and so that unpublished trials will not be ignored). Efforts to create data banks, in which original data can be

file drawer problem Many studies might be conducted but never reported, and they may include results that would substantially change the conclusions of a research synthesis.

obtained and results confirmed or added to a synthesis that is less subject to bias, may also be productive. Replication of study results is probably the simplest solution of all, and despite encouragement in many methods texts like this one, it is underutilized.

At the level of the individual researcher, reducing publication bias is a function of both ethics in honest reporting and competence in designing and conducting the most valid research possible. Therefore, good research communication depends on doing everything well, including telling the story. This chapter will provide guidance on some of the most common formats for written and oral presentation of a study.

13.1b General Considerations for Research Communication

There are several general considerations to keep in mind when producing a report:

- **The Audience.** Who is going to hear or read the report? What is their interest in your study? Reports will differ considerably depending on whether the audience will want technical detail, whether they are looking for a summary of results as in a poster presentation, or whether they are about to examine your research in a dissertation defense.
- **The Story.** We believe that every research project has at least one major story in it. Sometimes the story centers on a specific research finding. Sometimes it is based on a methodological problem or challenge. Other times it challenges or extends previous research. As you prepare your presentation or paper, you should attempt to tell your unique story to the reader. Even in formal journal articles where you will be required to be concise and detailed at the same time, a good story line can help make an otherwise dull report interesting to the reader.

The hardest part of telling the story in your research is finding the story in the first place. Usually when you come to writing up your research you have been steeped in the details for weeks or months (and sometimes even for years). You've been worrying about sampling responses, struggling with operationalizing your measures, dealing with the details of design, and wrestling with the data analysis. You're a bit like the ostrich that has its head in the sand. To find the story in your research, you have to pull your head out of the sand and look at the big picture. You have to try to view your research from your audience's perspective. You may have to let go of some of the details that you obsessed so much about and leave them out of the write-up or bury them in technical appendices or tables.

APA Format Also known as APA Style. The standard guide for formatting research write-ups in psychology and many other applied social research fields. The APA format is described in detail in the *Publication Manual of the American Psychological Association* (APA) and includes specifications for: how subsections of a research paper should be organized; writing style; proper presentation of results; research ethics; and how citations should be referenced.

13.2 The Written Report

The written report can take many forms, but you will find that in the social sciences and education, most journals, universities, and faculty will require what is known as "APA Style" or "**APA Format.**" The *Publication Manual of the American Psychological Association* (APA) provides the standard guide for formatting research write-ups. Its most recent version—the sixth edition (commonly known as "APA 6th"; American Psychological Association, 2010) includes specifications for how subsections should be organized, writing style, proper presentation of results, research ethics, and, of course, how citations should be referenced. If you are writing a term paper, most faculty members will require you to follow specific guidelines—often these are also the APA guidelines. Doing your thesis or dissertation? Every university typically has strict policies about formatting and style. There are legendary stories that circulate among graduate students about the dissertation that was rejected because the page margins were a quarter-inch off or the figures weren't labeled correctly! While the APA format guide itself is an excellent reference guide, and should be on the shelves of all serious researchers, students may find that a free online resource like the Purdue OWL Online Writing Lab's *APA Formatting and Style Guide* (Purdue Online Writing Lab, 2014, https://owl.english. purdue.edu) is a handy alternative.

To illustrate what a set of research report specifications might include, we present general guidelines for a class term paper and presentation. The presentation guidelines are intended to help you share your research proposal in class or at a conference. The paper guidelines are similar to the types of specifications you might be required to follow for publishing a journal article. However, you need to check the specific formatting guidelines for the report you are writing; the ones presented here are based on the APA format and may differ in some ways from guidelines required in other contexts. We include a sample research paper that reflects these guidelines in this text's appendix.

13.2a Key Elements and Formatting of a Research Paper

This section describes the elements and formatting considerations that you must typically address in a research write-up of an empirical study. Typically, a research paper includes these major sections: Title Page, Abstract, Introduction, Method, Results, Discussion, References, and possibly Appendices. Appendices might include IRB materials and measures. Often, tables and figures are either included as a separate section at the end of the paper (after references) or incorporated into the text.

The paper must have all the following sections in the order given, using the specifications outlined for each section (page numbers are our estimates of fairly typical proportions in a research article of about 15–25 pages. Of course, you should adjust these as appropriate for your context):

- Title Page
- Abstract (on a separate, single page)
- Introduction (2–3 pages)
- Method (7–10 pages)
- Results (2–3 pages)
- Discussion (2–3 pages)
- References
- Tables (one to a page)
- Figures (one to a page)
- Appendices

A "**running head**" is a brief version of the title (50 characters or less) that is included at the top-left of every page. To set up a running head, use your word processor's header function to insert a header. The running head should be left-justified on the top line of every page, and the page number should be right-justified on the same line (see example at the end of the chapter). The headings for each of these sections should be centered, using upper- and lowercase letters (Abstract, Method, etc.). Note that the heading for the introduction section is the title of the paper itself and not the word "Introduction." Subheadings (you will

running head A brief version of the title (50 characters or less) that is included at the top-left of every page of a research write-up.

find many in the "Methods" section) are usually bolded and left-aligned with the text beginning from the next line. See the APA manual for formatting different sections within subheadings.

Okay, now that you have an understanding of what goes inside your report, the next task is to think about how to proceed on writing the report itself. For this, there is no rule that you have to begin writing the sections in order, beginning with the "Introduction" section and ending with the "Conclusion" section. You might wish to write out the methods and the results sections first because they are typically more concrete and descriptive. You can then proceed to writing the "Introduction" and the "Discussion." For most people, the "Abstract" comes last, once they have a stronger sense of the full report.

Below, we describe key elements that go into each section.

Title Page

It's pretty obvious that the **title page** gets its own page! This page announces the title of the study, the author's name, and the institutional affiliation. This information should be centered and on separate lines. Try to place it around the middle of the page where the reader's eyes naturally fall.

In terms of choosing a title, you should aim to be as specific as you can without ending up with a really long title. Titles are typically limited to 12 words. The title should be interesting—it should not only convey a good first impression to your readers but also help you stay on track as you write your paper. If it's short enough, you might want to use your research question as your title. Alternatively, you might consider using your independent (predictor) and dependent variables (outcome) in the title.

At the top of the title page, you should have the words "Running Head": followed by an all-caps, short identifying title (two to four words) for the study. This running header should also appear on the top-left of every page of the paper (note that the words "Running Head" should only precede it on the title page).

Abstract

The **abstract** is limited to 150–200 words. It is also double-spaced. At the top of the page, centered, you should have the word "Abstract." Section 2.04 of the APA manual has additional information about formatting the abstract.

Many readers first review the abstract to determine if the entire article is of interest. For this reason, it is important to carefully draft the abstract. It should be written in paragraph form and should be a concise summary of all four major parts of an empirical paper (i.e., Introduction, Method, Results, and Discussion). Consider writing one or two sentences that capture the main points of each section. Most people find it useful to write the paper first, before writing the abstract. In order to shorten your abstract, you should eliminate nonessential information wherever possible—this includes transitions phrases (e.g., "the results demonstrate . . .").

Introduction. The first section in the body of the paper is the introduction. To reiterate, in APA format, you do not include a heading that says, "Introduction." You simply begin the paper in paragraph form following the title. The first sentence should have "grab"; that is, it should be memorable and engage the reader immediately (Gilgun, 2005). Keep in mind that the purpose of this section is to justify your study. That is, by the end of this section you'd like your reader to know enough of the background story to think "of course!" when you describe your research question and methods.

title page The initial page of a formal research write-up that includes the title of the study, the author's name, and the institutional affiliation.

abstract A written summary of a research project that is placed at the beginning of each research write-up or publication.

Every introduction will have the following (roughly in this order): a statement of the problem being addressed; a brief review of relevant literature (including citations and possibly some description of your search procedures); a description of the major constructs involved; and a statement of the research questions or hypotheses. The entire section should be in paragraph form, with the possible exception of the questions or hypotheses, which may be indented. Here are some details on the key components of this section:

- *Statement of the problem:* State the general problem area clearly and unambiguously. Discuss the importance and significance of the problem area.
- *Literature review and citations:* The literature review forms the empirical basis for conducting your research. Condense the literature in an intelligent fashion and include only the most relevant information. You might ask, what does "intelligent fashion" mean in this context? It is important to build a story—rather than simply citing previous work; try to weave a thread through past literature in order to build your case. It is also important that you review objectively and without bias, meaning you need to include studies that you agree with and those that you don't. Try to provide suggestions for the differences that underlie various studies. You might also want to organize the findings of a complex literature review in a table format if it would help the reader comprehend the material. You might even discover that a set of studies is ripe for a meta-analysis. Cite literature only from reputable and appropriate sources (such as peer-reviewed professional journals and books, and not sources like *Time*, *The Huffington Post*, *Yahoo News*, and so on). Ensure that all citations are in the correct format—consult the APA manual for in-text citation format.
- *Statement of constructs:* Explain each key construct in the research/evaluation project (e.g., in a causal study, describe both the cause and the effect). Ensure that explanations are readily understandable (that is, jargon-free) to an intelligent reader.
- *The Research Questions or Hypotheses:* Clearly state the major questions, objectives, or hypotheses for the research. For example, if you are studying a cause-effect relationship, describe the hypotheses and relate them sensibly to the statement of the problem. The relationship of the hypothesis to both the problem statement and literature review must be readily understood from reading the text.

Method. The Method section should begin immediately after the introduction (no page break) and should have the centered title, "Method." In APA format this section typically has four subsections: Participants, Measures, Design, and Procedures. Each of the four subsections should have a bold, left-justified section heading.

- *Participants:* This section should describe the population of interest, the sampling frame, ethical considerations and IRB review/approval, recruitment, eligibility, sample selection procedures, and characteristics of the sample itself. If the research participants are self-selected they should be described as such. Describe problems in contacting and obtaining complete measures of the sample. Whenever possible, it is good to include comments on how many participants were in the study and how that number was estimated. A power analysis is a method of determining how many participants would be needed to find a meaningful effect. A precision analysis provides expectations about how much accuracy will be provided by your measures with a particular sample size (typically, the 95 percent confidence interval (CI) around an outcome or effect size). A brief discussion of external validity is appropriate here; that is, you should state the degree to which you believe results will be generalizable from your sample to the population. One way to do this is by providing some details on subgroups in the sample (e.g., how many females and males). Sampling is covered in Chapter 4, "Sampling."

- *Measures:* This section should include a brief description of your constructs and all measures used to operationalize them. Short instruments can be presented in their entirety in this section. If you have more lengthy instruments, you may present some typical questions to give the reader a sense of what you did (and, depending on the context and the requirements of the publisher, include the full measure in an appendix).

  For preexisting instruments, you should cite any relevant information about reliability and validity, including evidence of cultural validity. For all instruments, you should briefly state how you determined reliability and validity, report the results, and discuss them. Measurement is covered in Chapter 5, "Introduction to Measurement." Note that the procedures used to examine reliability and validity should be appropriate for the measures and the type of research you are conducting. If you are using archival data, describe original data collection procedures and any indexes (for example, combinations of individual measures) in sufficient detail. For scales, you must describe briefly the scaling procedure you used and how you implemented it. You should also describe any issues with missing or anomalous data and how these issues were managed. If your study is qualitative, describe the procedures you used for collecting your measures in detail. Appendices that include measures are typically labeled by letter (for example, Appendix A) and cited appropriately in the body of the text.

- *Design:* Where appropriate, you should state the name of the design used and detail whether it is a true or quasi-experiment, survey, case study, and so on. Explain the composition of subgroups and the assignment to conditions. For nonstandard or tailored designs, you might also present the design structure (for instance, you might use the X and O notation used in this book). Describe major independent and dependent variables in this subsection. Ensure that the design is appropriate for the problem and that it addresses the hypothesis. You should also include a discussion of internal validity that describes the major plausible threats in your study and how the design accounts for them, if at all. Be your own study's critic here and provide enough information to show that you understand the threats to validity and whether you've been able to account for them all in the design or not.

- *Procedures:* Generally, this section ties together the sampling, measurement, and research design. In this section, you should briefly describe the overall sequence of steps or events from beginning to end (including sampling, measurement, and use of groups in designs), any procedures followed to assure that participants are protected and informed, and how their confidentiality will be protected (where relevant). Many journals now require a flow diagram giving specific details on every step from enrollment to analysis (often referred to as a "CONSORT-type" diagram after the international research group that developed the format for this kind of graphic; http://www.consort-statement.org).

An essential part of this subsection in a causal study is additional description of the program or independent variable that you are studying. Include sufficient information so that another researcher could replicate the essential features of the study.

Results. The heading for this section is centered with upper- and lowercase letters. You should indicate concisely what results you found in this research. Typically, a Results section begins with a descriptive report on characteristics of the participants. This is usually followed by psychometric or descriptive data on the measures, prior to reports of the primary analysis. It is always better to remind your reader of your initial hypothesis once again, before you start reporting the results of statistical analysis. You should tell the reader what kind of analysis was done—for example, if you used ANOVA or another type of analysis. It is important to state the alpha level that you used and to report effect sizes and confidence intervals. If possible, provide a statement on the practical or clinical significance of the results and how such a determination was made. State the implications of your results for your null and alternative hypotheses. Remember, your results don't have to support your hypotheses. In fact, the common experience in social research is the finding of no effect. Section 2.07 of the APA manual provides additional help in writing this section.

Discussion. In this section you interpret and discuss your results. In terms of your statistical analysis, try to explain what you found—is it along the lines of what you predicted? If not, why? Similarly, place your results in the context of previous research—do they agree with prior findings? If they are inconsistent, how can you explain this? Often, authors discuss whether they find a statistically significant effect but fail to account for its magnitude. In order to understand the implications of your findings, it is very important to discuss the magnitude of any effects you find.

Conclusion. In the last section, you should describe the conclusions you've reached. You should also describe the implications of your findings for theory and practice. You should discuss the overall strength of the research (for example, a discussion of the stronger and weaker validity areas) and should present some suggestions for possible future research that would be sensible, based on the results of this work.

References

There are really two components that are needed when "citing" a published research article in a research write-up. First, there is the **citation**, the way you mention or refer to the item in the text when you are discussing it. Second, there is the **reference** itself, the complete description of the name and location of the article, the authors, and the source. References are included in a separate section immediately following the body of the paper.

- *Reference Citations in the Text of Your Paper:* Cited references appear in the text of your paper and are a way of giving credit to the source of the information you quoted or described in your paper. The rules for citing references in texts differ, sometimes dramatically, from one format to another. In APA format, they generally consist of the following bits of information:

  - The author's last name (initials are included only if needed to distinguish between two authors with the same last name). If there are six or more

citation The mentioning of or referral to another research publication when discussing it in the text of a research write-up.

reference The complete description of the name and location of an article, the authors and the source (e.g., journal, book, or website). References are included in a separate section immediately following the body of the paper.

authors, the first author is listed followed by the term, "et al." (which means "and others" in Latin).
- Year of publication in parenthesis.
- Page numbers are given with a quotation or when only a specific part of a source was used:

"To be or not to be" (Shakespeare, 1660, p. 241)

For example, here is how you would cite a one author or multiple author reference:

One Work by One Author:

Rogers (1994) compared reaction times . . .

One Work by Multiple Authors:

If there are two authors, use both their names every time you cite them. If there are three to five authors, cite as follows:

Wasserstein, Zappulla, Rosen, Gerstman, and Rock (1994) [first time you cite in text]
Wasserstein et al. (1994) found [subsequent times you cite in text]

If there are six or more authors, use the "et al." citation approach every time you cite the article.

- *Reference List in Reference Section:* There are a wide variety of reference citation formats. Before submitting any research report, you should check to see which type of format is considered acceptable for that context. If there is no official format requirement, the most sensible thing is for you to select one approach and implement it consistently. Here, we'll illustrate by example some of the major reference items and how they might be cited in the reference section.

The reference list includes all the articles, books, and other sources used in the research and preparation of the paper and cited with a parenthetical (textual) citation in the text. References are listed in alphabetical order according to the authors' last names; if a source does not have an author, alphabetize according to the first word of the title, disregarding the articles *a*, *an*, and *the* if they are the first word in the title. Electronic references should have a digital object identifier or "doi" that is listed with the article you've downloaded. Websites and the date of access should be listed as well.

EXAMPLES

BOOK BY ONE AUTHOR:

Gardner, H. (2007). *Five minds for the future.* Cambridge, MA: Harvard Business Review Press.

BOOK BY TWO AUTHORS:

Freeman, R., & Freeman, S. (2004). *Take a paddle: Finger Lakes New York quiet water for canoes.* Englewood, FL: Footprint Press.

BOOK BY THREE OR MORE AUTHORS:

Heck, R. H., Thomas, S. L., & Tabata, L. N. (2010). *Multilevel and longitudinal modeling with SPSS.* New York: Routledge.

BOOK WITH NO GIVEN AUTHOR OR EDITOR:

Handbook of Korea (4th ed.). (1982). Seoul: Korean Overseas Information, Ministry of Culture & Information.

TWO OR MORE BOOKS BY THE SAME AUTHOR*:

Oates, J. C. (1990). *Because it is bitter, and because it is my heart.* New York: Dutton.

Oates, J. C. (1993). *Foxfire: Confessions of a girl gang.* New York: Dutton.

BOOK BY A GROUP OR ORGANIZATIONAL AUTHOR:

President's Commission on Higher Education. (1977). *Higher education for American democracy.* Washington, DC: U.S. Government Printing Office.

BOOK WITH AN EDITOR:

Bloom, H. (Ed.). (1988). *James Joyce's Dubliners.* New York: Chelsea House.

A TRANSLATION:

Dostoevsky, F. (1964). *Crime and punishment* (J. Coulson, Trans.). New York: Norton. (Original work published 1866).

AN ARTICLE OR READING IN A COLLECTION OF PIECES BY SEVERAL AUTHORS (ANTHOLOGY):

O'Connor, M. F. (1975). *Everything that rises must converge.* In J. R. Knott, Jr., & C. R. Raeske (Eds.), *Mirrors: An introduction to literature* (2nd ed., pp. 58–67). San Francisco: Canfield.

EDITION OF A BOOK:

Tortora, G. J., Funke, B. R., & Case, C. L. (1989). *Microbiology: An introduction* (3rd ed.). Redwood City, CA: Benjamin/Cummings.

A WORK IN SEVERAL VOLUMES:

Churchill, W. S. (1957). *A history of the English speaking peoples: Vol. 3. The age of revolution.* New York: Dodd, Mead.

ENCYCLOPEDIA OR DICTIONARY:

Cockrell, D. (1980). Beatles. In *The new Grove dictionary of music and musicians* (6th ed., Vol. 2, pp. 321–322). London: Macmillan.

ARTICLE FROM A WEEKLY MAGAZINE:

Jones, W. (1970, August 14). Today's kids. *Newsweek, 76,* 10–15.

ARTICLE FROM AN ONLINE MONTHLY MAGAZINE:

Howe, I. (2008, September). James Baldwin: At ease in apocalypse. *Harper's, 237,* 92–100, http://www.harpers.com/sept08/howe.html.

Note: Entries by the same author are arranged chronologically in the reference section by the year of publication, the earliest first. References with the same first author and different second and subsequent authors are listed alphabetically by the surname of the second author, and then by the surname of the third author. References with the same authors in the same order are entered chronologically by year of publication, the earliest first. References by the same author (or by the same two or more authors in identical order) with the same publication date are listed alphabetically by the first word of the title following the date; lowercase letters (a, b, c, and so on) are included after the year, within the parentheses.

ARTICLE FROM A NEWSPAPER:

Brody, J. E. (1976, October 10). Multiple cancers termed on increase. *New York Times (national ed.),* p. A37.

ARTICLE FROM A SCHOLARLY ACADEMIC OR PROFESSIONAL JOURNAL:

Barber, B. K. (2014). Cultural, family, and personal contexts of parent-adolescent conflict. *Journal of Marriage and the Family, 56,* 375–386, DOI: 12.23548.9090-456.

GOVERNMENT PUBLICATION:

U.S. Department of Labor. Bureau of Labor Statistics. (1980). *Productivity.* Washington, DC: U.S. Government Printing Office.

PAMPHLET OR BROCHURE:

Research and Training Center on Independent Living. (1993). *Guidelines for reporting and writing about people with disabilities* (4th ed.) [Brochure]. Lawrence, KS.

There are several major software programs available that can help you organize all of your literature references and automatically format them in the most commonly required formats, including APA format. These programs, which are essentially literature database management systems, typically integrate with your word processing software and can be used to insert citations correctly and produce reference lists. Many of them even allow you to conduct literature searches over the web and download literature citations and abstracts. If you will be doing serious research (anything from an honors thesis to a large, funded research grant), we highly recommend that you consider using a major bibliographic software program.

There are also increasingly good online and web-based sources for conducting literature searches. Many of these services are available through college or public libraries and allow you to search databases of journal articles and download formatted references into your bibliographic software (including the abstracts), and download or view the full text of research articles online. You should check with the reference librarians at your institution or local library to determine what electronic sources are available.

Tables

Any tables should have a heading with "Table #" (where # is the table number), followed by the title for the heading that describes concisely what is contained in the table. Tables and figures are usually included at the end of the paper after the references and before the appendices. In the text you should put a reference where each table or figure should be inserted using this form:

Insert Table 1 about here

Figures

Figures follow tables at the end of the paper, before the appendices with one figure on each page. In the text you should put a reference where each figure will be inserted using this form:

Insert Figure 1 about here

Appendices

Appendices should be used when necessary and depending on what kind of paper you are writing (e.g., thesis versus article for publication). Generally, you will only use them for presentation of measurement instruments, for detailed descriptions of the program or independent variable, and for any relevant supporting documents that you don't include in the body, such as consent forms. Many publications severely restrict the number of pages you are allowed and will not accept appendices. Even if you include such appendices, you should briefly describe the relevant material in the body and give an accurate citation to the appropriate appendix (for example, "see Appendix A").

Stylistic Elements

As you write your research paper, here are some general tips regarding stylistic considerations:

- *Professional Writing:* Avoid first-person and sex-stereotyped forms. Present material in an unbiased and unemotional (for example, no feelings about things), but not necessarily uninteresting, fashion.
- *Parallel Construction:* Keep tenses parallel within and between sentences (as appropriate).
- *Sentence Structure:* Use correct sentence structure and punctuation. Avoid incomplete and run-on sentences.
- *Spelling and Word Usage:* Make sure that spelling and word usage are appropriate. Avoid jargon and colloquialisms. Correctly capitalize and abbreviate words.
- *General Style:* Ensure that the document is neatly produced and reads well and that the format for the document has been correctly followed.

13.3 Other Forms of Research Communication

Students are often required to make class presentations and frequently have opportunities to share research in the form of conference posters and panels. In this section we'll provide some general guidance on effective presentations and posters.

13.3a Presentations

As difficult as writing can be, it may be safe to say that presentations produce more anxiety than just about any other part of the research process (right up there with statistics phobia). But if you stop to think about it, you almost certainly have a good knowledge base about presenting already. That is because you have observed countless teachers conducting lectures and other kinds of presentations over many years and can probably recognize a good presentation in the first few minutes. That expert judgment may include some basic elements that you can emulate in your presentation. Below you will see a summary of common elements of an effective presentation.

1. *Consider your audience.* Will you be talking with classmates and your instructor? If so, the main thing may be to be sure you follow any instructions about content and timing. Although it may not always be true, try to remember that you may be the class expert on the topic. This is your chance to tell your classmates the story of what you've been working on all semester. Think back to when you chose the topic and how enthusiastic you were about it then. If you can bring that enthusiasm to your presentation, there is a good chance your classmates will be interested. Your instructor may have some sort of rubric for evaluating your presentation. Most likely that comes directly from the instructions in your assignment, so review those carefully and be sure you hit every requirement. The professional audience at a conference is harder to predict, from the size of the audience to their background and focus. You may get some insight about this kind of presentation by attending and just observing. In the words of New York Yankee immortal Yogi Berra, "You can observe a lot just by looking." However, if you need to plan without the benefit of "looking," then focus your energy on the organization of your talk.

2. *Organize your presentation.* Most classroom and conference presentations will be in the 15–20-minute range. That may seem like an eternity, but if you break it down into components you will usually find it is challenging to fit everything into that kind of time frame. Time limits are usually strictly adhered to, and it is no fun to find yourself without time to complete your talk or address questions. Nor will an overly long presentation win you any positive feedback from fellow presenters, classmates, or your instructor. One way to manage this challenge is to structure your talk around the various sections, as you do in a paper. Below is one possible outline you might consider. If you are presenting a proposal rather than a completed study, you would be able to increase the time and attention to background literature and methods in order to help set up the justification for your study.

Introduction (about 1–2 minutes)

- Describe the origin of your interest
- Why is this topic important?
- What will you cover (review your outline)?

Literature Review (about 5 minutes) The review should give the audience a sense of:

- What are the major theories and controversies in this area?
- What are the key studies?
- Any special methodological issues?
- Summary and major unanswered questions.
- Now state your hypotheses or research questions.

Methods (about 4 minutes) The goal here is to communicate how you have chosen to study the issue. The usual components are the same as in a paper:

- Design
- Participants

- Ethical considerations
- Measures
- Analysis

Results (about 4 minutes)

- If there were any major issues with data (e.g., missing data, nonnormal data), describe what you did to manage the issue. Otherwise, stick to the primary analyses
- Consider using handouts if you have dense tables
- Use graphics but keep them clean and simple

Conclusions (about 1 minute)

- Briefly summarize the conclusions from your study. Provide a statement regarding relationship of your data to prior literature. Mention specific limitations in your study. Indicate whether the results were consistent with expectations or surprising in some way. What is the "take-home" message?

Questions/Discussion (about 5 minutes)

3. *Audiovisual Support.* Microsoft PowerPoint is the most common software program for making presentations of research results, but there are other options (including programs similar to PowerPoint, as well as very different ones like the nonlinear Prezi: http://prezi.com/). There is an abundance of commentary and research related to effective and poor use of PowerPoint. Here is a summary of the most frequently noted suggestions about use of slides, and some general presentation tips.

- Don't read your slides. First of all, they should be concise bullet points, not long passages of text. Second, your audience can probably read faster than you can talk, in which case reading generates annoyance and soon boredom.
- Simple is better. There are many available templates and options for color, animation, sound, and other effects. Effects should be used sparingly or not at all. If your goal is good communication and comprehension, aim for clean, clear, and simple. Fonts should be clear and relatively large (20-point or larger).
- Triple-check your spelling. Errors will stand out much more on the big screen, and you do not want them to distract from your talk or become the most memorable part of your presentation.
- Practice. If you can practice with at least one person in your audience, you will no doubt do better during the actual presentation. If you can even practice in the place where your presentation will take place (or some place similar), you are likely to feel even more comfortable when the time comes for your real presentation.
- Know your presentation so well that you could do it without slides. This should also help in anxiety management, and if, for some reason, the technology fails, the show will still go on.
- Aim for no more than one slide per minute of your talk.
- Last, but not least, try to have fun. That may not seem consistent with the conservative-sounding advice we've just given you, but if you think of the best teachers you have ever had, one commonality is likely to be that they seemed to be having fun. Also, having fun is pretty much incompatible with anxiety, so if you can generate enthusiasm and other positive emotions, thoughts, and behaviors, anxiety will likely be under control. One more bit of caution: Having

fun with your presentation might include some use of humor, but, of course, you aren't there to do a stand-up routine, so try to keep things on point.

Finally, here's a general goal, attributed to an anonymous Irish politician, about how to conduct a talk: "Tell them what you're going to tell them, then tell them, then tell them what you've told them." Find ways to reinforce the main points of your presentation to help everyone stay on track.

13.3b Posters

Many undergraduate and graduate students have their first experience in presenting a complete study as a poster presentation. This is often an ideal way to start your career in research presentations because it is typically a low-stress, friendly environment. The typical poster session is a large room with many posters and the presenter standing next to them. The conference attendees stroll by the posters with an occasional stop to ask a question or share a comment. Guidelines for constructing posters are typically very specific and produced for the particular conference or meeting. You can expect details related to the size of the poster as well as pointers about content (e.g., font size, use of color, suggestions about handout versions of your poster, and so on). In general, the layout of the poster follows a left-to-right format with the same sections as your paper: Abstract, Introduction/Background, Methods, Results, and Discussion. There is usually space for two or three tables and/or graphs. Many people use PowerPoint to construct their poster, on a single slide sized for printing to the required specifications. There are many templates and examples available on the Internet. There are numerous other poster design software packages that are excellent for producing very high-quality large text and images. Most colleges and universities are well equipped with large printers capable of producing very large posters (e.g., 72 inches tall or wide). **Figure 13.1** provides an example of a student poster presentation.

Figure 13.1 Doctoral student Francisco Lopez with a poster on his concept mapping study of quality of life of children with cancer. He won the conference Best Student Poster award.

Jim Donnelly

SUMMARY

This chapter discussed the last step in a typical research project—telling the story in a paper or presentation. We outlined the key elements that typically must be included somewhere in a standard research report—the Introduction, Methods, Results, and Discussion—and described what should be included in each of these sections. We described the major style issues you should watch out for when writing the typical report. And we presented one way to format a research paper appropriately, including how to cite other research in your report. Even though formatting rules can vary widely from one field to another, once you see how a set of rules guides report writing, it's a simple matter to change the formatting for a different audience or editorial policy.

So, with the write-up, the formal part of the research process is complete. You've taken the journey down the research road, from the initial plan for your trip, through all of the challenges along the way, and now with the write-up, on to your final destination. Now what? If you're a researcher, you don't stay in one place for very long. The thrill of the journey is just too much to resist. You begin pulling out the roadmaps (formulating research questions and hypotheses) and thinking about how you'd like to get to your next destination. There's a certain restlessness, a bit of research-based wanderlust that sets in. And there are all the lessons you learned from your previous research journeys. Now, if, on this next trip, you can only avoid the potholes! We hope that when you set out on your own research journey, you'll take this book along. Consider it a brief guidebook, a companion that might help point the way when you're feeling lost. Bon voyage!

Key Terms

abstract p. 331
APA format p. 329
citation p. 334

file drawer problem p. 328
reference p. 334
running head p. 330

title page p. 331

Suggested Websites

The CONSORT Statement Web Site

http://www.consort-statement.org/

The CONSORT group was initiated to develop standards for reporting randomized experiments. CONSORT stands for Consolidated Standards of Reporting Trials. The standards include a checklist of items that should always be reported, as well as templates for flowcharts showing participant progress. The goal is to make the reporting of results more transparent and to make sure that adequate information is included to enable critical review and synthesis of results.

Crossref

http://www.crossref.org/

Crossref is another organization designed to support consistency and accuracy in reporting of studies and retrieval of papers. At the time of this writing, the database included over 63 million records. It is a very handy place to locate a digital object identifier (DOI) when pulling your reference list together.

Review Questions

1. What publication would be most appropriate to consult for guidelines on writing and publishing research papers?

a. *Publication Manual* of the American Psychological Association
b. *Diagnostic* and *Statistical Manual of Mental Disorders (DSM-5)*
c. *The American Psychologist Guide to Good Writing*
d. Webster's *Encyclopedia of Writing*
(Reference: 13.2)

2. In which section of the research report should a statement of your research question or hypothesis first appear?

a. introduction
b. methods
c. results
d. conclusions
(Reference: 13.2a)

3. Professional writing is generally considered

a. unbiased and uninteresting.
b. unbiased and unemotional.
c. unemotional and uninteresting.
d. unemotional and unrelenting.
(Reference: 13.2a)

4. In the introduction section of a research paper, which of the following should be presented?

a. sample description
b. brief description of major constructs
c. description of procedures
d. construction of measures
(Reference: 13.2a)

5. Which of the following properly orders the sections of a typical research paper?

a. title page, abstract, introduction, methods, results, conclusions, references
b. title page, abstract, introduction, conclusions, methods, results, references
c. title page, abstract, introduction, design, discussion

d. title page, introduction, results, conclusions, references, abstract
(Reference: 13.2a)

6. In which part of the paper do you write about any threats to internal validity?

a. appendices
b. results
c. design
d. abstract
(Reference: 13.2a)

7. In your research project, you included a new measure of achievement orientation that is one-and-a-half pages in length. Where in your paper would it be most appropriate for you to present it?

a. measures
b. appendices
c. results
d. introduction
(Reference: 13.2a)

8. If you have six or more authors for a reference, do you need to list them all when citing the reference in the text?

a. Yes. List all six each time you cite the article.
b. Yes. List all six the first time you cite the article, and in subsequent references, list only the first author, followed by et al.
c. No. List only the first author.
d. No. List only the first author, followed by et al.
(Reference: 13.2a)

9. Formatting of a paper is consistent across disciplines and is also basically the same for term papers, theses, dissertations, and published studies.

a. True
b. False
(Reference: 13.2a)

10. If you are using previously published measures, you do not have to discuss reliability or validity.

a. True
b. False
(Reference: 13.2)

11. An abstract typically just gives a summary of the results of a study.

a. True
b. False
(Reference: 13.2a)

12. A paper that begins with this sentence "I wanted to study this issue because it make me very upset" is a good example of professional scientific writing.

a. True
b. False
(Reference: 13.2a)

13. Material obtained from online sources does not need a reference citation.

a. True
b. False
(Reference: 13.2a)

Appendix A

Sample Research Paper in APA Format

The following paper is a fictitious example of a research write-up. It is a made-up study that was created to look like a real research write-up. It is provided only to give you an idea of what a research paper write-up might look like. You should never copy any of the text of this paper in any other document and certainly should not cite this as a research paper in another paper.

Running head: SUPPORTED EMPLOYMENT   1

The Effects of a Supported Employment Program on Psychosocial

Indicators for Persons with Severe Mental Illness

Zeke DeAussie

Woofer College

SUPPORTED EMPLOYMENT   2

Abstract

This paper describes the psychosocial effects of a Supported Employment (SE) program for persons with severe mental illness. Employment specialists provided extended individualized supported employment for clients through a Mobile Job Support Worker. A 50% random sample was taken of all persons who entered the *Breakthroughs* Agency between 3/1/10 and 2/28/13 and who met study criteria. The 484 cases were randomly assigned to either the SE condition or the usual protocol, which consisted of life-skills training and employment in an in-house sheltered workshop. All participants were measured at intake and at three months after beginning employment on two measures of psychological functioning and two measures of self-esteem. Significant treatment effects were found on all four measures, but they were in the opposite direction from what was hypothesized. Effect sizes were negative and moderate, ranging from $-.48$ to $-.60$. Instead of functioning better and having more self-esteem, persons in SE had lower functioning levels and lower self-esteem. The most likely explanation is that people who work in low-paying service jobs in real-world settings experience significant job stress, whether they have severe mental illness or not.

Over the past quarter century a shift has occurred from traditional institution-based models of care for persons with severe mental illness (SMI) to more individualized, community-based treatments. The main focus of such efforts lies in rehabilitating persons with SMI toward lifestyles that more closely approximate those of persons without such illnesses. A central issue in this endeavor is the ability of a severely disabled person to hold a regular full-time job for a sustained period of time. Employment alongside others who do not have disabilities is thought to provide a concrete mechanism for broad community integration. Further, from a more general perspective, employment among individuals with SMI is also likely to weaken dependence on social disability programs as well as lead to an overall increase in national productivity.

Vocational Rehabilitation for Individuals with SMI

There have been several attempts to develop novel and radical program interventions designed to assist persons with SMI to sustain full-time employment while living in the community. Within this field of research, "Supported Employment (SE)" has emerged as one of the most effective approaches to vocational rehabilitation. Originally rooted in psychiatric rehabilitation (Cook, Jonikas, & Solomon, 2012), the key idea behind SE

SUPPORTED EMPLOYMENT   4

lies in helping people with disabilities participate as much as possible in the competitive labor market, working in jobs they prefer with the level of professional help they need (Bond et al., 2001).

SE programs typically provide individual placements in competitive employment – that is, community jobs paying at least minimum wage that any person can apply for – based on client choices and capabilities. These programs facilitate job acquisition by first locating a job in an integrated setting for minimum wage or above and then placing the person on the job - often sending staff to accompany clients on interviews. Further, these programs also provide one-on-one job coaching and ongoing support services (such as advocacy with co-workers and employers) even after the client is employed (Bond et al., 2001).

It is important to note that SE programs differ starkly from original models of vocational rehabilitation. The traditional models were based on the idea of sheltered workshop employment where clients were paid a piece rate and worked only with other individuals who were disabled. Sheltered work-shops tended to be "end points" for persons with severe and profound mental retardation since few ever moved from sheltered to competitive employment (Woest, Klein, & Atkins, 2006). There is widespread agreement among a

SUPPORTED EMPLOYMENT 5

network of scholars that sheltered workshops isolate people with disabilities from mainstream society (Wehman, 2010). Controlled studies of sheltered workshop performance of persons with mental illness suggested only minimal success (Griffiths, 2011) and other research indicated that persons with mental illness earned lower wages, presented more behavior problems, and showed poorer workshop attendance than workers with other disabilities (Ciardiello, 2012; Whitehead, 2007). The SE program, which was first introduced in the 1980s, was proposed as a less expensive and more normalizing alternative for persons undergoing rehabilitation (Wehman, 2005).

Evaluating Vocational Rehabilitation Approaches

The large majority of attempts to evaluate these two classes of rehabilitation programs (Supported Employment and Sheltered Employment) have naturally focused almost exclusively on employment outcomes. Evidence indicates that SE is more effective at increasing competitive employment among individuals with serious mental disorders as compared to traditional vocational rehabilitation. For instance, across 11 randomized studies, Crowther et al. (2001) found that about two-thirds of supported employment enrollees became competitively employed compared to less than one-fourth of those in traditional interventions.

SUPPORTED EMPLOYMENT   6

Social inclusion theory, however, suggests that vocational rehabilitation may have important therapeutic benefits that go beyond the
obvious monetary rewards. These latent rewards may include social
identity and status; social contacts and support; and a sense of personal
achievement (Shepherd, 2009). Being employed provides opportunities
to people with mental illnesses to participate in society as active citizens.
Work is also important both in maintaining mental health and in promot-
ing the recovery of those who have experienced mental health problems
(Boardman et al., 2003). Thus, enabling people to retain or gain employment likely has profound effects on a variety of different life domains.

Very few studies have analyzed non-vocational outcomes of participating in vocational rehabilitation programs. For example, Drake, McHugo,
Becker, Anthony, and Clark (2006) found no evidence of increased
hospitalization among SE program participants as compared to controls.
Similarly, Gold et al. (2006) found no substantial symptom change over
time in either group.

To date, however, no formal studies in the US have examined the effects of vocational rehabilitation programs on important social and clinical
outcomes such as self-esteem, anxiety and depression. This study seeks to

examine the effects of a new program of SE on psychosocial outcomes for persons with SMI.

SE Program at *Breakthroughs* site

A notable SE program was developed at *Breakthroughs* Agency in Chicago, the site for the present study, which created a new staff position called the mobile job support worker (MJSW) and removed the common six-month time limit for many placements. MJSWs provide ongoing, mobile support and intervention at or near the work site, even for jobs with high degrees of independence (Cook & Hoffschmidt, 2013). Time limits for many placements were removed so that clients could stay on as permanent employees if they and their employers wished. The suspension of time limits on job placements, along with MJSW support, became the basis of SE services delivered at *Breakthroughs*.

There are two key psychosocial outcome constructs of interest in this study. The first is the overall *psychological functioning* of the person with SMI. This would include the specification of severity of cognitive and affective symptoms as well as the overall level of psychological functioning. The second is the level of self-reported *self-esteem* of the person. This was measured both generally and with specific reference to employment.

The key hypothesis of this study is:

H_O: A program of supported employment will result in either *no change or negative effects* on psychological functioning and self-esteem.

which will be tested against the alternative:

H_A: A program of supported employment will lead to *positive effects* on psychological functioning and self-esteem.

Methods

Sample

The population of interest for this study is all adults with SMI residing in the United States in the early 2000s. The population that is accessible to this study consists of all persons who were clients of the *Breakthroughs* Agency in Chicago, Illinois between the dates of March 1, 2010 and February 28, 2013 who met the following criteria: 1) a history of severe mental illness (i.e., either schizophrenia, major depression, or bipolar disorder); 2) a willingness to achieve paid employment; 3) their primary diagnosis must not include chronic alcoholism or hard drug use; and 4) they must be 18 years of age or older. The sampling frame was obtained from records of the agency. Because of the large number of clients who pass through the agency each year (e.g., approximately 1000 who meet the criteria) a simple random sample of 50%

SUPPORTED EMPLOYMENT   9

was chosen for inclusion in the study. A power analysis was conducted

focused on the BPRS as a primary outcome measure. Alpha was set at .05

and a minimally important change in the BPRS set to one-half of one stan-

dard deviation (estimated as 2.5 scale points based on the test manual data).

These parameters indicated that power of .80 would be maintained with a fi-

nal sample size of 400. The actual final sample size was 484 persons over the

two-year course of the study. A CONSORT-type flow chart of participation at

each stage of the study is presented in Figure 1.

Insert Figure 1 about here

On average, study participants were 30 years old and high school grad-

uates (average education level = 13 years). The majority of participants

(70%) were male. Most had never married (85%), few (2%) were currently

married, and the remainder had been formerly married (13%). Just over

half (51%) are African American, with the remainder Caucasian (43%) or

other minority groups (6%). In terms of illness history, the members in the

sample averaged four prior psychiatric hospitalizations and spent a lifetime

average of nine months as patients in psychiatric hospitals. The primary

SUPPORTED EMPLOYMENT   10

diagnoses were schizphrenia (42%) and major depression (37%). Participants had spent an average of almost two and one-half years (29 months) at
the longest job they ever held.

The study sample cannot be considered representative of the original
population of interest. Generalizability was not a primary goal; the major
purpose of this study was to determine whether a specific SE program
could work in an accessible context. Any effects of SE evident in this
study can be generalized to urban psychiatric agencies that are similar to
Breakthroughs, have a similar clientele, and implement a similar program.

Measures

All but one of the measures used in this study are well-known
instruments in the research literature on psychosocial functioning. All
of the instruments were administered as part of a structured interview
that an evaluation social worker had with study participants at regular
intervals.

Two measures of psychological functioning were used. The Brief
Psychiatric Rating Scale (BPRS) (Overall & Gorham, 1962) is an 18-item
scale that measures perceived severity of symptoms ranging from "somatic
concern" and "anxiety" to "depressive mood" and "disorientation."

SUPPORTED EMPLOYMENT   11

Ratings are given on a 0-to-6 response scale where 0 = "not present" and 6 = "extremely severe" and the scale score is simply the sum of the 18 items. The Global Assessment Scale (GAS) (Endicott, Spitzer, Fleiss, & Cohen, 1976) is a single 1-to-100 rating on a scale where each ten-point increment has a detailed description of functioning (higher scores indicate better functioning). For instance, one would give a rating between 91-100 if the person showed "no symptoms, superior functioning . . ." and a value between 1-10 if the person "needed constant supervision . . .".

Two measures of self-esteem were used. The first is the Rosenberg Self Esteem (RSE) Scale (Rosenberg, 1965), a 10-item Likert-type scale rated on a 6-point response format where 1 = "strongly disagree" and 6 = "strongly agree" and there is no neutral point. The total score is simply the sum across the ten items, with five of the items being reversals. The second measure was developed explicitly for this study and was designed to measure the Employment Self-Esteem (ESE) of a person with SMI. This is a 10-item scale that uses a 4-point response format where 1 = "strongly disagree" and 4 = "strongly agree" and there is no neutral point. The final ten items were selected from a pool of 97 original candidate items, based upon high item-total score correlations and a judgment of face validity by a panel of three psychologists. This instrument was deliberately kept simple—a shorter response scale and no reversal items—because of the difficulties associated with measuring a population with SMI. The entire instrument is provided in Appendix A.

All four measures evidenced strong reliability and validity. Internal consistency reliability estimates using Cronbach's alpha ranged from .76 for ESE to .88 for RSE. Test-retest reliabilities were nearly as high, ranging from .72 for ESE to .83 for the BPRS. Inter-rater reliability for the GAS was .82 (Cohen's Kappa). Convergent validity was evidenced by the correlations within construct. For the two psychological functioning scales, the correlation was .68; while for the self-esteem measures it was somewhat lower at .57. Discriminant validity was examined by looking at the cross-construct correlations, which ranged from .18 (BPRS-ESE) to .41 (GAS-RSE).

Design

A pretest-posttest two-group randomized experimental design was used in this study. In notational form, the design can be depicted as:

$$R \quad O \quad X \quad O$$
$$R \quad O \quad \quad O$$

SUPPORTED EMPLOYMENT 13

where:

> R = the groups were randomly assigned
>
> O = the four measures (i.e., BPRS, GAS, RSE, and ESE)
>
> X = supported employment

The comparison group received the standard *Breakthroughs* protocol, which emphasized in-house training in life skills and employment in an in-house sheltered workshop. All participants were measured at intake (pretest) and at three months after intake (posttest).

This type of randomized experimental design is generally strong in internal validity. It rules out threats of history, maturation, testing, instrumentation, mortality, and selection interactions. Its primary weaknesses are in the potential for treatment-related mortality (i.e., a type of selection-mortality) and for problems that result from the reactions of participants and administrators to knowledge of the varying experimental conditions. In this study, the dropout rate was 4% ($n = 9$) for the control group and 5% ($n = 13$) in the treatment group. Because these rates are low and are approximately equal in each group, it is not plausible that there is differential mortality. There is a possibility that there were some deleterious effects due to participant knowledge of the other group's existence (e.g., compensatory rivalry, resentful demoralization). Staff members

SUPPORTED EMPLOYMENT   14

were debriefed at several points throughout the study and were explicitly asked

about such issues. There were no reports of any apparent negative feelings

from the participants in this regard. Nor is it plausible that staff might have

equalized conditions between the two groups. Staff members were given exten-

sive training and were monitored throughout the course of the study. Overall,

this study can be considered strong with respect to internal validity.

Procedures

Between 3/1/10 and 2/28/13, each person admitted to *Breakthroughs*

who met the study inclusion criteria was immediately assigned a random

number that gave him or her a 50/50 chance of being selected into the

study sample. For those selected, the purpose of the study was explained,

including the nature of the two treatments, and the need for and use of

random assignment. All participants were assured confidentiality and were

given an opportunity to decline to participate in the study. Only seven

people (out of 491) chose not to participate. At intake, each selected sam-

ple member was assigned a random number giving them a 50/50 chance

of being assigned to either the Supported Employment condition or the

standard in-agency sheltered workshop. In addition, all study participants

were given the four measures at intake. Informed consent was obtained by the social worker who collected all measures. The research protocol was reviewed and approved by the IRB of Woofer College.

All participants spent the initial two weeks in the program in training and orientation. This consisted of life skills training (e.g., handling money, getting around, cooking, and nutrition) and job preparation (employee roles and coping strategies). At the end of that period, each participant was assigned to a job site—at the agency sheltered workshop for those in the control condition, and to an outside employer if in the Supported Employment group. Control participants were expected to work full-time at the sheltered workshop for a three-month period, at which point they were posttested and given an opportunity to obtain outside employment (either Supported Employment or not). The Supported Employment participants were each assigned a case worker—called a Mobile Job Support Worker (MJSW)—who met with the person at the job site two times per week for an hour each time. The MJSW could provide any support or assistance deemed necessary to help the person cope with job stress, including counseling or working beside the person for short periods of time. In addition, the MJSW was always accessible by cellular telephone, and could be called by the participant or the employer at any time. At the end of three months, each participant was

SUPPORTED EMPLOYMENT   16

posttested and given the option of staying with their current job (with or without Supported Employment) or moving to the sheltered workshop.

Results

There were 484 participants in the final sample for this study, 242 in each condition. There were nine dropouts from the control group and 13 from the treatment group, leaving a total of 233 and 229 in each group, respectively, from whom both pretest and posttest were obtained. Due to unexpected difficulties in coping with job stress, 19 Supported Employment participants had to be transferred into the sheltered workshop prior to the posttest. In all 19 cases, no one was transferred prior to week 6 of employment, and 15 were transferred after week eight. In all analyses, these cases were included with the Supported Employment group (intent-to-treat analysis) yielding treatment effect estimates that are likely to be conservative.

The major results for the four outcome measures are shown in Figure 2. Note that the scores were standardized to T-scores to make them comparable across measures.

_____

Insert Figure 2 about here

_____

SUPPORTED EMPLOYMENT   17

It is immediately apparent that in all four cases the null hypothesis cannot be rejected. Contrary to expectations, Supported Employment cases did significantly *worse* on all four outcomes than did control participants.

The means, standard deviations, sample sizes, t-values (t-test for differences from pre to post), 95% confidence intervals, p values, and effect sizes (Cohen's d) are shown for the four outcome measures in Table 1.

Insert Table 1 about here

The results in the table confirm the impressions in the figures. Note that all t-values are negative except for the BPRS, where high scores indicate greater severity of illness. For all four outcomes, the t-values were statistically significant ($p < .05$). The effect sizes were all negative and moderate, ranging from $-.48$ to $-.60$.

Conclusions

The results of this study were clearly contrary to initial expectations. The alternative hypothesis suggested that SE participants would show improved psychological functioning and self-esteem after three months of employment. Exactly the reverse happened; SE participants showed significantly worse psychological functioning and self-esteem.

SUPPORTED EMPLOYMENT   18

There are two major possible explanations for this outcome pattern.
First, it seems reasonable that there might be a delayed positive or "boo-
merang" effect of employment outside of a sheltered setting. SE cases
may have to go through an initial difficult period of adjustment (longer
than three months) before positive effects become apparent. This "you
have to get worse before you get better" theory is commonly held in other
treatment-contexts like drug addiction and alcoholism. However, a second
explanation seems more plausible—that people working full-time jobs in
real-world settings are almost certainly going to be under greater stress and
experience more negative outcomes than those who work in the relatively
safe confines of an in-agency sheltered workshop. Put more succinctly,
the lesson here might very well be that work is hard. Sheltered workshops
are generally nurturing work environments where virtually all employees
share similar illness histories and where expectations about productivity are
relatively low. In contrast, getting a job at a local hamburger shop or as a
shipping clerk puts the person in contact with co-workers who may not be
sympathetic to their histories or forgiving with respect to low productivity.
This second explanation seems even more plausible in the wake of informal
debriefing sessions held as focus groups with the staff and selected research
participants. It was clear in the discussion that SE persons experienced

significantly higher job stress levels and more negative consequences. However, most of them also felt that the experience was a good one overall and that even their "normal" co-workers "hated their jobs" most of the time.

One lesson we might take from this study is that much of our contemporary theory in psychiatric rehabilitation is naive at best and, in some cases, may be seriously misleading. Theory led us to believe that outside work was a "good" thing that would naturally lead to "good" outcomes like increased psychological functioning and self-esteem. But, for most people (SMI or not) work is at best tolerable, especially for the types of low-paying service jobs available to study participants. While people with SMI may not function as well or have high self-esteem, we should balance this with the desire they may have to "be like other people," including struggling with the vagaries of life and work that others struggle with.

Future research in this area needs to address the theoretical assumptions about employment outcomes for persons with SMI. It is especially important that attempts to replicate this study also try to measure how SE participants feel about the decision to work, even if traditional outcome indicators suffer.

SUPPORTED EMPLOYMENT   20

References

Boardman, J., Grove, B., Perk, R., & Shepherd, G. (2003). Work and employment for people with psychiatric disabilities. *British Journal of Psychiatry, 182,* 467–468. doi:bmp.59.38500.7856-3-4.467.

Bond, G. R., Becker, D. R., Drake, R .E., Rapp, C. A., Miesler, N., Lehman, A. F., et al., (2001). Implementing Supported Employment as an Evidence Based Practice. *Psychiatric Services, 52,* 314–322. doi:35.89473276.

Chadsey-Rusch, J., & Rusch, F. R. (1986). *The ecology of the workplace.* In J. Chadsey-Rusch, C. Haney-Maxwell, L. A. Phelps, & F. R. Rusch (Eds.), *School-to-work transition issues and models* (pp. 59–94). Champaign, IL: Transition Institute at Illinois.

Ciardiello, J. A. (2012). Job placement success of schizophrenic clients in sheltered workshop programs. *Vocational Evaluation and Work Adjustment Bulletin, 14,* 125–128.

Crowther, R. E., Marshall, M., Bond, G. R., et al. (2001). Helping people with severe mental illness to obtain work: systematic review. *BMJ, 322,* 204–208. doi:21.9847.90594837058584.

Cook, J. A. (1992). Job ending among youth and adults with severe mental illness. *Journal of Mental Health Administration, 19,* 158–169. doi:95.948367285.938-302839.

Cook, J. A., & Hoffschmidt, S. (2013). Psychosocial rehabilitation programming: A comprehensive model for the 1990's. In R. W. Flexer & P. Solomon (Eds.), *Social and community support for people with severe mental disabilities: Service integration in rehabilitation and mental health.* Andover, MA: Andover Publishing.

Cook, J. A., Jonikas, J., & Solomon, M. (2012). Models of vocational rehabilitation for youth and adults with severe mental illness. *American Rehabilitation, 18,* 6–32. doi:78.9374.0347474038479.

Drake, R. E., McHugo, G. J., Becker, D. R., Anthony, W. A., & Clark, R. E. (2006). The New Hampshire Study of supported employment for people with severe mental illness. *Journal of Consulting and Clinical Psychology, 64,* 391–399. doi:70.844-9475-95.9588040.

Endicott, J. R., Spitzer, J. L., Fleiss, J. L., & Cohen, J. (1976). The Global Assessment Scale: A procedure for measuring overall severity of psychiatric disturbance. *Archives of General Psychiatry, 33,* 766–771.

Gold, P. B., Meisler, N., Santos A. B., Carnemolla, M. A., Williams, O. H., & Keleher, J. (2006). Randomized trial of supported employment integrated with assertive community treatment for rural adults with

SUPPORTED EMPLOYMENT   22

severe mental illness. *Schizophrenia Bulletin, 32,* 378–395. doi:87.0009-
943.00009837475783.

Griffiths, R. D. (2011). Rehabilitation of chronic psychotic patients.
Psychological Medicine, 4, 316–325. doi:12.038572-88831.2349.

Overall, J. E., & Gorham, D. R. (1962). The Brief Psychiatric Rating
Scale. *Psychological Reports, 10,* 799–812.

Rosenberg, M. (1965). *Society and adolescent self image.* Princeton,
NJ: Princeton University Press.

Shepherd, G. (2009). The value of work in the 1980s. *Psychiatric
Bulletin, 13,* 231–233.

Wehman, P. (2005). Supported competitive employment for persons
with severe disabilities. In P. McCarthy, J. Everson, S. Monn, & M. Barcus
(Eds.), *School-to-work transition for youth with severe disabilities*
(pp. 167–182). Richmond, VA: Virginia Commonwealth University.

Whitehead, C. W. (2007). *Sheltered workshop study: A nationwide
report on sheltered workshops and their employment of handicapped individuals (Workshop Survey, Volume 1).* U.S. Department of Labor Service
Publication. Washington, DC: U.S. Government Printing Office.

Woest, J., Klein, M., & Atkins, B. J. (2006). An overview of supported employment strategies. *Journal of Rehabilitation Administration,
10,* 130–135. doi:45.39283947-9383839.099992876.

SUPPORTED EMPLOYMENT   23

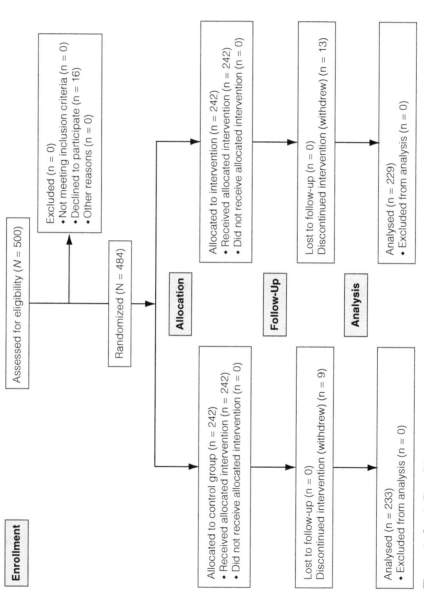

Figure 1. Study Flow Diagram.

SUPPORTED EMPLOYMENT   24

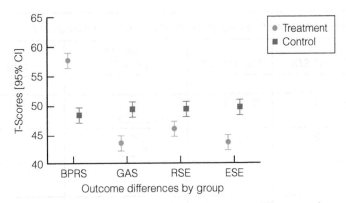

Figure 2. Standardized pre- to posttest mean differences for treatment and control groups for the four outcome measures. Note: Higher scores on BPRS indicate poorer status. Other measures scaled so that higher scores reflect better status.

SUPPORTED EMPLOYMENT   25

Table 1 *Descriptive and Inferential Statistics for the Outcome Measures*

Measure Group	Pre (*M, SD*)	Post (*M, SD*)	Change [95% CI]	*t (p)*	*d* [95%CI]
BPRS					
Treatment (*n* = 229)	3.2 (2.4)	5.1 (2.7)	1.9 [1.42, 2.38]		
Control (*n* = 233)	3.4 (2.3)	3.0 (2.5)	−0.4 [−.62, 2.0]		
Test statistics				−9.98 (.03)	−.60 [−.58, −.62]
GAS					
Treatment (*n* = 229)	59 (25.2)	43 (24.3)	−16 [−18.93, −13.69]		
Control (*n* = 233)	61 (26.7)	63 (22.1)	2 [.27, 4.27]		
Test statistics				−7.87 (.04)	−.58 [−.56, −.60]
RSE					
Treatment (*n* = 229)	42 (27.1)	31 (26.5)	−11 [−13.5, −8.5]		
Control (*n* = 233)	41 (28.2)	43 (25.9)	2 [−.49, 4.51]		
Test statistics				−5.19 (.049)	−.48 [−.46, −.50]
ESE					
Treatment (*n* = 233)	27 (18.6)	16 (20.3)	−11 [−12.8, −9.2]		
Control (*n* = 233)	25 (18.6)	24 (20.3)	−1 [−2.8, .8]		
Test statistics				−5.41 (.04)	−.53 [−.52, −.55]

SUPPORTED EMPLOYMENT   26

Appendix A

The Employment Self-Esteem Scale

Please rate how strongly you agree or disagree with each of the following statements.

1. I feel good about my work on the job.	Strongly Disagree	Somewhat Disagree	Somewhat Agree	Strongly Agree
2. On the whole, I get along well with others at work.	Strongly Disagree	Somewhat Disagree	Somewhat Agree	Strongly Agree
3. I am proud of my ability to cope with dif- ficulties at work.	Strongly Disagree	Somewhat Disagree	Somewhat Agree	Strongly Agree
4. When I feel uncomfortable at work, I know how to handle it.	Strongly Disagree	Somewhat Disagree	Somewhat Agree	Strongly Agree
5. I can tell that other people at work are glad to have me there.	Strongly Disagree	Somewhat Disagree	Somewhat Agree	Strongly Agree
6. I know I'll be able to cope with work for as long as I want.	Strongly Disagree	Somewhat Disagree	Somewhat Agree	Strongly Agree
7. I am proud of my relationship with my super- visor at work.	Strongly Disagree	Somewhat Disagree	Somewhat Agree	Strongly Agree
8. I am confident that I can handle my job without constant assistance.	Strongly Disagree	Somewhat Disagree	Somewhat Agree	Strongly Agree
9. I feel like I make a useful contri- bution at work.	Strongly Disagree	Somewhat Disagree	Somewhat Agree	Strongly Agree
10. I can tell that my co-workers respect me.	Strongly Disagree	Somewhat Disagree	Somewhat Agree	Strongly Agree

SUPPORTED EMPLOYMENT   27

Disclaimer: The above paper is a fictitious example of a research write-up. It is a made-up study that was created to look like a real research write-up. It is provided only to give you an idea of what a research paper write-up might look like. You should never copy any of the text of this paper in any other document and certainly should not cite this as a research paper in another paper.

Review Questions
Answer Key

Chapter 1

1. Answer: b

Rationale: Social research is empirical, meaning that it is based on what we perceive of the world around us. This is in contrast to social research being theoretical, meaning that much of it is concerned with developing, exploring, or testing the theories or ideas that social researchers have about how the world operates. While social research is both theoretical and empirical, a claim of empiricism asserts that the inquiry process is the observation and recording of data.

2. Answer: d

Rationale: Everyday life and research both involve observation, which makes them both empirical. However, research is systematic while everyday life is often a matter of happenstance. Formal social research also has the goal of contributing to public knowledge, that is, to our collective set of information about the world.

3. Answer: a

Rationale: Descriptive studies are designed primarily to describe what is going on or what exists in a particular context. The description of the proportion of people who hold various opinions—by such means as public opinion polls—is primarily descriptive in nature. By comparison, relational studies examine the relationship between two or more variables, while causal studies attempt to determine whether one or more variables cause or affect one or more outcomes.

4. Answer: b

Rationale: As opposed to cross-sectional studies (which take place at a single point in time), longitudinal studies occur over time, involving at least two (and often more) waves of measurement. Longitudinal studies can be classified as either a repeated measures design (if we have two or a few waves of measurement) or a time series design (if we have many waves of measurement—20 or more—over time).

5. Answer: c

Rationale: A meta-analysis uses statistical methods to combine the results of similar studies quantitatively

to reach general conclusions. A systematic review is a research synthesis approach that focuses on a specific question or issue and uses specific preplanned methods to identify, select, assess, and summarize the findings of multiple research studies. Often, a systematic review involves a panel of experts who discuss the research literature and reach conclusions about how well a discovery works to address a problem or issue. So, while a meta-analysis is always a quantitative synthesis, a systematic review may be a judgmental expert-driven synthesis, a meta-analysis, or both.

6. Answer: c

Rationale: A relationship refers to the correspondence between two variables. A negative (or inverse) relationship implies that high values on one variable are associated with low values on the other. This pattern is distinguished from positive relationships (in which high values on one variable are associated with high values on the other, and low values on one are associated with low values on the other) and curvilinear relationships (in which a relationship changes over the range of both variables), as well as no relationship (in which one variable is not correlated with the other variable).

7. Answer: d

Rationale: The dependent variable is the effect or outcome of a program or treatment. The independent variable is the cause of that effect—the program or treatment itself that we or nature manipulates in the hope of producing an effect on the dependent variable. For example, if we are researching the effects of a teacher certification program on teacher retention, the program is the independent variable, and our measures of teacher retention are the dependent variables.

8. Answer: b

Rationale: The EBP movement represents a major attempt of the research enterprise to achieve a better integration of research and practice. The EBP movement started in medicine and quickly moved to other fields such as psychology and education.

9. Answer: a

Rationale: A hypothesis is a specific statement of prediction that describes in concrete (rather than theoretical) terms what we expect will happen in our study. Hypothesis tests in empirical research involve formulating two hypothesis statements—one that describes our prediction and one that describes all the other possible outcomes with respect to the hypothesized relationship. Typically, we call the hypothesis that we support (our prediction) the "alternative hypothesis," while the hypothesis that describes the remaining possible outcomes in our study is called the "null hypothesis."

10. Answer: b

Rationale: Data are typically referred to as "quantitative" if they are in numerical form and "qualitative" if they are not. All quantitative data are based upon qualitative judgments, and all qualitative data can be described and manipulated numerically.

11. Answer: c

Rationale: The unit of analysis is the major entity that we are analyzing in the study. In this question, the unit is the individual student because we would have final exam scores for each student in the class.

12. Answer: c

Rationale: Research begins with broad questions, then narrows down by focusing in on specific questions or hypotheses that can reasonably be studied in a research project. After operationalizing the research variables, the researcher engages in direct measurement or observation of the question of interest. From this narrowest point of the hourglass, the researcher then analyzes the data, reaches conclusions, and generalizes back to the research questions from the data.

13. Answer: a

Rationale: In causal studies, at least two major variables are of interest: the cause and the effect. Usually, the cause is some type of event, program, or treatment that is the independent variable.

14. Answer: a

Rationale: This is one of the most frequently repeated comments in research but is still a source of confusion. Correlation implies a systematic association between two things, so as one goes up or down, the other moves in a predictable way. But this synchronization does not directly tell us how or why those observations are generated.

15. Answer: b

Rationale: Actually, "translational research" refers to the process of moving research findings from the basic discovery phase to practice in real life. We can think of the research enterprise as encompassing a research-practice continuum within which translation occurs. If a discovery survives this applied research testing, there is usually a process of seeing how well it can be implemented in and disseminated to a broad range of contexts that extend beyond the original controlled studies. The process of translating measures to other languages is a complex but less encompassing kind of translation, often referred to as cross-cultural validation.

Chapter 2

1. Answer: b

Rationale: The modern history of human subject protections also began at the Nuremberg trial. At the beginning of the trial, the judges had no prior basis in law by which to judge the Nazi physicians. They developed 10 principles for this purpose, and these principles formed the basis of what came to be known as the Nuremberg Code for research involving human subjects. The Code has been extremely important as a reference point for all regulations related to the protection of human subjects. Among other things, it established principles of informed consent, voluntary participation without coercion, clear scientific justification for research, and most important, limits on the risk of harm.

2. Answer: d

Rationale: In hindsight, Milgram's studies provided important stimuli that are now reflected in improved assessment of risk of harm, restrictions on the use of deception, and requirements for informing participants of the complete purpose of the study as well as follow-up appropriate to their interests and well-being. Conflict of interest has been a focus of ethics review in many other studies, but not Milgram's.

3. Answer: d

Rationale: The *Belmont Report* established these three universal principles with the intent of providing reference points for all future studies with human participants. They have been incorporated into the guidelines that all IRBs follow in reviewing proposals for research.

REVIEW QUESTIONS ANSWER KEY

Rationale: The elements of informed consent are supposed to enable every potential participant to understand what they will be involved in and to have the opportunity to ask questions and withdraw at any time. Informed consent is not merely "the consent form."

5. Answer: a

Rationale: Anonymity means that no information is collected that could reveal the identity of the individual participant. Confidentiality means that identifying information will be collected but kept private by the researcher.

6. Answer: c

Rationale: The National Research Act of 1974 lead to the federally regulated system of local controls represented by the IRB. Each IRB is responsible for review and monitoring of research with human participants at their local institutions in order to make sure that research protocols protect their participants according to all relevant guidelines. This includes colleges, universities, hospitals, and other organizations actively conducting research.

7. Answer: a

Rationale: Advocacy slogans like "stop protecting us to death" helped communicate the message of urgency that patients and caregivers wanted researchers and medical professionals to understand. The result was a change in the more conservative patient protection guidelines so that access to treatment could be made available sooner rather than later.

8. Answer: c

Rationale: IRBs are required to maintain an Institutional Animal Care and Use Committee or IACUC, but the legal and ethical framework governing the use of animals in research is much less well developed than the system for human participants.

9. Answer: b

Rationale: This is the federal definition of research misconduct. It occurs when researchers intentionally compromise the integrity of the research process. Research misconduct has a major impact on all aspects of the research enterprise, including public trust.

10. Answer: a

Rationale: In this case, a patient participant (Jesse Gelsinger) died as a result of participation in a clinical study. This situation represented a conflict of interest because the researchers' judgment about their primary interest (Jesse's well-being and the integrity of their study) was likely compromised by a secondary interest (personal gain) because they had a financial interest in a company that would have profited from the study.

11. Answer: b

Rationale: Piecemeal and duplicate publication is basically misrepresentation of research because they may give a false impression of the original study by packaging and repackaging the same data in various formats. There may be cases in which presentation of study results to vastly different audiences is justified, or when a large data set is split into substudies because they are not duplicate but, in fact, represent different questions with a particular sample and are better published separately.

12. Answer: a

Rationale: Although most researchers can cite personal examples where this did not happen, appropriate credit for contributions to a paper is part of honest reporting and fairness in research. Many journals have now published guidelines for determining authorship.

13. Answer: b

Rationale: We cannot promise anonymity *and* confidentiality because if data are collected anonymously, we have no idea who the data came from and therefore cannot keep confidentiality. We must choose one or the other.

14. Answer: b

Rationale: The point of informed consent is not to get a form signed but to inform each participant of what the study will involve, including procedures, risk of harm, any anticipated benefits, management of the data (especially regarding data security and confidentiality or anonymity), and his or her rights as a participant. Quite often, hurried researchers mistake the documentation of the process ("getting the form signed") with the importance of good communication with participants.

15. Answer: a

Rationale: There are ethical implications of every choice a researcher can consider, from conceptualization of a research question to sampling, design, analysis, and reporting. Therefore, it is best to take a proactive approach to ethical considerations and think through as many of your choices as you can, and then be very active in monitoring the implementation and reporting phases of your study to be sure that all ethical obligations have been met.

Chapter 3

1. Answer: b
Rationale: First, all qualitative data can be assigned meaningful numerical values without necessarily detracting from the qualitative information. Second, numbers in and of themselves cannot be interpreted without understanding the assumptions that underlie them.

2. Answer: c
Rationale: While ethnography emphasizes the study of entire cultures, field research requires the researcher to go "into the field" to observe the phenomenon as a participant observer. The purpose of grounded theory is to develop theories about phenomena of interest by way of observation.

3. Answer: a
Rationale: Grounded theory is a complex iterative process that tends to evolve toward one or a few core categories that are central. Eventually, one approaches conceptually dense theory as new observation leads to new linkages that lead to revisions in the theory and more data collection. Although the core concepts or categories are identified and fleshed out in detail, grounded theory does not have a clearly demarcated point for ending a study, instead aiming for a point at which the data do not contribute further to understanding (aka "saturation").

4. Answer: c
Rationale: While one of the most common methods for qualitative data collection, participant observation is demanding. This is due in large part to the months or years of intensive work required for the researcher to become accepted as a natural part of the culture. This acceptance is necessary to ensure that the observations are of the natural phenomenon, rather than of socially desirable responses.

5. Answer: a
Rationale: Sometimes conducted prior to a quantitative study, qualitative research may provide the ability to approach the existing literature on the topic with a fresh perspective born of direct experience, enhancing the ability to generate new theories and hypotheses. In addition, qualitative measures can help develop a deep understanding of how people think about complicated issues. In doing so,

qualitative researchers commonly develop information that is much more detailed that that of quantitative researchers. However, the ability to generate detailed information comes at a cost: The determination of generalizable themes is difficult (though many qualitative researchers aren't interested in generalizing).

6. Answer: a
Rationale: Whereas "internal validity" represents one of the traditional criteria for judging quantitative research, "credibility" provides an alternative criterion for judging qualitative research. From this perspective, the participants are the only ones who can legitimately judge the credibility of the results, since qualitative research serves to describe or understand the phenomena of interest from the participant's eyes.

7. Answer: b
Rationale: Confirmability refers to the degree to which others can confirm or corroborate the results.

8. Answer: d
Rationale: Even though qualitative research tends to assume that each researcher brings a unique perspective to the study, "confirmability" can be enhanced by checking and rechecking the data, having another researcher take a "devil's advocate" role, actively searching for and describing negative instances that contradict prior observations, and conducting a data audit.

9. Answer: c
Rationale: Indirect measure, content analysis, and secondary analysis are all unobtrusive measures because they do not interfere with ongoing natural activities, while participant observation is a method of qualitative data collection that places the participant directly into the stream of activities in some context or situation.

10. Answer: b
Rationale: Content analysis is typically used to identify patterns in the analysis of text documents and other narratives (e.g., online discussion forums). The analysis can be quantitative, qualitative, or both.

11. Answer: a
Rationale: This is the most basic characteristic of a qualitative measure as contrasted with a quantitative one, but as the text notes, this definition is very limited.

12. Answer: b
Rationale: Generally speaking, quantitative research methods are better for testing theories than for

generating them, relative to qualitative methods. However, it should be noted that when quantitative findings are quite different from what was expected, the theory that led to the study hypothesis will be challenged and possibly revised.

13. Answer: a
Rationale: This is one of the text's most fundamental lessons and bears repeating.

14. Answer: a
Rationale: If a researcher was concerned about biasing the results of the study by the act of measuring them, then unobtrusive measures would be helpful if an alternative can be identified. Contrasting the findings one obtains with standard measures versus unobtrusive ones is very informative as a mixed-methods approach.

15. Answer: a
Rationale: Secondary analysis allows a researcher to attempt to replicate original findings, a very important and underused scientific method. There are many publicly available data sets that allow researchers to reexamine questions and pose new ones.

Chapter 4

1. Answer: c
Rationale: The Proximal Similarity Model is an approach to generalization that involves developing a theory about which contexts (persons, places, and times) are more like the study, and which are less so. When different contexts are placed in terms of their relative similarities, we can call this implicit theoretical dimension a gradient of similarity and can generalize the results of the study to other persons, places, or times that are more like (that is, more proximally similar to) the study.

2. Answer: b
Rationale: Whereas the theoretical population refers to the group to which you would ultimately like to generalize your results, the accessible population is the subset of the larger population to which you would actually be able to gain access. Researchers may have difficulty developing a reasonable sampling plan for the population of interest (the theoretical population) and so may have to sample from a smaller portion of that population—the accessible population.

3. Answer: d
Rationale: We sample so that we can get an estimate of a variable without measuring the entire population directly. In sampling, the units that are sampled—typically people—supply one or more responses for each of our variables. To look across the responses for an entire sample, we use a statistic. If we could measure the entire population and calculate a value like a mean or average, this would not be referred to as a statistic, but rather as a parameter of the population.

4. Answer: b
Rationale: A standard deviation is the spread of the scores around the average in a single sample. But the sample we draw is just one of a potentially infinite number of samples that we could have taken. To get from our sample statistic to an estimate of the population parameter, the sampling distribution provides a theoretical distribution of an infinite number of samples of the same size as the sample in our study. The standard error is the spread of the average of averages in a sampling distribution.

5. Answer: b
Rationale: In sampling contexts, the standard error is called sampling error. A low sampling error means that we had relatively less variability or range in the sampling distribution—in other words, our statistical estimates are more precise. Larger sample sizes are closer to the size of the actual population itself. So, since larger samples are less variable than smaller ones (all things being equal), the larger our sample size, the smaller the standard error.

6. Answer: a
Rationale: To have a random selection method, we must set up some process or procedure that assures that the different units in our population have equal probabilities of being chosen. While simple methods of probability sampling (such as drawing a name from a hat) can be simple, they can also be cumbersome when sample sizes are large. For this reason, we tend to use computers as the mechanism for generating random numbers as the basis for random selection.

7. Answer: a
Rationale: Simple random sampling—the simplest form of random sampling—is a method of sampling that involves drawing a sample from a population so that every possible sample has an equal probability of being selected. Because simple random sampling is a form of probability sampling, it is reasonable to generalize the results from the sample back to the population.

8. Answer: c
Rationale: Stratified random sampling, also sometimes called proportional or quota random sampling, involves dividing the population into homogeneous

subgroups and then taking a simple random sample in each subgroup. In addition, stratified random sampling has more statistical precision (less variance) than simple random sampling, better assuring that both the overall population and key subgroups are represented.

9. Answer: c
Rationale: Systematic random sampling is a probability sampling method in which we determine randomly where we want to start selecting in the sampling frame, and then follow a rule to select every xth element in the sampling frame list (which must be randomly ordered, at least with respect to the characteristics we are measuring). Systematic random samples are generally easier to draw and more precise than simple random samples.

10. Answer: d
Rationale: Cluster (area) random sampling is a probability sampling method that involves dividing the population into groups called clusters (usually along geographic boundaries), randomly selecting clusters, and then measuring all units within sampled clusters. This method is useful when sampling a population that is spread across a wide area geographically so as to reduce travel required between sites.

11. Answer: b
Rationale: The reference to the bell reflects the shape of the distribution of scores, with a large number in the middle and a gradual decline in the number of cases as we go toward the edge (or "tail") of the distribution, which corresponds to the edge of the bell shape.

12. Answer: a
Rationale: The standard error of measurement is referred to as the sampling error when calculated in a sample survey. Often, we hear comments like: "These results are considered accurate within plus or minus three percentage points" In an instance like that, we can guess that the standard error, or sampling error, is 1.5 points because, as you remember, in a normal distribution approximately two standard errors marks the 95% Confidence Interval.

13. Answer: b
Rationale: A nonprobability sample *might be* representative of the population, but we cannot use probability theory to support such an inference. In some studies, a nonprobability sampling strategy might be our only choice, and careful consideration of our options will allow us to do the best we can in trying to represent the population.

14. Answer: b
Rationale: Actually, the correct term is "snowball sampling," and this item is intended to allow you to check your snow-related research methods knowledge! The experienced researcher will realize that only rarely do you get an avalanche. Most of the time, you have to work hard to get a decent-sized snowball sample.

15. Answer: a
Rationale: Table 3-1 reviews all of the major strategies and identifies the strengths and weaknesses of each one.

Chapter 5

1. Answer: a
Rationale: A nominal level of measurement describes a category or kind of something. Nominal categories like party affiliation can be coded with numeric values to make analysis easier, but the difference is a difference in kind, but not in quantity, as with ordinal, interval, and ratio level measures.

2. Answer: c
Rationale: This coefficient has been very widely used because of its ease of calculation and interpretation. Sometimes, it seems that the ease of use prevents researchers from going further in assessing the reliability of a scale, and in turn, they are left with a limited view of reliability.

3. Answer: b
Rationale: Construct validity is the approximate truth of the conclusion that our operationalization accurately reflects its construct. It involves generalizing from our program or measures to the concept of our program or measures. Two general forms of construct validity include translation validity (whether the operationalization is a good reflection of the construct) or criterion-related validity (the degree to which an operationalization should function in predictable ways in relation to other operationalizations, based upon our theory of the construct).

4. Answer: a
Rationale: Content validity is an approach to translation validity (itself a form of construct validity), which involves deciding on the criteria that constitute a good description of the content domain and using those criteria as a type of checklist when examining our program or intervention.

5. Answer: c
Rationale: Criterion-related validity involves the development of evidence of the validity of a measure

based on its relationship to another independent measure as predicted by our theory of how the measures should behave. As opposed to translation validity, criterion-related validity usually involves making a prediction about how the operationalization will perform based on our theory of the construct.

6. Answer: b
Rationale: As a type of criterion-related validity (itself a type of construct validity), predictive validity involves an operationalization's ability to distinguish between groups that it should theoretically be able to distinguish between. As in any discriminating test, the results are more powerful if you are able to show that you can discriminate between two groups that are very similar.

7. Answer: a
Rationale: In convergent validity, measures of constructs that theoretically should be related to each other correspond or "converge" with similar constructs. In discriminant validity, measures of constructs that theoretically should not be related to each other discriminate between dissimilar constructs. For this reason, construct validity (specifically, criterion-related validity) is enhanced when the correlation between a measure and like constructs is high and the correlation with unlike constructs is low.

8. Answer: c
Rationale: A threat to construct validity known as a mono-method bias occurs when we use only a single method of measurement (and, in turn, cannot provide multiple sources of evidence that we are really measuring all of the construct that we intend to). Mono-method bias refers to our measures or observations, not to our programs or causes.

9. Answer: c
Rationale: Experimenter expectancy is a "social" threat to construct validity in which the researcher can knowingly or unwittingly bias the study by letting research participants know of her or his hypotheses. This threat can best be countered by making use of multiple experimenters who differ in their characteristics and/or through measuring experimenter's expectations before a study and then adjusting findings according to those expectations.

10. Answer: a
Rationale: Remember that external validity (Chapter 3) is most concerned with generalizing to people, places,

and contexts, while construct validity is concerned with the generalization to related aspects of the construct being studied.

11. Answer: b
Rationale: Reliance on single forms of measurement can limit the validity of our observations by giving us only one view. Mono-method and mono-operation bias can best be overcome by the use of multiple measures, particularly if they are different forms of measurement.

12. Answer: a
Rationale: This is the most complex sounding of the threats to validity, but breaking it down into component parts will remind you that the essence of it is quite simple (and unfortunately, too common).

13. Answer: a
Rationale: The organizational scheme for validity presented in this chapter should help you tie the concepts together, as well as remember them more easily. It might even help you draw a model of them with icons or graphics to help you remember which is which.

14. Answer: b
Rationale: Actually, it is just the opposite. While some kind of face validity evidence might be the easiest to obtain and is, in fact, important in a practical sense, it is the weakest form of evidence of construct validity.

15. Answer: a
Rationale: In predictive validity, you assess the operationalization's ability (your measure of some ability in this example) to predict something it should theoretically be able to predict.

Chapter 6

1. Answer: a
Rationale: All three methods attempt to scale a construct that is thought to be best represented as a single dimension.

2. Answer: c
Rationale: One of the main objectives of the chapter is to help you understand the distinctions among similar terms in measurement. It may help you to remember one of the famous indexes like the Consumer Price Index as an example of something that reflects a general construct (the cost of living) as a function of a number of variables (key consumer items).

3. Answer: d
Rationale: The important thing here is to learn that construction of an index follows an orderly systematic process that is useful no matter what the particular construct is to be measured.

4. Answer: b
Rationale: The use of weighted indexes is quite common and straightforward. As you can imagine, identifying the weights can require considerable thought. For example, if you were going to determine a formula for admission to a college or university, how would you identify the key variables and the amount of weight that each should have in determining eligibility for admission?

5. Answer: c
Rationale: This comment boils all of the complexity of scale development down to its essence.

6. Answer: a
Rationale: Samuel Messick's career was devoted to developing better tests and being a critical observer of how they were used and sometimes misused. He invented the idea of consequential validity to illustrate the importance of the real-world impact of tests, particularly when their imperfections led to unfair negative consequences for test takers.

7. Answer: c
Rationale: Thurstone's techniques are all general methods of scale construction.

8. Answer: b
Rationale: The study of the item-total correlations allows the scale developer to identify the poor items (that can be eliminated), as well as the good items.

9. Answer: c
Rationale: This scaling method allows the analyst to locate items in relation to one another based on the scalogram analysis.

10. Answer: d
Rationale: As the text advises, be sure to carefully define the focus, or all of the methods will prove to be inadequate.

11. Answer: b
Rationale: The terms are often used as if they mean the same thing, but one of the main lessons of this chapter is that there are substantial differences between the terms, including purpose and procedures.

12. Answer: a
Rationale: Tests are generally used in situations in which an evaluation is to be made based on

an objective measure. Scales are generally used when an estimate of a more subjective aspect of a construct is needed (e.g., knowledge of research methods versus attitude toward research).

13. Answer: a
Rationale: In this important step, you identify how each of the components in your index will be represented or measured.

14. Answer: b
Rationale: The set of questions on a survey cannot be considered a scale unless a scaling process was followed to identify the questions and determine how the responses would be combined. We need to follow a process and test it along the way to assert that the items form a valid and reliable scale.

15. Answer: b
Rationale: A scale is only a Likert scale if the procedures in scale development that Rensis Likert described have been properly followed.

Chapter 7

1. Answer: d
Rationale: Questionnaires and interviews comprise the two broad categories of surveys. Questionnaires are usually completed by the respondent and may include the mail survey, the group-administered questionnaire, and the household drop-off survey. Interviews are completed by the interviewer based on what the respondent says, and may include the personal interview and the telephone interview, among others.

2. Answer: d
Rationale: Before asking questions about the topic you are interested in, sometimes it is necessary to first ensure that respondents are qualified or have relevant experience to provide a valid response.

3. Answer: a
Rationale: The chance of obtaining a socially desirable response from a respondent can be reduced by avoiding embarrassing questions. In addition, interviewers can lessen the risk of eliciting socially desirable answers by not asking questions that they themselves find difficult, that make them uncomfortable, or for which they have strong opinions. When sensitive questions are important, then making sure that the interview is conducted in a private setting is very important. Lastly, if a researcher is concerned with the possibility of someone other than the intended respondent

providing answers to questions, then personal interviews rather than phone interviews or mail surveys should be employed.

4. Answer: b
Rationale: Dichotomous response formats are a type of structured response that has only two possible values, such as True/False, Yes/No, and gender (Male/Female).

5. Answer: d
Rationale: To keep multiple-filter questions from becoming too complex, try to have three or fewer levels (two jumps) for any questions, use graphics such as an arrow and box to jump, and if possible, jump to a new page.

6. Answer: b
Rationale: If several questions are required to elicit the response desired, you should consider splitting each of the questions into two or more separate ones. Double-barreled questions can often be spotted by looking for the conjunction "and" in the question.

7. Answer: a
Rationale: By posing questions in terms of a hypothetical respondent, you might get a reasonable estimate of the respondent's point of view (for example, you could ask respondents how much money "people you know" typically give in a year to charitable causes).

8. Answer: b
Rationale: In "check the answer" type questions, the respondent places a check (or "X") next to the responses. By convention, the check-mark format is typically used when it is possible for respondents to check more than one response.

9. Answer: b
Rationale: Multioption variables are used when we want to allow respondents to select multiple items. Because respondents can select any or all of the options, this type of variable must be treated in the analysis as though each option is a separate dichotomous (yes/no) variable.

10. Answer: d
Rationale: Unstructured response formats consist of space for written text. These formats can vary from short comment boxes to the transcript of an interview. While the former are relatively simple to solicit (for example, by instructing in a survey to "Please add any other comments in the box below"), transcription involves decisions regarding whether to transcribe verbatim or only major ideas, whether it's

important to distinguish between speakers, as well as the handling of nonconversational events that take place and of the thoughts of the interviewer.

11. Answer: a
Rationale: The key here is really just understanding the term *dichotomous,* a common and important term in scale development and statistics.

12. Answer: b
Rationale: The open-ended comment card is probably the most widely used *unstructured* response format.

13. Answer: b
Rationale: This is not a matter that can be so easily decided. You have to consider how long the survey is and how fatigued respondents might be at the end, especially if you are studying people who might in some way have difficulty with a long survey (e.g., cancer patients). At the same time, you should ensure that the questions asked at the beginning of the questionnaire pique the respondent's interest and motivate them to continue on with the questionnaire.

14. Answer: b
Rationale: The "Golden Rule" means that you can treat respondents with care and respect, as you would like to be treated yourself in a research (or other) situation.

15. Answer: a
Rationale: The most likely bad outcome of the use of tape or other direct recording devices in interview research is that people will be more likely to provide socially desirable responses as a result of their perceived loss of anonymity. Therefore, practice with immediate recording and in getting comfortable with the questions is very important in establishing and keeping rapport, and in getting the data you are after.

Chapter 8

1. Answer: c
Rationale: Internal validity is the approximate truth about inferences regarding cause and effect and is only relevant in studies that try to establish a causal relationship.

2. Answer: c
Rationale: No other plausible alternative explanation means that the presumed cause is the only reasonable explanation for changes in the outcome measures.

3. Answer: b
Rationale: Plausible alternative explanations for a hypothesized cause-effect relationship can be

reduced by arguing that the threat in question is not a reasonable one, by measurement of a threat and demonstration that either it does not occur at all or occurs minimally; by design (adding treatment or control groups, waves of measurement, and the like); by use of statistical analyses; and/or by preventive action (such as managing the attitudes of participants in treatment-comparison group quasi-experiments).

4. Answer: a
Rationale: It is possible to have internal validity in a study and not have construct validity. Internal validity means that you have evidence that what you did in the study (i.e., the program) caused what you observed (i.e., the outcome) to happen. It does not tell you whether what you did for the program was what you wanted to do or whether what you observed was what you wanted to observe—those are construct validity concerns.

5. Answer: b
Rationale: You cannot tell which way an individual's score will move based on the regression to the mean phenomenon. Even though the group's average will move toward that of the population's, some individuals in the group are likely to move in the other direction.

6. Answer: c
Rationale: Adding additional groups may be added to a design in order to rule out specific threats to validity. For example, the addition of a nonequivalent group consisting of persons who are from a similar setting as the treatment and control or comparison groups, but are unaware of the original two groups, can help rule out social threats such as intergroup communication, group rivalry, or demoralization of a group that is denied a desirable treatment or receives a less desirable one.

7. Answer: d
Rationale: Expanding across observations means collecting multiple measurements in order to determine whether the scores converge or discriminate as expected.

8. Answer: c
Rationale: If changes in scores are not due to the treatment effect, but instead are due to selective loss of participants from the study, our results are biased by "mortality" (loss of participants).

9. Answer: c
Rationale: If you can minimize plausible alternative explanations, your study by definition is well designed. Multiple measures are desirable but they address only one kind of alternative explanation. Having a guarantee of statistical significance does not necessarily imply good design, and there is almost always a need for further research, even though we all hope to do research that is definitive in some way.

10. Answer: b
Rationale: Consistent with the overall theme of the chapter, redundancy in a design strategy allows the researcher to account for some of the alternative explanations for observed results. You will have more confidence if multiple measures provide convergent results.

11. Answer: a
Rationale: Like most things in life, an ounce of prevention beats a pound of cure. However, each study will have unique aspects that need to be considered, and adding multiple strategies to your design will often allow you to see if your preventive action worked and will provide additional sources of control.

12. Answer: b
Rationale: The main reason for using this design-building strategy is to consider ways of controlling *all* of the possible threats to validity.

13. Answer: b
Rationale: Covariation is one of the necessary conditions for causality but is not the same thing. Sometimes, things vary over time together, without one causing the other.

14. Answer: b
Rationale: In fact, ruling out a potential threat to validity by argument alone is weaker than using the other strategies. Therefore, you should generally plan to begin with a thoughtful analysis of potential threats and include arguments, as well as measurement, design, analysis, and preventive action.

15. Answer: a
Rationale: The term *causal relationship* implies that some time has elapsed between the occurrence of the cause and the effect. Therefore, a design that attempts to measure a cause and its

effect or effects must include sufficient time for the process to occur. It helps to have a theory of how the process works, in order to do the best job of figuring out exactly when the measurements should occur.

Chapter 9

1. Answer: b
Rationale: When you are interested in determining whether a program or treatment causes an outcome or outcomes to occur, then you are interested in having strong internal validity. In the simplest type of experiment, we create two groups that are "equivalent" to each other, except that one group receives the program and the other does not. In such a design, if differences are observed in outcomes between these two groups, then the differences must be due to the only thing that differs between them—that one got the program and the other didn't.

2. Answer: b
Rationale: Probabilistic equivalence does not mean that we expect that the groups we create through random assignment will be exactly the same. Rather, probabilistic equivalence means that we know perfectly the odds that we will find a difference between two randomly assigned groups.

3. Answer: c
Rationale: In most real-world contexts, experimental design is intrusive and difficult to carry out. You are to some extent setting up both an intrusion and an artificial situation through experimental designs so that you can assess your causal relationship with high internal validity. By doing so, you are limiting the degree to which you can generalize your results to real contexts where you have not set up an experiment. You have reduced your external validity to achieve greater internal validity.

4. Answer: a
Rationale: Factorial designs are signal-enhancing experimental designs in which a number of different variations of a treatment are examined. In the null case, treatments have no effect, and the lines in a graph of means overlap each other. A main effect is a treatment effect that is a consistent difference between levels of one factor. An interaction effect exists when differences on one factor depend on the level of another factor.

5. Answer: b
Rationale: Probabilistic equivalence is achieved through the mechanism of random assignment to groups. The chance that the two groups will differ just because of the random assignment (i.e., by chance alone) can be calculated when subjects are randomly assigned to groups.

6. Answer: b
Random selection provides representativeness, which is the essence of external validity. Random assignment provides probabilistic equivalence, which is essential to internal validity.

7. Answer: a
Rationale: Random selection concerns sampling, since it concerns how you draw the sample of people for your study from a population. On the other hand, random assignment is how you assign the sample that you draw to different groups or treatments in your study.

8. Answer: a
Rationale: Factorial design has great flexibility for exploring or enhancing the "signal" (treatment) in research studies. Whenever you are interested in examining treatment variations, factorial designs should be strong candidates as the designs of choice.

9. Answer: d
Rationale: While factorial designs enhance the signal in experimental designs, covariance design and blocking design are commonly used to remove variability or noise. In these designs, we typically use information about the makeup of the sample or about pre-program variables to remove some of the noise in the study.

10. Answer: b
Rationale: In factorial designs, a factor is an independent variable. A level is a subdivision of a factor (like female or male if one of the factors was gender). The number of numbers in the notation of a factorial design tells you how many factors there are and the number of values tells you how many levels. So, since there are two numbers in the design (2 × 3), we know there are two factors, and from the number values (2 and 3), we know that one factor has two levels and the other three.

11. Answer: a
Rationale: By allowing everyone who participates in the study to be in both the treatment and control groups (at different times), the switching-replications design provides fairness as well as the ability to compare the effects of the treatment (or more generally, the independent variable).

12. Answer: b
Rationale: Random selection is how you draw the sample of people for your study from a population. Random assignment is how you assign the sample that you draw to different groups or treatments in your study. This is confused sometimes, even by experienced researchers, so be sure you have this very clear in your mind.

13. Answer: b
Rationale: A main effect is a consistent difference, graphically portrayed when the experimental and control group pre- and posttest scores are represented by parallel lines that are at significantly different levels on a scale. Interactions suggest that different levels of the independent variables produce different-strength effects and appear as non-parallel or crossed lines in a graph.

14. Answer: a
Rationale: The null case is the one that we assume when developing our null hypothesis. We test this against the alternative hypothesis that there is a significant effect.

15. Answer: a
Rationale: This design has two levels of one independent variable and three levels of the other. To fully test this model, you would need subjects assigned to all of the possible combinations of these variables, which would be six. Reread the cocaine abuse treatment study example in this section and write out all of the possible combinations, if this is not clear.

Chapter 10

1. Answer: b
Rationale: Probably the most commonly used quasi-experimental design is the nonequivalent-groups design.

2. Answer: b
Rationale: In quasi-experimental designs, assignment to groups is not random; therefore, the groups may be different prior to the study. Under the worst circumstances, this can lead us to conclude that our program didn't make a difference when, in fact, it did, or that it did make a difference when, in fact, it didn't.

3. Answer: c
Rationale: In a nonequivalent-groups design, research participants are not randomly assigned to groups; therefore, any prior differences between the groups may affect the outcome of the study. This design was named the nonequivalent-groups design to remind us that it is unlikely that the groups are equivalent.

4. Answer: a
Rationale: Selection of participants is an issue whenever random assignment is not used (i.e., whenever you have a nonexperimental design). The discussion considers all of the possible ways that selection might interact with other threats to internal validity to undermine a study.

5. Answer: c
Rationale: An observation that one group's pretest means in a quasi-experimental design are higher than those of another group suggests that there is a potential selection threat. However, such a difference may or may not explain any differences in posttest means. Even given the initial advantage of one group, there may still be a legitimate treatment effect.

6. Answer: b
Rationale: Internal validity focuses not on our ability to generalize but on whether a causal relationship can be demonstrated for the immediate context of research designs that address causal questions. Depending on how well the analyst can model the true pre-post relationship, only factors that would naturally induce a discontinuity in the pre-post relationship can be considered threats to the internal validity of inferences from the regression-discontinuity design.

7. Answer: b
Rationale: From an ethical perspective, regression-discontinuity designs are compatible with the goal of getting the program to those most in need. Since participants are assigned to program or comparison groups solely on the basis of a cutoff score on a pre-program measure, regression-discontinuity designs do not require us to assign potentially needy individuals to a no-program comparison group in order to evaluate the effectiveness of a program.

8. Answer: d

Rationale: A discontinuity in regression lines indicates a program effect in the regression-discontinuity design. But the discontinuity in and of itself is not sufficient to tell us whether the effect is positive or negative. To make this determination, we must know who received the program and how to interpret the direction of scale values on the outcome measures.

9. Answer: a

Rationale: In a "recollection" proxy-pretest design, participants are asked to estimate where their pretest level would have been, had they taken one. Alternately, an "archived" proxy-pretest design involves the review of archived records to stand in for the pretest. Use of the proxy-pretest design is typically born of necessity in situations where you have to evaluate a program that has already begun.

10. Answer: b

Rationale: By including two measures prior to the program, the double-pretest design is a very strong quasi-experimental design with respect to internal validity. If the program and comparison groups are maturing at different rates, this should be detected as a change from pretest 1 to pretest 2.

11. Answer: b

Rationale: The term *nonequivalent* means that assignment to groups was not random. That is, the researcher did not control the assignment to groups through the mechanism of random assignment, so the groups may be different on some or all of the measures prior to the study.

12. Answer: a

Rationale: Examining graphs of your data before conducting statistical tests is generally an excellent way to start an analysis, because a picture is worth a thousand words. You will immediately get an idea of any patterns in your data as well as any oddities like strange distributions and extreme values that should be double-checked for accuracy.

13. Answer: a

Rationale: This may be a major issue in designing a study because more participants necessitate more resources. However, as noted, there may be many situations where randomization is impossible, and so the regression-discontinuity design is the strongest practical alternative.

14. Answer: a

Rationale: The nonequivalent dependent variables (NEDV) design makes great use of theory in that, if you have a theory that is strong enough to allow you to make specific predictions, then you are in a good position to make a causal argument with regard to the pattern of results you obtain.

15. Answer: b

Rationale: In this case, the threat would be labeled a selection-maturity threat because the difference in the groups becomes apparent over time due to differential maturation. A selection-mortality threat might be a case when missing data from dropouts are related to something about the study (e.g., maybe the treatment is unpleasant for some).

Chapter 11

1. Answer: b

Rationale: Data preparation involves the cleaning and organizing of data for analysis. Generation of descriptive statistics consists of describing the data, while the testing of hypotheses and models is the function of inferential statistics.

2. Answer: a

Rationale: Of the four types of validity (internal, construct, external, and conclusion), conclusion validity may be considered the most important because it is relevant whenever we are trying to decide if there is a relationship in our observations. Conclusion validity is the degree to which the conclusion we reach is credible or believable.

3. Answer: c

Rationale: Conclusion validity was originally developed in relation to statistical inference, but it also pertains to qualitative research. Although conclusions or inferences in qualitative research may be based entirely on impressionistic data, we can ask whether there is conclusion validity by considering whether it is a reasonable conclusion about a relationship in our observations.

4. Answer: a

Rationale: whenever you investigate a relationship using NHST, you essentially have two possible conclusions: accepting or rejecting the null hypothesis. However, you might conclude that there is a relationship, when, in fact, there is not, or you might infer that there is not a relationship, when, in fact, there is.

Nevertheless, even though it also pertains to causal relationships, conclusion validity is only concerned with whether or not there is a relationship, not if that relationship is causal.

5. Answer: b
Rationale: Reliability is related to the idea of noise or "error" that obscures your ability to see a relationship. In general, you can improve reliability by doing a better job of constructing measurement instruments, by increasing the number of questions on a scale, or by reducing situational distractions in the measurement context.

6. Answer: b
Rationale: The four interrelated components that influence the conclusions you might reach from a statistical test in a research project include: (1) sample size (the number of units [e.g., people] accessible to the study), (2) effect size (the salience of the treatment relative to the noise in measurement), (3) alpha level (or significance level—the odds that the observed result is due to chance), and (4) power (the odds that you will observe a treatment effect when it occurs).

7. Answer: a
Rationale: Type I error is defined as concluding that there is a relationship when, in fact, there is not, while Type II error is defined as inferring that there isn't a relationship when, in fact, there is. Type I error concerns the odds of confirming our theory incorrectly, while Type II error deals with the odds of not confirming our theory when it is true. Ideally, Type I error should be kept small when we cannot afford/risk wrongly concluding that our program works.

8. Answer: d
Rationale: A threat to conclusion validity is a factor that can lead you to reach an incorrect conclusion about a relationship in your observations. Conclusion validity can be improved by increasing statistical power (through increased sample size, raising the alpha level, and/or increasing the effect size), bettering the reliability of your measures, and/or by assuring good implementation of your program.

9. Answer: b
Rationale: "Fishing" and the error rate problem occur when we conduct multiple analyses and treat each one as though it were independent. For example, suppose you conducted 20 analyses of the same data at the 0.05 level of significance. Treating each analysis as independent, you would expect to find one of them significant by chance alone (20 analyses × 0.05 = 1.0), even if there is no real relationship in the data.

10. Answer: b
Rationale: This is the definition of a Type II error.

11. Answer: b
Rationale: This is a good conceptual definition and a good literal definition. That is because if you look back at the formula and the example, you'll see that the *SD* is essentially determined by averaging the deviations of each score from the mean of the sample.

12. Answer: a
Rationale: The tradition of using the 0.05 level goes back to Sir Ronald Fisher, the father of analysis of variance and a widely respected authority on statistics, who suggested that this level of confidence in a result was convenient and adequate.

13. Answer: a
Rationale: These terms are used to indicate the rule that will be used in a study to determine at what point a result is considered statistically significant.

14. Answer: b
Rationale: "Fishing" in data until you find a "whopper" is not good practice because quite often, there will be something wrong with the result due to Type I error. In other words, you will end up with a "fish story" that will not be very believable.

15. Answer: a
Rationale: If you remember that power refers to our ability to find effects or results that are "really there," then the weak signal in the noisy background clearly reflects a problem of power.

16. Answer: a
Rationale: This rule of thumb implies that we want to have at least 80 chances out of 100 of finding a relationship when there is one.

Chapter 12

1. Answer: b
Rationale: The general linear model is an inferential statistic technique based on a system of equations that is used as the mathematical framework for most of the statistical analyses used in applied social research. The term *linear* refers to the fact that

we are fitting a line. The term *model* refers to the equation that summarizes the line that we fit.

2. Answer: b
Rationale: A dummy variable is a numerical variable used in regression analysis to represent subgroups of the sample in your study. In regression analysis, a dummy variable is often used to distinguish different groups of participants.

3. Answer: a
Rationale: The *t*-test assesses whether the means of two groups are statistically different from each other. Mathematically, a *t*-test is the difference between group means divided by the variability of groups. Statistical significance, however, is not the same as practical importance—the former concerns the level of certainty with which researchers make conclusions regarding differences between groups, while the latter relates to the question of whether the magnitude of such differences is meaningful.

4. Answer: c
Rationale: Given the alpha level, the degrees of freedom, and the *t*-value, you can look the *t*-value up in a standard table of significance (available as an appendix in the back of most statistics texts) to determine whether the *t*-value is large enough to be significant. If it is, you can conclude that the difference between the means for the two groups is different (even given the variability). Computer programs routinely produce the exact *p*-value associated with a result.

5. Answer: d
Rationale: When there is relatively little overlap of scores (low variability) and groups have different means (larger effect size), it is more likely that a statistically significant difference will be discovered. From a signal-to-noise point of view, statistically significant differences between groups are more likely when there is less noise (variability) and more signal (effect size).

6. Answer: c
Rationale: The regression line describes the relationship between two or more variables when using regression analysis as an inferential statistic technique. In the analysis of data using the general linear model, although the regression line does not perfectly describe any specific point, it does accurately describe the pattern in the data.

7. Answer: b
Rationale: The error term (*e*) describes the vertical distance from the straight line to each point. This term is called "error" because it is the degree to which the line is in error in describing each point.

8. Answer: c
Rationale: The key idea with the *t*-test is that you can see how big the difference between the means of two groups is relative to their *variability*. As the graphs in this section show, the absolute difference is much less meaningful than the difference relative to the degree of dispersion of the scores.

9. Answer: a
Rationale: All three statistical tests yield mathematically equivalent results, providing the exact same answer. First, an independent *t*-test describes the difference between the groups relative to the variability of the scores in the groups. Second, you could compute a one-way analysis of variance (ANOVA) between two independent groups. Finally, regression analysis can be used to regress the posttest values onto a dummy-coded treatment variable.

10. Answer: c
Rationale: Dummy variables—typically represented in regression equations by the letter *Z*—are useful because, like switches that turn various parameters on and off in an equation, they enable the use of a single regression equation to represent multiple groups. In addition, another advantage of a 0,1 dummy-coded variable is that, even though it is a nominal-level variable, you can treat it statistically like an interval-level variable.

11. Answer: a
Rationale: It may help you to think about the meaning of the words *descriptive* and *inferential*. A descriptive statistic will give us some sense of how things appear to be on the surface, while an inferential statistic will go beyond this level to *infer* the probable veracity of a statement about a broader or deeper question.

12. Answer: b
Rationale: Again, the name is informative because it tells us that there is a general (applies to many situations) and linear (based on a straight line)" with "(based on a straight line) model (an abstract representation).

13. Answer: a
Rationale: Graphs can provide you with important insights about the relationships in your data, including the possibility that there are nonlinear aspects to the pattern you see. For example, you can check to see if there is a curve rather than a straight line, or maybe a cluster of data points apart from most of the others. Such issues may violate the assumption of linearity in the general linear model (GLM) and cause you to consider data transformations or alternative analytic strategies.

14. Answer: a
Rationale: When we think of a line describing a pattern of data, then the slope of that line gives us a key piece of information. As the text shows you, this line corresponds to the strength of the relationship, so that the stronger the relationship, the greater the slope. That is, if the relationship is strong, then the change in one variable will be relatively easily determined by knowing the level of the other variable.

15. Answer: a
Rationale: Although in many cases you might want to examine the means of your experimental and control groups in an analysis of variance or covariance, it is possible to code your groups with a dummy variable (e.g., 1 = treatment, 0 = control) and enter them into a regression equation.

Chapter 13

1. Answer: a
Rationale: The *Publication Manual of the American Psychological Association,* 6th edition, documents the APA's instructions for the formatting of research articles or research reports. Other resources may be useful to help improve your writing, and publishers may also have their own requirements.

2. Answer: a
Rationale: The hypothesis (or hypotheses) should be clearly stated in the introduction section of a research report and should be specific about what is predicted. The relationship of the hypothesis to both the problem statement and literature review should be readily understood.

3. Answer: b
Rationale: So as to present arguments in as unbiased a fashion as possible, the professional writing style is characterized by avoidance of first-person and sex-stereotyped forms of writing. In addition,

the professional writing style can be described as unemotional (e.g., no "feelings" about things), although this should not come at the expense of writing an uninteresting paper.

4. Answer: b
Rationale: The introduction section of a research paper should state and explain each key construct in the research/evaluation project. The explanations should be readily understandable (i.e., jargon-free) to an intelligent reader.

5. Answer: a
Rationale: The title page provides the title of the study, the author's name, and the institutional affiliation. The abstract should be limited to one page, written in paragraph form, and should be a concise summary of the entire paper. Following this, the body of the paper contains the introduction, methods, results, and conclusions sections. The references section lists all the articles, books, and other sources used in the research and preparation of the paper and cited with a parenthetical (textual) citation in the text.

6. Answer: c
Rationale: In the design and procedures section of your methods section, you should discuss both threats to internal validity and how they are addressed by the design. In addition, any threats to internal validity that are not well controlled should also be considered.

7. Answer: b
Rationale: Generally, appendices are only used for presentation of extensive measurement instruments, for detailed descriptions of the program or independent variable, and for any relevant supporting documents that you do not include in the body of the paper.

8. Answer: d
Rationale: If there are six or more authors, the first author is listed, followed by et al., and then the year of the publication is given in parentheses. Only the author's last name is provided, unless first initials are necessary to distinguish between two authors with the same last name. Page numbers are provided if a direct quotation was used, or when only a specific part of a source was used.

9. Answer: b
Rationale: Following the available guidelines is one of the most important aspects of good writing. The conventions presented in this chapter might be found quite commonly across many research

reports, but always pay close attention to the instructions you have been given about how to format a paper—before you begin to write.

10. Answer: b
Rationale: Although it helps if you do not have to develop your own measures, you should still justify their use in your study by showing the reader that they are appropriate to your population and that they have sufficient reliability and validity to use as you intend.

11. Answer: b
Rationale: The abstract is a key piece of the research report because it concisely summarizes the entire study from introduction to discussion. Writing an abstract takes practice, especially when given tight word limits.

12. Answer: b
Rationale: Although, in fact, many people do want to study things that upset them so they can understand and perhaps change them, scientific writing is more objective than a personal essay would be.

13. Answer: b
Rationale: Every idea, finding, or other thing you include in your paper should have a reference citation. Internet sources should be as specifically referenced as possible.

Glossary

.05 level of significance This is the probability of a Type I error. It is selected by the researcher before a statistical test is conducted. The .05 level has been considered the conventional level since it was identified by the statistician R. A. Fisher. *See also alpha level.*

abstract A written summary of a research project that is placed at the beginning of each research write-up or publication.

accessible population A group that reflects the theoretical population of interest and that you can get access to when sampling. This is usually contrasted with the theoretical population.

alpha level The *p* value selected as the significance level. Specifically, alpha is the Type I error, or the probability of concluding that there is a treatment effect when, in reality, there is not.

alternative hypothesis A specific statement of prediction that usually states what you expect will happen in your study.

analysis of covariance (ANCOVA) An analysis that estimates the difference between groups on the posttest after adjusting for differences on the pretest.

anonymity The assurance that no one, including the researchers, will be able to link data to a specific individual. This is the strongest form of privacy protection in research because no identifying information is collected on participants.

APA Format Also known as APA Style. The standard guide for formatting research write-ups in psychology and many other applied social research fields. The APA format is described in detail in the *Publication Manual of the American Psychological Association* (APA) and includes specifications for: how subsections of a research paper should be organized; writing style; proper presentation of results; research ethics; and how citations should be referenced.

applied research Research where a discovery is tested under increasingly controlled conditions in real-world contexts.

assent Assent means that a child has affirmatively agreed to participate in a study. It is not the mere absence of an objection. The appropriateness of an assent procedure depends on the child's developmental level and is determined by the researcher in consultation with the IRB.

attribute A specific value of a variable. For instance, the variable *sex* or *gender* has two attributes: male and female.

average inter-item correlation An estimate of internal consistency reliability that uses the average of the correlations of all pairs of items.

average item-total correlation An estimate of internal consistency reliability where you first create a total score across all items and then compute the correlation of each item with the total. The average inter-item correlation is the average of those individual item-total correlations.

basic research Research that is designed to generate discoveries and to understand how the discoveries work.

bell curve Also known as a normal curve. A type of distribution where the values of a variable have a smoothed histogram or frequency distribution that is shaped like a bell. In a normal distribution, approximately 68 percent of cases occur within one standard deviation of the mean or center, 95 percent of the cases fall within two standard deviations, and 99 percent are within three standard deviations.

Belmont Report The report includes basic standards that should underlie the conduct of any biomedical and behavioral research involving human participants. It emphasizes universal principles that are unbounded by time or technology. The three core principles described in the *Belmont Report* are 1) respect for persons 2) beneficence, and 3) justice.

beneficence The expected impact on a person's well-being that may result from participation in research. Researchers should attempt to maximize the benefits of participation and take

391

steps to identify and limit the potential for harm. This is typically done when planning a study by a careful risk/benefit assessment.

bias A systematic error in an estimate. A bias can be the result of any factor that leads to an incorrect estimate. When bias exists, the values that are measured do not accurately reflect the true value. For instance, in sampling bias can lead to a result that does not represent the true value in the population.

blocking designs A type of randomized experimental design that helps minimize noise through the grouping of units (e.g., participants) into one or more classifications called blocks that account for some of the variability in the outcome measure or dependent variable.

case study An intensive study of a specific individual or specific context.

causal Pertaining to a cause-effect question, hypothesis, or relationship. Something is causal if it leads to an outcome or makes an outcome happen.

causal relationship A cause-effect relationship. For example, when you evaluate whether your treatment or program causes an outcome to occur, you are examining a causal relationship.

causal studies A study that investigates a causal relationship between two variables.

cause construct The abstract idea or theory of what the cause is in a cause-effect relationship you are investigating.

central tendency An estimate of the center of a distribution of values. The most usual measures of central tendency are the mean, median, and mode.

citation The mentioning of or referral to another research publication when discussing it in the text of a research write-up.

cluster random sampling or area random sampling A sampling method that involves dividing the population into groups called clusters, randomly selecting clusters, and then sampling each element in the selected clusters. This method is useful when sampling a population that is spread across a wide area geographically.

codebook A written description of the data that describes each variable and indicates where and how it can be accessed.

coding The process of categorizing qualitative data.

Cohen's Kappa A statistical estimate of inter-rater agreement or reliability that is more robust than percent agreement because it adjusts for the probability that some agreement is due to random chance.

comparison group In a comparative research design, like an experimental or quasi-experimental design, the control or comparison group is compared or contrasted with a group that receives the program or intervention of interest.

compensatory equalization of treatment A social threat to internal validity that occurs when the control group is given a program or treatment (usually by a well-meaning third party) designed to make up for or "compensate" for the treatment the program group gets. By equalizing the group's experiences, this threat diminishes the researcher's ability to detect the program effect.

compensatory rivalry A social threat to internal validity that occurs when one group knows the program another group is getting and, because of that, develops a competitive attitude with the other group. Often it is the comparison group knowing that the program group is receiving a desirable program (e.g., new computers) that generates the rivalry.

conclusion validity The degree to which conclusions you reach about relationships in your data are reasonable.

concurrent validity An operationalization's ability to distinguish between groups that it should theoretically be able to distinguish between.

confidence intervals (CIs) A confidence interval is used to indicate the precision of an estimate of a statistic. The CI provides the lower and upper limits of the statistical estimate at a specified probability level. For instance, a 95% confidence interval for an estimate of a mean or average (the statistic) is the range of values within which there is a 95% chance that the true mean is likely to fall.

confidentiality An assurance made to study participants that identifying information about them acquired through the study will not be released to anyone outside of the study.

conflict of interest A conflict of interest exists in research when a researcher's primary interest in the integrity of a study is compromised by a secondary interest such as personal gain (e.g., financial profit).

consequential validity The approximate truth or falsity of assertions regarding the intended or unintended consequences of test interpretation and use.

construct validity The degree to which inferences can legitimately be made from the operationalizations in your study to the theoretical constructs on which those operationalizations are based.

content analysis The analysis of text documents. The analysis can be quantitative, qualitative, or both. Typically, the major purpose of content analysis is to identify patterns in text.

content validity A check of the operationalization against the relevant content domain for the construct.

control group In a comparative research design, like an experimental or quasi-experimental design, the control or comparison group is compared or contrasted with a group that receives the program or intervention of interest.

convergent validity The degree to which the operationalization is similar to (converges on) other operationalizations to which it should be theoretically similar.

correlation A single number that describes the degree of relationship between two variables. A correlation always ranges from -1 to $+1$.

correlation matrix A table of all inter-correlations for a set of variables.

covariance designs A type of randomized experimental design that helps minimize noise through the inclusion of one or more variables (called covariates) that account for some of the variability in the outcome measure or dependent variable.

covariate A measured variable that is correlated with the dependent variable. It can be included in an analysis to provide a more precise estimate of the relationship between an independent and dependent variable.

criterion-related validity The validation of a measure based on its relationship to another independent measure as predicted by your theory of how the measures should behave.

Cronbach's Alpha One specific method of estimating the internal consistency reliability of a measure. Although not calculated in this manner, Cronbach's Alpha can be thought of as analogous to the average of all possible split-half correlations.

cross-sectional studies A study that takes place at a single point in time.

data audit The process of systematically assessing the quality of data in a qualitative study.

de-identification The process of removing identifying information from data sets in order to assure the anonymity of individuals.

debriefing The process of providing participants with full information about the purpose of a study once a person has completed participation. Participants are typically offered a standard written description of the study and why it was conducted, given the opportunity to ask questions, and given the opportunity to have their data withdrawn if they are not satisfied with what has happened in the study.

deception The intentional use of false or misleading information in study procedures. Researchers must justify the use of deception on scientific grounds and be certain to provide complete debriefing about the actual nature of the study once it has been completed.

Declaration of Helsinki The World Medical Association adopted the Declaration of Helsinki in 1964 in order to provide a set of principles to guide the practice of medical research. The principles include such statements as "research protocols should be reviewed by an independent committee prior to initiation."

deductive Top-down reasoning that works from the more general to the more specific.

degrees of freedom (df) A statistical term that is a function of the sample size. In the *t*-test formula, for instance, the df is the number of persons in both groups minus 2.

dependent variable The variable affected by the independent variable; for example, the outcome.

descriptive statistics Statistics used to describe the basic features of the data in a study.

descriptive studies A study that documents what is going on or what exists.

dichotomous response A measurement response that has two possible options (e.g., true/false or yes/no).

dichotomous response format A question response format that allows the respondent to choose between only two possible responses.

diffusion or imitation of treatment A social threat to internal validity that occurs because a comparison group learns about the program either directly or indirectly from program group participants.

direct observation The process of observing a phenomenon to gather information about it. This process is distinguished from participant observation, in that a direct observer doesn't typically try to become a participant in the context and does strive to be as unobtrusive as possible so as not to bias the observations.

discriminant validity The degree to which the operationalization is not similar to (or diverges from) other operationalizations that it theoretically should not be similar to.

dispersion The spread of the values around the central tendency. The two common measures of dispersion are the range and the standard deviation.

distribution The manner in which a variable takes different values in your data.

double-barreled question A question in a survey that asks about two issues but only allows the respondent a single answer. For instance, the question "What do you think of proposed changes in benefits and hours in your workplace?" asks simultaneously about two issues but treats it as though they are one.

double-pretest design A design that includes two waves of measurement prior to the program.

double entry An automated method for checking data-entry accuracy in which you enter data once and then enter them a second time, with the software automatically stopping each time a discrepancy is detected until the data enterer resolves the discrepancy. This procedure assures extremely high rates of data-entry accuracy, although it requires twice as long for data entry.

dual-media surveys A survey that is distributed simultaneously in two ways. For instance, if you distribute a survey to participants as an attachment they can print, complete, and fax back, or they can complete directly on the web as a web form, you can describe this as a dual-media survey.

dummy variable A variable that uses discrete numbers, usually 0 and 1, to represent different groups in your study. Dummy variables are often used in equations in the General Linear Model (GLM).

effect construct The abstract idea or theory of what the outcome is in a cause-effect relationship you are investigating.

electronic survey A survey that is administered via a computer program, typically distributed via email and/or a web site.

email survey Any survey that is distributed to respondents via email. Generally, the survey is either embedded in the email message and the respondent can reply to complete it, is transmitted as an email attachment that the respondent can complete and return via email, or is reached by providing a link in the email that directs the respondent to a website survey.

empirical Based on direct observations and measurements of reality.

error term A term in a regression equation that captures the degree to which the regression line is in error (that is, the residual) in describing each point.

ethnography Study of a culture using qualitative field research.

evidence-based practice (EBP) A movement designed to encourage or require practitioners to employ practices that are based on research evidence as reflected in research syntheses or practice guidelines.

evolutionary epistemology The branch of philosophy that holds that ideas evolve through the process of natural selection.

exception dictionary A dictionary that includes all nonessential words like "is," "and," and "of," in a content analysis study.

exhaustive The property of a variable that occurs when you include all possible answerable responses.

expert sampling A sample of people with known or demonstrable experience and expertise in some area.

external validity The degree to which the conclusions in your study would hold for other persons in other places and at other times.

face validity A validity that checks that "on its face" the operationalization seems like a good translation of the construct.

factor A major independent variable.

factor analysis A multivariate statistical analysis that uses observed correlations (variability) as input and identifies a fewer number of unobserved variables, known as factors, that describe the original data more efficiently.

factorial designs Designs that focus on the program or treatment, its components, and its major dimensions, and enable you to determine whether the program has an effect, whether different subcomponents are effective, and whether there are interactions in the effects caused by subcomponents.

fairness in publication credit Authorship credit should be established on the basis of the quality and quantity of one's contributions to a study rather than on status, power, or any other factor. Many research journals have now adopted guidelines to help authorship groups determine credit.

field research A research method in which the researcher goes into the field to observe the phenomenon in its natural state.

file drawer problem Many studies might be conducted but never reported, and they may include results that would substantially change the conclusions of a research synthesis.

filter or contingency question A question you ask the respondents to determine whether they are qualified or experienced enough to answer a subsequent one.

fishing and the error rate problem A problem that occurs as a result of conducting multiple analyses and treating each one as independent.

focus group A qualitative measurement method where input on one or more focus topics is collected from participants in a small-group setting where the discussion is structured and guided by a facilitator.

forced-choice response scale A response scale that does not allow for a neutral or undecided value. By definition, a forced-choice response scale has an even number of response options. For example, the following would be a forced-choice response scale: 1 = strongly disagree; 2 = disagree; 3 = agree; 4 = strongly agree.

frequency distribution A summary of the frequency of individual values or ranges of values for a variable.

fully crossed factorial design A design that includes the pairing of every combination of factor levels.

general linear model (GLM) A system of equations that is used as the mathematical framework for many of the statistical analyses used in applied social research.

generalizing, generalizability The process of making an inference that the results observed in a sample would hold in the population of interest. If such an inference or conclusion is valid we can say that it has generalizability.

gradient of similarity The dimensions along which your study context can be related to other potential contexts to which you might wish to generalize. Contexts that are closer to yours along the gradient of similarity of place, time, people, and so on can be generalized to with more confidence than ones that are further away.

grounded theory A theory rooted in observation about phenomena of interest. Also, a method for achieving such a theory.

group-administered questionnaire A survey that is administered to respondents in a group setting. For instance, if a survey is administered to all students in a classroom, we would describe that as a group-administered questionnaire.

group interview An interview that is administered to respondents in a group setting. A focus group is a structured form of group interview.

guideline A systematic process that leads to a specific set of research-based recommendations for practice that usually includes some estimates of how strong the evidence is for each recommendation.

Guttman or cumulative scaling A method of scaling in which the items are assigned scale values that allow them to be placed in a cumulative ordering with respect to the construct being scaled.

heterogeneity sampling Sampling for diversity or variety.

hierarchical modeling A statistical model that allows for the inclusion of data at different levels, where the unit of analysis at some levels is nested within the unit of analysis at others (e.g., student within class within school within school district)

history threat A threat to internal validity that occurs when some historical event affects your study outcome.

household drop-off survey A paper-and-pencil survey that is administered by dropping it off at the respondent's household and, either picking it up at a later time, or having the respondent return it directly. The household drop-off method assures a direct personal contact with the respondent while also allowing the respondent the time and privacy to respond to the survey on their own.

hypothesis A specific statement of prediction.

hypothesis guessing A threat to construct validity and a source of bias in which participants in a study guess the purpose of the study and adjust their responses based on that.

hypothetico-deductive model A model in which two mutually exclusive hypotheses that together exhaust all possible outcomes are tested, such that if one hypothesis is accepted, the second must therefore be rejected.

impact research Research that assesses the broader effects of a discovery or innovation on society.

implementation and dissemination research Research that assesses how well an innovation or discovery can be distributed in and carried out in a broad range of contexts that extend beyond the original controlled studies.

incomplete factorial design A design in which some cells or combinations in a fully crossed factorial design are intentionally left empty.

independent variable The variable that you manipulate. For instance, a program or treatment is typically an independent variable.

index A quantitative score that measures a construct of interest by applying a formula or a set of rules that combines relevant data.

indirect measure An unobtrusive measure that occurs naturally in a research context.

inductive Bottom-up reasoning that begins with specific observations and measures and ends up as general conclusion or theory.

inferential statistics Statistical analyses used to reach conclusions that extend beyond the immediate data alone.

informed consent A policy of informing study participants about the procedures and risks involved in research that ensures that all participants must give their consent to participate.

Institutional Review Boards (IRBs) A panel of people who review research proposals with respect to ethical implications and decide whether additional actions need to be taken to assure the safety and rights of participants.

instrumentation threat A threat to internal validity that arises when the instruments (or observers) used on the posttest and the pretest differ.

inter-rater or inter-observer reliability The degree of agreement or correlation between the ratings or codings of two independent raters or observers of the same phenomenon.

interaction effect An effect that occurs when differences on one factor depend on which level you are on another factor.

internal consistency reliability A correlation that assesses the degree to which items on the same multi-item instrument are interrelated. The most common forms of internal consistency reliability are the average inter-item correlation, the average item-total correlation, the split half correlation and Cronbach's Alpha.

internal validity The approximate truth of inferences regarding cause-effect or causal relationships.

interquartile range The difference between the 75th (upper quartile) and 25th (lower quartile) percentile scores on the distribution of a variable. The interquartile range is an estimate of the spread or variability of the measure.

interval The interval distance between values on a measure or variable.

interval level of measurement Measuring a variable on a scale where the distance between numbers is interpretable. For instance, temperature in Fahrenheit or Celsius is measured on an interval level.

interval level response format A response measured using numbers spaced at equal intervals where the size of the interval between potential response values is meaningful. An example would be a 1-to-5 response scale.

interval response scale A measurement response format with multiple response options that are set at equal intervals (e.g., 1-to-5 or 1-to-7).

item analysis The systematic statistical analysis of items on a scale or test undertaken to determine the properties of the items, especially for purposes of selecting final items or deciding which items to combine. An item-to-total-score correlation is an example of an item analysis.

justice Justice means that participation in research should be based on fairness and not on circumstances that give researchers access to or control of a population based on status. The principle of justice also means that recruiting should be done in a way that respects the rights of individuals and does not take advantage of their status as patients, students, prisoners, children, or others not fully able to make independent decisions.

Kefauver-Harris Amendments After the Thalidomide tragedy, these amendments to the Food, Drug and Cosmetic Act were passed to ensure greater drug safety. For the first time, drug manufacturers were legally required to present evidence on the safety and effectiveness of their products to the FDA before marketing them. It also established the requirement that participants be fully informed about potential risks or harm, and that based on this information, they voluntarily agree to participate in clinical trials.

level A subdivision of a factor into components or features.

level of measurement The relationship between numerical values on a measure. There are different types of levels of measurement (nominal, ordinal, interval, ratio) that determine how you can treat the measure when analyzing it. For instance, it makes sense to compute an average of an interval or ratio variable but does not for a nominal or ordinal one.

Likert-type response scale A response format where responses are gathered using numbers spaced at equal intervals.

Likert or summative scaling A method of scaling in which the items are assigned interval-level scale values and the responses are gathered using an interval-level response format.

linear model Any statistical model that uses equations to estimate lines.

literature review A systematic compilation and written summary of all of the literature published in scientific journals that is related to a research topic of interest. A literature review is typically included in the introduction section of a research write-up.

longitudinal studies A study that takes place over time.

low reliability of measures A threat to conclusion validity that occurs because measures that have low reliability by definition have more noise than measures with higher reliability, and the greater noise makes it difficult to detect a relationship (i.e., to see the signal of the relationship relative to the noise).

mail survey A paper-and-pencil survey that is sent to respondents through the mail.

main effect An outcome that shows consistent differences between all levels of a factor.

maturation threat A threat to internal validity that occurs as a result of natural maturation between pre- and post-measurement.

mean A description of the central tendency in which you add all the values and divide by the number of values.

meaning units In qualitative data analysis, a small segment of a transcript or other text that captures a concept that the analyst considers to be important.

median The score found at the exact middle or 50th percentile of the set of values. One way to compute the median is to list all scores in numerical order and then locate the score in the center of the sample.

meta-analysis A type of research synthesis that uses statistical methods to combine the results of similar studies quantitatively in order to allow general conclusions to be made.

mixed methods research Research that uses a combination of qualitative and quantitative methods.

modal instance sample Sampling for the most typical case.

mode The most frequently occurring value in the set of scores.

model specification The process of stating the equation that you believe best summarizes the data for a study.

mono-method bias A threat to construct validity that occurs because you use only a single method of measurement.

mono-operation bias A threat to construct validity that occurs when you rely on only a single implementation of your independent variable, cause, program, or treatment in your study.

mortality threat A threat to internal validity that occurs because a significant number of participants drop out.

multi-option or multiple-response variable A question format in which the respondent can pick multiple variables from a list.

multiple-group threats An internal validity threat that occurs in studies that use multiple groups, for instance, a program and a comparison group.

multistage sampling The combining of several sampling techniques to create a more efficient or effective sample than the use of any one sampling type can achieve on its own.

mutually exclusive The property of a variable that ensures that the respondent is not able to assign two attributes simultaneously. For example, gender is a variable with mutually exclusive options if it is impossible for the respondents to simultaneously claim to be both male and female.

National Research Act An Act passed by the US Congress in 1974. It represents the first serious attempt to build a comprehensive system of research ethics in the U.S. It created a national commission to develop guidelines for human subjects research and to oversee and regulate the use of human experimentation in medicine.

negative relationship A relationship between variables in which high values for one variable are associated with low values on another variable.

nominal level of measurement Measuring a variable by assigning a number arbitrarily in order to name it numerically so that it might be distinguished from other objects. The jersey numbers in most sports are measured at a nominal level.

nominal response format A response format that has a number beside each choice where the number has no meaning except as a placeholder for that response.

nonequivalent-groups design (NEGD) A pre-post two-group quasi-experimental design structured like a pretest-posttest randomized experiment, but lacking random assignment to group.

nonequivalent dependent variables (NEDV) design A single-group pre-post quasi-experimental design with two outcome measures where only one measure is theoretically predicted to be affected by the treatment and the other is not.

nonprobability sampling Sampling that does not involve *random* selection.

nonproportional quota sampling A sampling method where you sample until you achieve a specific number of sampled units for each subgroup of a population, where the proportions in each group are not the same.

normal curve A common type of distribution where the values of a variable have a bell-shaped histogram or frequency distribution. In a normal distribution, approximately 68 percent of cases occur within one standard deviation of the mean or center, 95 percent of the cases fall within two standard deviations, and 99 percent are within three standard deviations.

null case A situation in which the treatment has no effect.

null hypothesis The hypothesis that describes the possible outcomes other than the alternative hypothesis. Usually, the null hypothesis predicts there will be no effect of a program or treatment you are studying.

Nuremberg Code This code was developed following the trial of Nazi doctors after World War II. It includes 10 principles to guide research involving human subjects. The Code has been extremely important as a reference point for all regulations related to the protection of human subjects. Among other things, it established the principles of informed consent, voluntary participation without coercion, clear scientific justification for research, and most important, limits on the risk of harm.

one-tailed hypothesis A hypothesis that specifies a direction; for example, when your hypothesis predicts that your program will increase the outcome.

operationalization The act of translating a construct into its manifestation—for example, translating the idea of your treatment or program into the actual program, or translating the idea of what you want to measure into the real measure. The result is also referred to as an *operationalization*; that is, you might describe your actual program as an *operationalized program*.

ordinal level of measurement Measuring a variable using rankings. Class rank is a variable measured at an ordinal level.

ordinal response format A response format in which respondents are asked to rank the possible answers in order of preference.

parallel-forms reliability The correlation between two versions of the same test or measure that were constructed in the same way, usually by randomly selecting items from a common test question pool.

participant observation A method of qualitative observation where the researcher becomes a participant in the culture or context being observed.

pattern-matching NEDV design A single group pre-post quasi-experimental design with multiple outcome measures where there is a theoretically specified pattern of expected effects across the measures.

Pearson product moment correlation A particular type of correlation used when both variables can be assumed to be measured at an interval level of measurement.

peer review A system for reviewing potential research publications where authors submit potential articles to a journal editor who solicits several reviewers who agree to give a critical review of the paper. The paper is sent to these reviewers with no identification of the author so that there will be no personal bias (either for or against the author). Based on the reviewers' recommendations, the editor can accept the article, reject it, or recommend that the author revise and resubmit it.

personal interview A one-on-one interview between an interviewer and respondent. The interviewer typically uses an interview guide that provides a script for asking questions and follow-up prompts.

Phase I study A research study designed to test a new drug or treatment in a small group of people for the first time to evaluate its safety, determine a safe dosage range, and identify potential side effects.

Phase II study A research study designed to test a drug or treatment that is given in a larger group of people than in a Phase I study to see if it is effective and to further evaluate its safety.

Phase III study A research study designed to test a drug or treatment that is given to large groups of people using a highly controlled research design, in order to confirm the intervention's effectiveness, monitor side effects, compare it to commonly used treatments, and collect information that will allow the drug or treatment to be used safely.

phenomenology A philosophical perspective as well as an approach to qualitative methodology that focuses on people's subjective experiences and interpretations of the world.

piecemeal and duplicate publication An ethical issue in the dissemination of research referring to the possibility that, in order to maximize personal reward (status, promotion, pay, and such) from publication, a researcher would essentially repackage the results of a single study in multiple articles.

placebo The use of a treatment that may look like the real treatment, but is inert. The use of placebos in research is allowed if careful assessment of the scientific need and care of participants can be provided.

plagiarism The use of another person's work without proper credit.

point-of-experience survey A survey that is delivered at or immediately after the experience that the respondent is being asked about. A customer-satisfaction survey is often a point-of-experience survey.

policy research Research that is designed to investigate existing policies or develop and test new ones.

poor reliability of treatment implementation A threat to conclusion validity that occurs in causal studies because treatments or programs with more inconsistent or unreliable implementation introduce more noise than ones that are consistently carried out, and the greater noise makes it difficult to detect a relationship (i.e., to see the signal of the relationship relative to the noise).

population The group you want to generalize to and the group you sample from in a study.

population parameter The mean or average you would obtain if you were able to sample the entire population.

positive relationship A relationship between variables in which high values for one variable are associated with high values on another variable, and low values are associated with low values on the other variable.

posttest-only nonexperimental design A research design in which only a posttest is given. It is referred to as nonexperimental because no control group exists.

posttest-only randomized experiment An experiment in which the groups are randomly assigned and receive only a posttest.

posttest-only single-group design A design that has an intervention and a posttest where measurement of outcomes is only done within a single group of program recipients.

pre-post nonequivalent groups quasi-experiment A two-group quasi-experimental design structured like a randomized experiment, but lacking random assignment to group.

predictive validity A type of construct validity based on the idea that your measure is able to predict what it theoretically should be able to predict.

pretest-posttest Any research design that uses measurement both before and after an intervention or program.

privacy Privacy in research means protection of personal information about participants. It is typically accomplished through the use of confidentiality procedures that specify who will have access to personally identifying data and the limits of that access. It can also include the use of anonymous data in which no personally identifying information is ever collected.

probabilistic Based on probabilities.

probabilistically equivalent The notion that two groups, if measured infinitely, would on average perform identically. Note that two groups that are probabilistically equivalent would seldom obtain the exact same average score.

probability sampling Method of sampling that utilizes some form of *random selection*.

program group In a comparative research design, like an experimental or quasi-experimental design, the program or treatment group receives the program of interest and is usually contrasted with a no-treatment comparison or control group or a group receiving another treatment.

proportional quota sampling A sampling method where you sample until you achieve a specific number of sampled units for each subgroup of a population, where the proportions in each group are the same.

protocol A detailed document summarizing the purpose and procedures of a study. It provides a reader such as an IRB reviewer with enough information to know how participants are recruited, consented, and treated in all phases of a study. It typically includes details on all measures, the handling and storage of data, consent and assent forms, debriefing materials,

and any other item that might have an impact on a participant.

proximal similarity model A model for generalizing from your study to other contexts based upon the degree to which the other context is similar to your study context.

proxy-pretest design A post-only design in which, after the fact, a pretest measure is constructed from preexisting data. This is usually done to make up for the fact that the research did not include a true pretest.

qualitative The descriptive nonnumerical characteristic of some object. A qualitative variable is a descriptive nonnumerical observation.

qualitative data Data in which the variables are not in a numerical form, but are in the form of text, photographs, sound bites, and so on.

qualitative measures Data not recorded in numerical form.

quantitative The numerical representation of some object. A quantitative variable is any variable that is measured using numbers.

quantitative data Data that appear in numerical form.

quasi-experimental design Research designs that have several of the key features of randomized experimental designs, such as pre-post measurement and treatment-control group comparisons, but lack randomized assignment to groups.

quota sampling Any sampling method where you sample until you achieve a specific number of sampled units for each subgroup of a population.

random assignment The process of assigning your sample into two or more subgroups by chance. Procedures for random assignment can vary from flipping a coin to using a table of random numbers to using the random-number capability built into a computer.

random error A component or part of the value of a measure that varies entirely by chance. Random error adds noise to a measure and obscures the true value.

random heterogeneity of respondents A threat to statistical conclusion validity. If you have a very diverse group of respondents, they are likely to vary more widely on your measures or observations. Some of their variety may be related to the phenomenon you are looking at, but at least part of it is likely to just constitute individual differences that are irrelevant to the relationship being observed.

random irrelevancies in the setting A threat to conclusion validity that occurs when factors in the research setting that are irrelevant to the relationship being studied add noise to the environment, which makes it harder to detect a relationship between key variables, even if one is present.

random selection Process or procedure that assures that the different units in your population are selected by chance.

randomized block design Experimental designs in which the sample is grouped into relatively homogeneous subgroups or blocks within which your experiment is replicated. This procedure reduces noise or variance in the data.

range The highest value minus the lowest value.

ratio level of measurement Measuring a variable on a scale where the distance between numbers is interpretable and there is an absolute zero value. For example, weight is a ratio measurement.

reference The complete description of the name and location of an article, the authors and the source (e.g., journal, book, or website). References are included in a separate section immediately following the body of the paper.

regression-discontinuity (RD) design A pretest-posttest program-comparison-group quasi-experimental design in which a cutoff criterion on the preprogram measure is the method of assignment to group.

regression analysis A general statistical analysis that enables us to model relationships in data and test for treatment effects. In regression analysis, we model relationships that can be depicted in graphic form with lines that are called *regression lines*.

regression line A line that describes the relationship between two or more variables.

regression point displacement (RPD) design A pre-post quasi-experimental research design where the treatment is given to only one unit in the sample, with all remaining units acting as controls. This design is particularly useful for studying the effects of community-level interventions, where outcome data are routinely collected at the community level.

regression threat or regression artifact or regression to the mean A statistical phenomenon that causes a group's average performance on one measure to regress toward or appear closer to the mean of that measure, more than anticipated or predicted. Regression occurs whenever you have a nonrandom sample from a population and two measures that are imperfectly correlated. A regression threat will bias your estimate of the group's posttest performance and can lead to incorrect causal inferences.

relational studies A study that investigates the connection between two or more variables.

relationship An association between two variables such that, in general, the level on one variable is related to the level on the other. Technically, the term correlational relationship is redundant: a correlation by definition always refers to a relationship. However the term correlational relationship is used to distinguish it from the specific type of association called a causal relationship.

repeated measures Two or more waves of measurement over time.

replicate, replication A study that is repeated in a different place, time, or setting.

requests for proposals (RFPs) A document issued by a government agency or other organization that, typically, describes the problem that needs addressing, the contexts in which it operates, the approach the agency would like you to take to investigate the problem, and the amount the agency would be willing to pay for such research.

research A type of systematic investigation that is empirical in nature and is designed to contribute to public knowledge.

research-practice continuum The process of moving from an initial research idea or discovery to practice, and the potential for the idea to influence our lives or world.

research enterprise The macro-level effort to accumulate knowledge across multiple empirical systematic public research projects.

research misconduct Fabrication, falsification, or plagiarism in proposing, performing, or reviewing research, or in reporting research results, are all prime examples of research misconduct. Fabrication is making up data or results and recording or reporting them. Falsification is manipulating research materials, equipment, or processes, or changing or omitting data or results such that the research is not accurately represented in the research record.

research question The central issue being addressed in the study, typically phrased in the language of theory.

research synthesis A systematic study of multiple prior research projects that address the same research question or topic and that summarizes the results in a manner that can be used by practitioners.

resentful demoralization A social threat to internal validity that occurs when the comparison group knows what the program group is getting, and instead of developing a rivalry, control group members become discouraged or angry and give up.

respect for persons This principle means that people are to be treated as independent and autonomous individuals. It also means that the well-being of those who are not in a position to be fully autonomous should also be protected. See also "vulnerable populations."

respondent-driven sampling A nonprobability sampling method that combines chain-referral or snowball sampling, with a statistical weighting system that helps compensate for the fact that the sample was not drawn randomly.

response A specific measurement value that a sampling unit supplies.

response brackets A question response format that includes groups of answers, such as between 30 and 40 years old, or between $50,000 and $100,000 annual income.

response format The format you use to collect the answer from the respondent.

response scale A sequential-numerical response format, such as a 1-to-5 rating format.

reversal items Items on a multi-item scale whose wording is in the opposite direction of the construct of interest. Reversal items must have their scores reversed prior to computing total scale scores. For example, if you had a multi-item self-esteem scale, an item such as "I generally feel bad about myself as a person" would be a reversal item.

right to service The right of study participants to the best available and most appropriate services relevant to a condition that is part of a study.

running head A brief version of the title (50 characters or less) that is included at the top-left of every page of a research write-up.

sample The actual units you select to participate in your study.

sampling The process of selecting units (e.g., participants) from the population of interest.

sampling distribution The theoretical distribution of an infinite number of samples of the population of interest in your study.

sampling error Error in measurement associated with sampling.

sampling frame The list from which you draw your sample. In some cases, there is no list; you draw your sample based upon an explicit rule.

sampling model A model for generalizing in which you identify your population, draw a fair sample, conduct your research, and finally generalize your results from the sample to the population.

scaling The branch of measurement that involves the construction of an instrument that associates qualitative constructs (i.e., objects) with quantitative metric units.

scalogram analysis A method of analysis of a set of scale items used when constructing a Guttman or cumulative scale. In scalogram analysis, one attempts to determine the degree to which responses to the set of items allows the items to be ordered cumulatively in one dimension with respect to the construct of interest.

secondary analysis Quantitative analysis of existing data that is done either to verify or extend previously accomplished analyses or to explore new research questions. Using existing data can be an efficient alternative to collecting original primary data and can extend the validity, quality and scope of what can be accomplished in a single study.

selection-history threat A threat to internal validity that results from any other event that occurs between pretest and posttest that the groups experience differently.

selection-instrumentation threat A threat to internal validity that results from differential changes in the test used for each group from pretest to posttest.

selection-maturation threat A threat to internal validity that arises from any differential rates of normal growth between pretest and posttest for the groups.

selection-mortality threat A threat to internal validity that arises when there is differential non-random dropout between pretest and posttest.

selection-regression threat A threat to internal validity that occurs when there are different rates of regression to the mean in the two groups between pretest and posttest.

selection-testing threat A threat to internal validity that occurs when a differential effect of taking the pretest exists between groups on the posttest.

selection threat or selection bias Any factor other than the program that leads to pretest differences between groups.

semantic differential A scaling method in which an object is assessed by the respondent on a set of bipolar adjective pairs.

separate pre-post samples design A design in which the people who receive the pretest are not the same as the people who take the posttest.

simple random sampling A method of sampling that involves drawing a sample from a population so that every possible sample has an equal probability of being selected.

single-group design Any research design that involves only a single group in measuring outcomes.

single-group threats A threat to internal validity that occurs in a study that uses only a single program or treatment group and no comparison or control.

single-option variable A question response list from which the respondent can check only one response.

slope The change in y for a change in x of one unit.

snowball sampling A sampling method in which you sample participants based upon referral from prior participants.

social threats to internal validity Threats to internal validity that arise because social research is conducted in real-world human contexts where people will react to not only what affects them, but also to what is happening to others around them.

split-half reliability An estimate of internal consistency reliability that uses the correlation between the total score of two randomly selected halves of the same multi-item test or measure.

standard deviation An indicator of the variability of a set of scores in a sample around the mean of that sample.

standard error The spread of the averages around the average of averages in a sampling distribution.

standard error of the difference A statistical estimate of the standard deviation one would obtain from the distribution of an infinite number of estimates of the difference between the means of two groups.

statistic A value that is estimated from data.

statistical power The probability that you will conclude there is a relationship when in fact there is one. We typically want statistical power to be at least 0.80 in value.

stratified random sampling A method of sampling that involves dividing your population into homogeneous subgroups and then taking a simple random sample in each subgroup.

structured response format Provides a specific format for the respondent to choose their answer. For example, a checkbox question lists all of the possible responses.

survey A measurement tool used to gather information from people by asking questions about one or more topics.

switching-replications design A two-group design in two phases defined by three waves of measurement. The implementation of the treatment is repeated in both phases. In the repetition of the treatment, the two groups *switch* roles; the original control group in phase 1 becomes the treatment group in phase 2, whereas the original treatment group acts as the control. By the end of the study, all participants have received the treatment.

symmetric matrix A square (as many rows as columns) table of numbers that describes the relationships among a set of variables, where each variable represents a row or column. Each value in the table represents the relationship between the row and column variable for that cell of the table. The table is "symmetric" when the relationship between a specific row and column variable is identical to the relationship between the same column and row. A correlation matrix is a symmetric matrix.

systematic error A component of an observed score that consistently affects the responses in the distribution.

systematic random sampling A sampling method where you determine randomly where you want to start selecting in the sampling frame and then follow a rule to select every xth element in the sampling frame list (where the ordering of the list is assumed to be random).

systematic review A type of research synthesis that focuses on a specific question or issue and uses preplanned methods to identify, select, assess, and summarize the findings of multiple research studies.

***t*-value** The estimate of the difference between the groups relative to the variability of the scores in the groups.

telephone interview A personal interview that is conducted over the telephone.

temporal precedence The criterion for establishing a causal relationship that holds that the cause must occur before the effect.

test-retest reliability The correlation between scores on the same test or measure at two successive time points.

testing threat A threat to internal validity that occurs when taking the pretest affects how participants do on the posttest.

tests Measurement instruments designed to assess a respondent's knowledge, skill, or performance.

Thalidomide tragedy This event involved the occurrence of very serious birth defects in children of pregnant women who had been given Thalidomide as a sedative. The drug side effects should have been known and available to doctors and patients, but were not until much harm had been done.

theoretical Pertaining to theory. Social research is theoretical, meaning that much of it is concerned with developing, exploring, or testing the theories or ideas that social researchers have about how the world operates.

theoretical population A group which, ideally, you would like to sample from and generalize to. This is usually contrasted with the accessible population.

third-variable or **missing-variable problem** An unobserved variable that accounts for a correlation between two variables.

threats to conclusion validity Any factors that can lead you to reach an incorrect conclusion about a relationship in your observations.

threats to construct validity Any factors that cause you to make an incorrect conclusion about whether your operationalized variables (e.g., your program or outcome) reflect well the constructs they are intended to represent.

threats to external validity Any factors that can lead you to make an incorrect generalization from the results of your study to other persons, places, times, or settings.

threats to internal validity Any factors that lead you to draw an incorrect conclusion that your treatment or program causes the outcome.

threats to validity Reasons your conclusion or inference might be wrong.

Thurstone scaling A class of scaling methods (the method of equal-appearing intervals, the method of successive intervals, and the method of paired comparisons) that were designed to yield unidimensional, interval-level, multi-item scales.

time series Many waves of measurement over time.

title page The initial page of a formal research write-up that includes the title of the study, the author's name, and the institutional affiliation.

translation validity A type of construct validity related to how well you translated the idea of your measure into its operationalization.

translational research The systematic effort to move research from initial discovery to practice and ultimately to impacts on our lives.

treatment group In a comparative research design, like an experimental or quasi-experimental design, the program or treatment group receives the program of interest and is usually contrasted with a no-treatment comparison or control group or a group receiving another treatment.

triangulate Combining multiple independent measures to get at a more accurate estimate of a variable.

true score theory A theory that maintains that an observed score is the sum of two components: true ability (or the true level) of the respondent; and random error.

Tuskegee Syphilis Study This was a 40-year observational study of the impact of untreated syphilis on men. The participants in the study were low-income African American men who were led to believe they were receiving treatment when in fact they were not. This was one of the major stimuli of the Belmont Conference.

two-group posttest-only randomized experiment A research design in which two randomly assigned groups participate. Only one group receives the program and both groups receive a posttest.

two-tailed hypothesis A hypothesis that does not specify a direction. For example, if your hypothesis is that your program or intervention will have an effect on an outcome, but you are unwilling to specify whether that effect will be positive or negative, you are using a two-tailed hypothesis.

Type I Error An erroneous interpretation of statistics in which you falsely conclude that there is a significant relationship between variables when in fact there is none. In other words, you reject the null hypothesis when it is true.

Type II Error The failure to reject a null hypothesis when it is actually false.

unit of analysis The entity that you are analyzing in your analysis; for example, individuals, groups, or social interactions.

unitizing In content analysis, the process of dividing a continuous text into smaller units that can then be analyzed.

unobtrusive measures Methods used to collect data without interfering in the lives of the respondents.

unstructured interviewing An interviewing method that uses no predetermined interview protocol or survey and where the interview questions emerge and evolve as the interview proceeds.

unstructured response format A response format that is not predetermined and that allows the respondent or interviewer to determine how to respond. An open-ended question is a type of unstructured response format.

validity The best available approximation of the truth of a given proposition, inference, or conclusion.

variability The extent to which the values measured or observed for a variable differ.

variable Any entity that can take on different values. For instance, age can be considered a variable

because age can take on different values for different people at different times.

variance A statistic that describes the variability in the data for a variable. The variance is the spread of the scores around the mean of a distribution. Specifically, the variance is the sum of the squared deviations from the mean divided by the number of observations minus 1. The standard deviation and variance both measure dispersion, but because the standard deviation is measured in the same units as the original measure and the variance is measured in squared units, the standard deviation is usually more directly interpretable and meaningful.

violated assumptions of statistical tests The threat to conclusion validity that arises when key assumptions required by a specific statistical analysis are not met in the data.

voluntary participation The right of voluntary participation is one of the most basic and important of all research participant rights. It means that fully informed individuals have the right to decide whether or not to participate in a study without coercion. It also means that they can terminate participation at any point without negative consequences.

vulnerable populations This is the term used to designate those who may not be fully in control of their decision making. There are several specific groups considered vulnerable, including children, prisoners, and people with impaired cognitive capacity (e.g., with mental retardation or dementia). Children under 18 are not of legal age to provide consent. Instead, researchers must seek permission from parents or legal guardians.

web survey A survey that is administered over a website (either Intranet or Internet). Respondents use their web browser to reach the website and complete the survey.

weighted index A quantitative score that measures a construct of interest by applying a formula or a set of rules that combines relevant data where the data components are weighted differently.

References

Chapter 1

Campbell, D. T. (1988). Evolutionary epistemology. In E. S. Overman (Ed.), *Methodology and epistemology for social science: Selected papers of Donald T. Campbell*. Chicago: University of Chicago Press.

Cook, T. D., & Campbell, D. T. (1979). *Quasi-experimentation: Design and analysis for field settings*. Boston: Houghton Mifflin Company.

Dougherty, D., & Conway, P. H. (2008). The "3T's" road map to transform US health care. *JAMA, 299*, 2319–2321.

Foster, G. D., Sherman, S., Borradaile, K. E., Grundy, K. M., Vander Veur, S. S., Nachmani, J., Karpyn, A., Kumanyika, S., & Shults, J. (2008). A policy-based school intervention to prevent overweight and obesity. *Pediatrics, 121*, 794–802.

Gibbs, L. E. (2003). *Evidence-based practice for the helping professions*. Pacific Grove, CA: Thomson, Brooks-Cole.

Institute of Medicine. (2008). *Knowing what works in health care*. Washington, DC: National Academy Press.

Khoury, M. J., Gwinn, M., Yoon, P. W., Dowling, N., Moore, C. A., & Bradley, L. (2007). The continuum of translation research in genomic medicine: How can we accelerate the appropriate integration of human genome discoveries into health care and disease prevention? *Genetics in Medicine, 9*, 665–674.

Marriott, F. H. C. (1990). *A dictionary of statistical terms*. New York: Longman Scientific and Technical.

McDonald, P. W., & Viehbeck, S. (2007). From evidence-based practice making to practice-based evidence making: Creating communities of (research) and practice. *Health Promotion Practice, 8*, 140–144.

Nagel, E. (1979). *The structure of science: Problems in the logic of scientific explanation* (2nd ed.). Indianapolis, IN: Hackett Publishing Company.

Popper, K. (1985). Evolutionary epistemology. In D. M. Miller (Ed.), *Popper selections* (pp. 78–86). Princeton, NJ: Princeton University Press.

Popper, K. R. (1959). *The logic of scientific discovery*. New York: Basic Books.

Sackett, D. L. (1997). Evidence-based medicine. *Seminars in Perinatology, 21*, 3–5.

Sung, N., Crowley, W., Genel, M., Salber, P., Sandy, L., Sherwood, L., . . . Rimoin, D. (2003). Challenges facing the National Clinical Research Enterprise. *Journal of Investigative Medicine, 51*, S385–S385.

Sung, N. S., Crowley, W. F., Jr, Genel, M., Salber, P., Sandy, L., Sherwood, L. M., . . . Rimoin, D. (2003). Central challenges facing the National Clinical Research Enterprise. *JAMA, 289*, 1278–1287.

Trochim, W., Kane, C., Graham, M., & Pincus, H. (2011). Evaluating translational research: A process marker model. *Clinical and Translational Sciences, 4*, 153–162.

Urban, J. B., & Trochim, W. (2009). The role of evaluation in research-practice integration working toward the "Golden Spike." *American Journal of Evaluation, 30*, 538–553. doi:10.1177/1098214009348327.

Westfall, J. M., Mold, J., & Fagnan, L. (2007). Practice-based research—"Blue Highways" on the NIH roadmap. *JAMA, 297*, 403–406.

Chapter 2

Amdur, R. J., & Bankert, E. A. (2007). *Institutional review board member handbook*. Sudbury, MA: Jones & Bartlett.

American Psychological Association. (2013). *Ethical principles of psychologists and code of conduct*. Accessed at http://www.apa.org/ethics/code/index.aspx?item=11.

Bartlett, T. (2012). Former Harvard psychologist fabricated and falsified, report says. *The Chronicle of Higher Education*, September 5, 2012. Accessed 9-13 from: http://chronicle.com/blogs/percolator/report-says-former-harvard-psychologist-fabricated-falsified/30748.

Beecher, H. K. (1966). Ethics and clinical research. *The New England Journal of Medicine, 274,* 1354–1360.

Bhattacharjee, Y. (2013). The psychology of lying. *The New York Times Magazine,* April 26, 2013. Accessed from: http://www.nytimes .com/2013/04/28/magazine/diederik-stapels -audacious-academic-fraud.html?pagewanted=1& _r=2&smid=tw-share&.

Blass, T. (2004). *The man who shocked the world.* New York: Basic Books.

Carey, B. (2011). Fraud case seen as red flag for psychology research. *The New York Times,* November 2, 2011. Accessed 9-13 from http:// www.nytimes.com/2011/11/03/health /research/noted-dutch-psychologist-stapel -accused-of-research-fraud.html.

Dunn, C. M., & Chadwick, G. L. (2004). *Protecting study volunteers in research.* Boston, MA: Thomson.

Epstein, S. (1996). *Impure science: AIDS, activism, and the politics of knowledge.* Berkeley, Los Angeles, and London: University of California Press.

Festinger, D. S., Dugosh, K. L., Marlowe, D. B., & Clements, N. T. (2014). Achieving new levels of recall in consent to research by combining remedial and motivational techniques. *Journal of Medical Ethics, 40,* 264–268.

Gardner, H., Csikszentmihalyi, M., & Damon, W. (2002). *Good work.* New York: Basic Books.

Kaiser, D. (2009, November–December). *Physics and pixie dust.* Retrieved from: http://www .americanscientist.org/bookshelf/pub/physics -and-pixie-dust.

Killen, J. Y., Jr. (2008). HIV research. In E. J. Emanuel (Ed.), *The Oxford textbook of clinical research Ethics.* New York, NY: Oxford University Press.

McNeil, D. G., Jr. (2010). U.S. apologizes for syphilis tests in Guatemala. *The New York Times,* October 1, 2010. Accessed 12/4 from http://www.nytimes.com/2010/10/02/health /research/02infect.html?_r=0.

Meinert, C. (2012). *Clinical trials: Design, conduct and analysis* (2nd ed.). New York, NY: Oxford University Press.

Messner, D. A. (2006). *Rulemaking under fire: Accelerated approval as a case study of crisis-mode rulemaking.* Atlanta, GA: School of Public Policy, Georgia Institute of Technology.

Milgram, S. (1965, 2008). *Obedience.* University Park, PA: Penn State Media Sales.

National Institutes of Health. (April 18, 2008). *Clinical trial phases.* Retrieved from National Institutes of Health website: http://www.nlm.nih. gov/services/ctphases.html.

Resnik, D. B. (2011). *What is ethics in research & why is it important?* Retrieved from the National Institute of Environmental Health Sciences website: http://www.niehs.nih.gov/research /resources/bioethics/whatis/.

Scott-Jones, D. (2000). Recruitment of research participants. In B. D. Sales and S. Folkman (Eds.), *Ethics in research with human participants.* Washington, DC: American Psychological Association.

Tyson, P. (2000). *The experiments.* Retrieved from the Public Broadcasting Service (PBS) website: http://www.pbs.org/wgbh/nova/holocaust /experiside.html.

Chapter 3

Bluebond-Langner, M. (1978). *The private worlds of dying children.* Princeton, NJ: Princeton University Press.

Cameron, W. B. (1963). *Informal sociology.* New York: Random House.

Campbell, D. T. (1975). Degrees of freedom and case study. *Comparative Political Studies, 8,* 178–193.

Dart, J., & Davies, R. (2003). A dialogical, story-based evaluation tool: The most significant change technique. *American Journal of Evaluation, 24,* 137–155.

Fine, M., Torre, M. E., Boudin, I. B., Clark, J., Hylton, D., Martinez, M., …Upegui, D. (2003). Participatory action research: From within and beyond prison bars. In P. M. Camic, J. E. Rhodes, and L. Yardley (Eds.), *Qualitative research in psychology.* Washington, DC: American Psychological Association.

Finger, S. (1975). Child-holding patterns in Western art. *Child Development, 46,* 267–271.

Glaser, B., & Strauss, A. (1967). *The discovery of grounded theory: Strategies for qualitative research.* Chicago, IL: Aldine De Gruyther.

Greene, J. C., & Caracelli, V. J. (1997). Advances in mixed-method evaluation: The challenges and benefits of integrating new paradigms. *New Directions for Evaluation, 74.*

Lincoln, Y. S., & Guba, E. (1985) *Naturalistic enquiry.* Beverly Hills, CA: Sage.

Robinson, P., Giorgi, B., & Ekman, S-L. (2012). The lived experience of early-stage Alzheimer's disease: A three-year longitudinal phenomenological case study. *Journal of Phenomenological Psychology, 43,* 216–238.

Sofaer, S. (1999). Qualitative methods: What are they and why use them? *Health Services Research, 34* (5, Part II), 1101–1118.

Wilson, H. S., Hutchinson, S. A., & Holzemer, W. L. (2002). Reconciling incompatibilities: A grounded theory of HIV medication adherence and symptom management. *Qualitative Health Research, 12,* 1309–1322.

Chapter 4

Aitken, M. E., Graham, C. J., Killingsworth, J. B., Mullins, S. H., Parnell, D. N., & Dick, R. M. (2004). All-terrain vehicle injury in children: Strategies for prevention. *Injury Prevention, 10,* 303–307.

Belluck, P. (2009, January 17). Test subjects who call the scientist Mom or Dad. *The New York Times. Retrieved* from: http://www.nytimes.com/2009/01/18/science/18kids.html?pagewanted=all.

Campbell, D. T. (1986). Relabeling internal and external validity for applied social scientists. In W. Trochim (Ed.), *Advances in quasi-experimental design and analysis, 31,* 67–77. San Francisco: Jossey-Bass.

Centers for Disease Control and Prevention. National home and hospice care survey. Retrieved July 7, 2012, from http://www.cdc.gov/nchs/nhhcs.htm.

Ding, Y., Detels, R., Zhao, Z., Zhu, Y., Zhu, G., Zhang, B., et al. (2005). HIV infection and sexually transmitted diseases in female commercial sex workers in China. *Journal of Acquired Immune Deficiency Syndrome, 38,* 314–319.

Fishkin, J. S. (2006). Strategies of public consultation. *The Integrated Assessment Journal, 6,* 57–72.

Heckathorn D. D. (1997). Respondent driven sampling: A new approach to the study of hidden populations. *Social Problems, 44,* 174–199.

Henderson, R. H., & Sundaresan T. (1982) Cluster sampling to assess immunization coverage: A review of experience with a simplified sampling method. *Bulletin of the World Health Organization, 60,* 253–260. Available at: http://whqlibdoc.who.int/bulletin/1982/Vol60-No2/bulletin_1982_60(2)_253-260.pdf.

Henry, G. T. (1990). *Practical sampling.* Newbury Park, CA: Sage.

Knol, A., et al. (2009). Expert elicitation on ultrafine particles: Likelihood of health effects and causal pathways. *Part Fibre Toxicology, 6,* 19.

Magin, P., Adams, J., Heading, G., Pond, D., & Smith, W. (2006). The causes of acne: A qualitative study of patient perceptions of acne causation and their implications for acne care. *Dermatology Nursing, 18,* 344–349.

Simon, G. E., Ludman, E. J., et al. (2008). Association between obesity and depression in middle-aged women. *Gen Hosp Psychiatry, 30,* 32–39.

Walsh, W., Jones, L. M., Cross, T. P., & Lippert, T. (2010). Prosecuting child sexual abuse: The importance of evidence type. *Crime & Delinquency, 56,* 436–454.

Chapter 5

Cohen, J. (1960). A coefficient of agreement for nominal scales. *Educational and Psychological Measurement, 20,* 37–46.

Cook, T. D., & Campbell, D. T. (1979*). Quasi-experimentation: Design & analysis for field settings.* Boston: Houghton Mifflin Company.

Stevens, S. S. (1946). On the theory of scales of measurement. *Science, 103,* 677–680.

Wood, J. M. (2007, October 3). Understanding and computing Cohen's kappa: A tutorial. *WebPsychEmpiricist.* Retrieved May 6, 2014 from http://wpe.info/papers_table.html.

Chapter 6

APA. (1999). *Standards for educational and psychological testing.* Washington, DC: The American Psychological Association.

Duncan, O. D. (1981). A socioeconomic index for all occupations. In A. J. Reiss, Jr. (Ed.), *Occupations and social status* (pp. 139–161). New York: Free Press.

Fabiano, G. A., Hulme, K., Linke, S., Nelson-Tuttle, C., Pariseau, M., Gangloff, B., ...Buck, M. The Supporting a Teen's Effective Entry to the Roadway (STEER) program: Feasibility and preliminary support for a psychosocial intervention for teenage drivers with ADHD. *Cognitive and Behavioral Practice, 18,* 267–280.

Hauser, R. M., & Warren, J. R. (1996). *Socioeconomic indexes for occupations: A review, update, and critique.* Madison, WI: Center for Demography and Ecology.

Messick, S. (1996). Validation of psychological assessment: Validation of inferences from persons' responses and performances as scientific inquiry into score meanings. *American Psychologist, 50,* 741–749.

Messick, S. (1998). Test validity: A matter of consequence. *Social Indicators Research, 45,* 35–44.

Provenzo, E. F. (Ed.). (2008). *Encyclopedia of the social and cultural foundations of education.* Thousand Oaks, CA: Sage.

Rosenberg, M. (1979). Conceiving the self. *New York: Basic Books.*

Stevens, G., & Cho, J. H. (1985). Socioeconomic indices and the new 1980 census occupational classification scheme. *Social Science Research, 14,* 142–168.

Stevens, S. S. (1959). Measurement, psychophysics, and utility. In C. W. Churchman and P. Ratoosh (Eds.), *Measurement: Definitions and theories.* New York: John Wiley.

Stevens, S. S. (1946). On the theory of scales of measurement. *Science, 103,* 677–680.

U.S. Department of Labor. (2004). Consumer price index, 2004, from http://www.bls.gov/cpi/home.htm.

Chapter 7

Allred, S. B., & Ross-Davis, A. (2010). The dropoff and pickup method: An approach to reduce nonresponse bias in natural resource surveys. *Small-scale Forestry, 10,* 305–318.

Hager, M. A., Wilson, S., Thomas, P., & Rooney, P. M. (2003). Response rates for mail surveys of nonprofit organizations: A review and empirical test. *Nonprofit and Voluntary Sector Quarterly, 32,* 252–267.

Pew Research Center for the People and the Press. (2013). *Cell phone surveys.* Retrieved March 4, 2013, from http://www.people-press.org/methodology/collecting-survey-data/cell-phone-surveys/.

Chapter 9

Armitage, P. (2003). Fisher, Bradford Hill, and randomization. *International Journal of Epidemiology, 32,* 925–928.

Campbell, D. T., & Stanley, J. C. (1963). *Experimental and quasi-experimental designs for research.* Chicago: Rand McNally.

Dollahite, J., Pijai, E., Scott-Pierce, M., Parker, C., & Trochim, W. (2014). A randomized controlled trial of a community-based nutrition education program for low-income parents. *Journal of Nutrition Education and Behavior, 46,* 102–109.

Salsburg, D. (2001). *The lady tasting tea: How statistics revolutionized science in the twentieth century.* New York: W. H. Freeman.

Spoth R. L., Redmond, C., & Shin, C. (2000). Reducing adolescents' aggressive and hostile behaviors: Randomized trial effects of a brief family intervention 4 years past baseline. *Archives of Pediatric Adolescent Medicine, 154,* 1248–1257.

Chapter 10

Cook, T. D., & Campbell, D. T. (1979). *Quasi-experimentation: Design & analysis issues for field settings.* Boston: Houghton Mifflin.

Hazan, C., & Shaver, P. (1987). Romantic love conceptualized as an attachment process. *Journal of Personality & Social Psychology, 52,* 511–524.

Spoor, E., de Jonge, J. E., & Hamers, J. P. H. (2010). Design of the DIRECT-project: Interventions to increase job resources and recovery opportunities to improve job-related health, well-being, and performance outcomes in nursing homes. *BMC Public Health, 10*(293). http://www.biomdecentral.com/1471-2458/10/293.

Chapter 11

Cumming, G. (2012). *Understanding the new statistics: Effect sizes, confidence intervals, and meta-analysis.* New York: Routledge.

Kline, R. (2013). *Beyond significance testing: Statistics reform in the behavioral sciences.* Washington, DC: American Psychological Association.

Chapter 12

Smith, M.L. & Glass, G.V. (1977). Meta-analysis of psychotherapy outcome studies. *American Psychologist, 32,* 752-760.

Chapter 13

American Psychological Association. (2010). *Publication manual of the American Psychological Association* (6th ed.). Washington, DC: Author.

Gilgun, J. (2005). "Grab" and good science: Writing up the results of qualitative research. *Qualitative Health Research, 15,* 256–262. doi:10.1177/1049732304268796.

Ioannidis, J.P.A. (2005). Why most published research findings are false. *PLoS Medicine, 2(8):* e124.

Purdue Online Writing Lab. (2013). *APA formatting and style guide.* Retrieved December 15, 2013, from https://owl.english.purdue.edu/owl/resource/560/01/.

Rosenthal, R. (1979). The file drawer problem and tolerance for null results. *Psychological Bulletin, 86,* 638–641. doi:10.1037/0033-2909.86.3.638.

Index